A Voyage Through Turbulence

Turbulence is widely recognized as one of the outstanding problems of the physical sciences, but it still remains only partially understood despite having attracted the sustained efforts of many leading scientists for well over a century.

In *A Voyage Through Turbulence*, we are transported through a crucial period of the history of the subject via biographies of twelve of its great personalities, starting with Osborne Reynolds and his pioneering work of the 1880s. This book will provide absorbing reading for every scientist, mathematician and engineer interested in the history and culture of turbulence, as background to the intense challenges that this universal phenomenon still presents.

A Voyage Through Turbulence

Edited by

PETER A. DAVIDSON
University of Cambridge

YUKIO KANEDA
Nagoya University

KEITH MOFFATT
University of Cambridge

KATEPALLI R. SREENIVASAN
New York University

CAMBRIDGE
UNIVERSITY PRESS

University Printing House, Cambridge CB2 8BS, United Kingdom

One Liberty Plaza, 20th Floor, New York, NY 10006, USA

477 Williamstown Road, Port Melbourne, VIC 3207, Australia

314-321, 3rd Floor, Plot 3, Splendor Forum, Jasola District Centre, New Delhi - 110025, India

79 Anson Road, #06-04/06, Singapore 079906

Cambridge University Press is part of the University of Cambridge.

It furthers the University's mission by disseminating knowledge in the pursuit of education, learning and research at the highest international levels of excellence.

www.cambridge.org
Information on this title: www.cambridge.org/9780521149310

© Cambridge University Press 2011

First published 2011

A catalogue record for this publication is available from the British Library

Library of Congress Cataloging in Publication data
A voyage through turbulence / [edited by] P.A. Davidson . . . [et al.].
p. cm.
Includes bibliographical references.
ISBN 978-0-521-19868-4 (hardback)
1. Turbulence. I. Davidson, P. A. (Peter Alan), 1957– II. Title.
QA913.V69 2011
532´.0527 – dc23 2011022992

ISBN 978-0-521-19868-4 Hardback
ISBN 978-0-521-14931-0 Paperback

Contents

Contributors

Brian Launder *School of Mechanical, Aerospace and Civil Engineering, University of Manchester, Manchester M13 9PL, UK*

Derek Jackson *Professor Emeritus, University of Manchester, Manchester M13 9PL, UK*

Eberhard Bodenschatz *Max Planck Institute for Dynamics and Self-Organization (MPIDS), Am Fassberg 17, 37077 Göttingen, Germany*

Michael Eckert *Forschungsinstitut, Deutsches Museum, Museumsinsel 1, 80538 München, Germany*

A. Leonard *Graduate Aerospace Laboratories, California Institute of Technology, Pasadena, CA, 91125, USA*

N. Peters *Institut für Technische Verbrennung, RWTH Aachen, Templergraben 64, 52056 Aachen, Germany*

K. R. Sreenivasan, *Courant Institute of Mathematical Sciences, and Department of Physics, New York University, NY 10012, USA*

Roberto Benzi *Dip. di Fisica, Univ. Roma Tor Vergata, via della Ricerca Scientifica 1, 00133, Roma, Italy*

Gregory Falkovich *Department of Physics of Complex Systems, Faculty of Physics, Weizmann Institute of Science, Rehovot, 76100 Israel*

Charles Meneveau *Department of Mechanical Engineering and Center for Environmental and Applied Fluid Mechanics, Johns Hopkins University, Baltimore, MD, USA*

James J. Riley *Department of Mechanical Engineering, Box 352600, University of Washington, Seattle, WA 98195, USA*

H.K. Moffatt *Department of Applied Mathematics and Theoretical Physics, University of Cambridge, Wilberforce Road, Cambridge, UK*

Ivan Marusic *Department of Mechanical Engineering, University of Melbourne, Victoria, 3010, Australia*

Timothy B. Nickels *Emmanuel College, Cambridge*

Gregory Eyink   *Department of Applied Mathematics and Statistics, The Johns Hopkins University. Baltimore, MD 21218, USA*

Uriel Frisch   *UNS, CNRS, OCA, Lab. Lagrange, B.P. 4229, 06304 Nice Cedex 4, France*

Roddam Narasimha   *Jawaharlal Nehru Centre for Advanced Scientific Research, Bangalore, 560064, India*

D. I. Pullin   *Graduate Aerospace Laboratories, California Institute of Technology, Pasadena CA 91125, USA*

Daniel I. Meiron   *Graduate Aerospace Laboratories, California Institute of Technology, Pasadena CA 91125, USA*

Preface

> I have dream'pt of bloudy turbulence, and this whole night
> hath nothing seen but shapes and forms ...
> Shakespeare (1606): *Troilus and Cressida*, V, iii, 11

"Will no-one rid me of this turbulent priest?" So, according to tradition, cried Henry II, King of England, in the year 1170, even then conveying a hint of present frustration and future trouble. The noun form 'la turbulenza' appeared in the Italian writings of that great genius Leonardo da Vinci early in the 16th century, but did not appear in the English language till somewhat later, one of its earliest appearances being in the quotation above from Shakespeare. In his "Memorials of a Tour in Scotland, 1803", William Wordsworth wrote metaphorically of the turmoil of battles of long ago: "Yon foaming flood seems motionless as ice; its dizzy turbulence eludes the eye, frozen by distance ...". Perhaps we might speak in similar terms of long-past intellectual battles concerning the phenomenon of turbulence in the scientific context.

Turbulence in fluids, or at least its scientific observation, continued to elude the eye until Osborne Reynolds in 1883 conducted his brilliant 'flow visualisation' experimental study "of the circumstances which determine whether the motion of water shall be direct or sinuous, and of the law of resistance in parallel channels". Although the existence and potential importance of 'eddying' as opposed to steady streamlined flow had been recognized previously, notably by the great 19th-century French pioneers of hydrodynamics, Barré de Saint-Venant and his follower Joseph Boussinesq, the study of turbulence as a recognizable branch of fluid mechanics may be said to date from this famous 1883 investigation of Reynolds, who correctly identified the competing roles of fluid inertia and viscosity in promoting hydrodynamic instability and the transition from smooth to irregular flow. He did not use the word 'turbulent', opting rather for the phrase 'sinuous flow'; but just four years later, William Thomson (Lord Kelvin) introduced[1] the phrase 'turbulent flow', and (in a later paper the same year) the abstraction 'turbulence', to the literature of fluid mechanics.

---

[1] 'On the propagation of laminar motion through a turbulently moving inviscid liquid', *Phil. Mag.* **24**, 342–353 (1887).

Some decades elapsed before the word gained acceptance in the scientific literature. Even in 1897, Boussinesq used the more eloquent phrase "écoulement tourbillonnant et tumultueux des liquides" within the title of a book[2] devoted essentially to the phenomenon of turbulent flow as then understood. One is reminded of the song from the 1970s of Guy Béart:

> Tourbillonnaire, tourbillonaire,
> Deux pas en avant, quatre en arrière!

which we might perhaps facetiously translate with regard to the history of the subject, and with some degree of poetic license:

> Turbulence toiler, on the rack,
> For each step forward, two steps back!

In this book, we propose to explore the development of ideas in turbulence over the 100-year period 1880–1980. We describe this as a 'voyage' through turbulence, rather than a 'history', because we make no claims to the completeness that a history would demand. Rather we invite the reader to join this voyage in the company of a group of twelve great scientists who contributed to the development of the subject over this period, during which its intense challenge and difficulty came to be increasingly appreciated. The problem of turbulence has challenged mathematicians, physicists and engineers alike, and our choice of voyagers reflects this span of disciplines:

Osborne Reynolds (1842–1912)	Scientist and Engineer
Ludwig Prandtl (1875–1953)	Aerodynamicist and Engineer
Theodore von Kármán (1881–1963)	Aerodynamicist and Engineer
Geoffrey Ingram Taylor (1886–1975)	Physicist, Applied Mathematician and Engineer
Lewis Fry Richardson (1881–1953)	Meteorologist and Mathematician
Andrej Nicolaevich Kolmogorov (1903–1987)	Mathematician and Statistician
Stanley Corrsin (1920–1986)	Fluid Dynamicist
George Keith Batchelor (1920–2000)	Fluid Dynamicist
Alan Townsend (1917–2010)	Physicist and Fluid Dynamicist
Robert Kraichnan (1928–2008)	Mathematical Physicist
Satish Dhawan (1920–2002)	Aerodynamicist and Engineer
Philip Saffman (1931–2008)	Mathematician and Fluid Dynamicist

---

[2] *Théorie de l'écoulement tourbillonnant et tumultueux des liquides dans les lits rectilignes a grande section (vol. 1)*, Gauthier–Villars, 1897.

Some among these (e.g. Prandtl, von Kármán, Taylor) have the status of great founder-figures who interacted during the inter-war years through copious correspondence as well as through the International Congresses of the period. Others (e.g. Kolmogorov, Corrsin, Batchelor, Dhawan) were pivotal figures in the development of post-war schools of turbulence, radiating outwards from their centres of activity (the Russian school, the Johns Hopkins school, the Cambridge school, and the school of the Indian Institute of Science, Bangalore, respectively). Yet others (e.g. Richardson, Townsend, Kraichnan, Saffman) were individualists, whose brilliant contributions made a profound impact upon the subject.

Many names of other departed colleagues come to mind, for whom separate chapters could well have been justified – J.M. Burgers, Kampé de Fériet, Klebanoff, S.J. Kline, Kovasznay, Laufer, Liepmann, Lighthill, Loitsianski, Monin, Obukhov, Perry, O.M. Phillips, W.C. Reynolds, Tani, Yaglom, P.Y. Zhou, . . . , to name but a few. Their contributions are referred to in chapters of this book. We beg the indulgence of the reader in the choices we have made, in the interest of providing a reasonably compact yet balanced picture[3].

Why, it may be asked, should the problem of turbulence exert such enduring fascination within the scientific community? First perhaps because it is recognized as a prototype of problems in the physical sciences exhibiting both strong nonlinearity and irreversibility, a combination of circumstances that leads to great irregularity in both space and time of the fields considered. This is also why its resolution has eluded the best minds of the 20th century. The role of vortex structures is seen as of central importance, while a statistical approach is needed to cope with the irregularity of turbulent flow at all scales. No fully satisfactory treatment combining these aspects has yet been found. The remark that "Turbulence is the most important unsolved problem of classical physics" attributed to Nobel Laureate Richard Feynman (and perhaps originating with Einstein) remains true to this day. Horace Lamb, author of the great classic treatise *Hydrodynamics*, is alleged to have said "When I meet my Creator, one of the first things I shall ask of Him is to reveal to me the solution to the problem of turbulence" (or words to that effect – see Sidney Goldstein[4]). Certainly, von Kármán repeated this sentiment at the meeting *Mécanique de la Turbulence* in Marseille (1961)! Meanwhile, Robert Kraichnan, Einstein's last postdoc, was mounting a massive theoretical attack on the problem, importing techniques from quantum field theory and developing these techniques in

---

[3] In partial mitigation, we provide in Table 13.1 a chronologically ordered table of 'events' in the history of turbulence up to the mid 1970s, with focus on the emergence of new ideas and papers of seminal importance.

[4] 'Fluid mechanics in the first half of this century', *Ann. Rev. Fluid Mech.* **1**, 1–28 (1969).

entirely original ways; nevertheless, despite his efforts, turbulence has remained impervious to purely theoretical onslaught even after the lapse of another half-century.

Second, the great span of applications of fluid mechanics has generated an ever-growing need to achieve a better fundamental understanding of the origins and effects of turbulence in practical circumstances. This need was first fuelled by the rapid development of aerodynamics in the early part of the 20th century. We tend to take air-transport for granted nowadays, but it is salutary to recall that mastery of flight, arguably the greatest engineering accomplishment of the 20th century, first required an understanding of flow in the viscous boundary layer on an aircraft wing and of the conditions leading to instability and turbulence in such boundary layers. Soon, the relevance of turbulence in meteorology and oceanography came to be recognized, here with the additional factors (sometimes complicating, sometimes simplifying!) of density stratification and Coriolis effects due to the Earth's rotation. Then at the planetary, stellar and inter-stellar levels, the relevance of turbulence for the generation and evolution of magnetic fields as observed in the cosmos came to be similarly recognized in the post-war years. And of course, turbulence remained all along of key importance in Mechanical and Chemical Engineering, in which it is the essential requirement for the effective mixing of fluid ingredients to promote chemical or combustive interactions.

The authors of the 12 chapters of this volume are all experts in various aspects of turbulence, and have detailed (and in some cases personal) knowledge of the personalities of whom they write, and of their impact on the field. Although influenced by editorial comment in some cases, the opinions expressed remain those of the authors themselves, and we, as editors of the volume, are deeply grateful to them all for the care and effort that they have devoted to their task. We hope that this volume, incomplete though it may be, will give a balanced perspective of the development of ideas and research in turbulence over what was in many ways an exceedingly turbulent century!

The original idea for this book arose during the programme on *The Nature of High Reynolds Number Turbulence* held at the Isaac Newton Institute for Mathematical Sciences, August–December 2008. We wish to express our warm thanks to the Director and the staff of the Institute for their unfailing encouragement and support, and for providing an ideal environment for the initiation of a project of this kind. By happy chance, the book will be published just before the *European Turbulence Conference (ETC13)* to be held in Warsaw in September 2011. At the suggestion of Konrad Bajer, this conference will be followed by a symposium *Turbulence – the Historical Perspective*, based on the chapters of this volume. We wish to thank Konrad for taking this most timely initiative. Finally, we wish to thank David Tranah of Cambridge University Press, who has taken a close personal interest in the work, and has steered it from initial conception all the way through to final publication; without his guidance and encouragement, we would not have been able to bring the project to completion.

Peter A. Davidson
Yukio Kaneda
Keith Moffatt
Katepalli R. Sreenivasan

1

Osborne Reynolds: a turbulent life

Brian Launder and Derek Jackson

1.1 Introduction

1.1.1 Scope

Articles on Osborne Reynolds' academic life and published works have appeared in a number of publications beginning with a remarkably perceptive anonymous obituary notice published in *Nature* within eight days of his death (on 21 February 1912) and a more extensive account written by Horace Lamb, FRS, and published by the Royal Society (Lamb, 1913) about a year later. More recent reviews have been provided by Gibson (1946), a student of Reynolds and later an academic colleague, by Allen (1970), who provided the opening article in a volume marking the passage of 100 years from Reynolds taking up his chair appointment at Manchester in 1868, and by Jackson (1995), in an issue of *Proc. Roy. Soc.* celebrating the centenary of the publication of Reynolds' 1895 paper on what we now call the Reynolds decomposition of the Navier–Stokes equations, about which more will be said later in the present chapter. A significant portion of the present account is therefore devoted to Reynolds' family and background and to hitherto unreported aspects of his character to enable his contributions as a scientist and engineer to be viewed in the context of his life as a whole. While inevitably some of what is presented here on his academic work will be known to those who have read the articles cited above, archive material held by the University of Manchester and The Royal Society and other material brought to light in the writers' personal enquiries provide new perspectives on parts of his career.

1.1.2 Family background

Osborne Reynolds came from a well-established Suffolk family with strong clerical connections (Crisp, 1911), which between 1800 and 1880 owned some

1

500 acres of land and much of the property around the small village of Debach located about 5 miles NNW of Woodbridge (White, 1844). Starting in 1779, three consecutive rectors of the parish of Debach-with-Boulge came from the Reynolds family. The Rev. Robert Reynolds was instituted on 6 September 1779 on his own petition. When he retired in September 1817 his son, the Rev. Osborne Shribb Reynolds, became Rector. Then, on *his* death in December 1848, his eldest son, the Rev. Osborne Reynolds, took over for a while. This last named, father of the main subject of this chapter, entered Cambridge University in 1832 as a fee-paying pensioner. After matriculating from Trinity College in 1833, he transferred to Queens' College from whence he graduated in 1837 as 13th Wrangler (Venn, 1954). At that point it seemed that he was destined to follow a clerical career like his father and grandfather before him for he was ordained a deacon in Ely Cathedral the following year and became a priest a year later.

The Rev. Osborne Reynolds married Jane Bryer, née Hickman, the 22-year-old widow of the late Rev. Thomas Bryer, at Hampstead Church on June 25th, 1839 (*The Times*, 29 June 1839). Their first child, Jane, was born in 1840 and, soon afterwards, they moved to Ireland where the Rev. Reynolds had obtained a position as principal of the First Belfast Collegiate College, in Donegall Place (Martins Belfast Directory, 1842–43, p. 82). Their second child, Osborne, was born on 23 August 1842 (Crisp, 1911).

It seems, however, that the Rev. Reynolds still saw a career in the church as his goal for in 1843 he returned to England with his family to take up an appointment as curate at the parish church in Chesham, Buckinghamshire. However, his tenure of this post proved to be short-lived. On February 6th, 1844, his wife died as a result of complications following the birth, three weeks earlier, of their second son Edward (*The Times*, 12 February 1844), leaving the Rev. Reynolds with the responsibility of bringing up his three small children alone. That task, allied with the financial limitations of his post as curate at Chesham, provided the incentive for him to seek an alternative position. In October 1845 he was appointed headmaster of Dedham Grammar School in Essex.

The Rev. Reynolds took up his post at the end of 1845 and held it for almost eight years (Jones, 1907). Besides carrying out his duties as headmaster he provided the personal tuition of his children, who lived with him at Dedham (1851 Census of Great Britain). It seems that he was also working on inventions, for while in post he took out the first two of the six patents that would be registered in his name (Ramsey, 1949).

In fact, the small market town of Dedham is located only some 25 miles south-west of Debach. This relative proximity meant that the Rev. Reynolds was able to keep very much in touch with his father, the Rev. Osborne Shribb

Reynolds, and with the family's farming interests in Debach (White, 1844). On visits there he occupied a farmhouse on the family estate. Moreover, when his father died in post in 1848, the Rev. Osborne Reynolds was able to take over as rector, which he did on his own petition (White, 1844), while remaining headmaster of Dedham Grammar School (Clergy List, 1850, p. 14). This arrangement lasted until May 1850 when a replacement was appointed rector by the Church of England (Crockfords Clerical Directory, 1850). Family misfortune still seemed to stalk the Rev. Reynolds, for the following year his ten-year-old daughter, Jane, died at Dedham.

In 1854, having inherited much of the land and property in Debach, he resigned as headmaster at Dedham Grammar School (or 'was persuaded to resign', as one contemporary account (Jones, 1907) seems to imply) to take on what amounted to the life of a gentleman farmer, managing the family estate in Debach which then employed some 30 staff (1861 Census of Great Britain). He lived at Debach House with his two sons whom he continued to educate, concentrating, it seems, on mathematics and mechanics. As will be seen later, his elder son Osborne warmly acknowledged his father's role in stimulating his own interest in mechanics. By the time of the next census (1871) his sons had both gone up to Cambridge University leaving Osborne Reynolds Senior free to concentrate on managing his estate and farming interests which, apparently, continued to flourish. Later, however, in the agricultural depression of 1879, it appears that Osborne Reynolds Senior encountered financial difficulties. In 1880, at the age of 66, he relinquished the estate to take up the post of rector of Rockland St Mary in Norfolk (a placement arranged under the auspices of his Cambridge College, Queens' (Crockfords Clerical Directory, 1880, p. 839)). He finally retired in 1889 and moved to Clipston, Northamptonshire, where his younger son, Edward, was then rector (Crockfords Clerical Directory, 1890, p. 1069) and he died there on June 7th the following year (*The Ipswich Journal*, 14 June 1890; *The Manchester Guardian*, 18 June 1890).

In summary, the talented but ill-fated Rev. Reynolds had an exceptionally strong influence on the formation and development of his elder son, Osborne Reynolds, who forms the subject of the remainder of this article. Not only was he directly responsible for Osborne's primary and secondary education but he stimulated in him a fascination for mechanics that was to be the bedrock of his life's work. He went on to play a major role in shaping the path and covering the not inconsiderable cost of the further five years which his son spent receiving practical training in mechanical engineering and a university education in mathematics which were so pivotal to his subsequent career. Moreover, the manner in which the Rev. Reynolds coped with the serious domestic misfortunes he faced must have provided a source of inspiration for his son

Osborne when strangely parallel developments later threatened to derail his own life.

1.1.3 Osborne Reynolds: education and first professional steps

As noted above, Osborne Reynolds Junior was born in August 1842 in Belfast where the Rev. Reynolds was briefly Principal of the Collegiate School. Despite the brevity of the family's stay there, numerous Irish history websites include Osborne Reynolds in their listings of famous Irishmen, lists that also include Sir George Stokes and Lord Kelvin, the former of whom plays a significant role later in this account.

At the age of 19, having acquired from his father not just his schooling but also a fascination for mechanics, Osborne Reynolds entered the engineering workshop of Edward Hayes of Stony Stratford, a well-known trainer (and teacher) of mechanical engineers. There, typically, a dozen 'privileged apprentices' (or, more transparently, since fees of up to £300 per annum were charged, the sons of wealthy families) would be taking their first steps in learning the rudiments of engineering manufacture, management and science (*The Engineer*, 14 September 1877, p. 183). The nature of the training is well brought out by the following:

> The pupils... who stood out at the works in their white overalls... were instructed in every aspect of the business, turning, fitting, erecting and other trades, and were trained not by the workmen, but by the Hayes foremen, and to a large extent by Edward Hayes himself. Whilst at the Works they had to conform to all the general rules and hours worked by the labour force. Most evenings were also spent with Edward Hayes being instructed in the technicalities of drawing, planning and estimating. Other subjects taught were mathematics, mechanics and natural philosophy... The pupils received... a thorough grounding in the basis of engineering theory as well as shop-floor experience, to enable them to qualify as professional engineers, who could undertake all types of design work...
>
>   (from Neil Loudon, *Boats without Water*, unpublished ms, personal communication).

Reynolds' object in taking this placement, as explained later in a testimonial provided by Mr Hayes, was "to learn in the shortest time possible how work should be done and, as far as time would admit, to be made a working Mechanic before going to Cambridge to work for Honours" (University of Manchester Archive). Indeed, in 1863 Osborne Reynolds was duly admitted as a pensioner to Queens' College at the comparatively late age of 21 to read mathematics – too old to be eligible for the recently established entrance scholarships. Moreover, his father's tuition had failed to include instruction in classical Greek on which, to his annoyance, he thus had to devote many hours of private study. On passing the requisite examination early in his second year he ceremonially demonstrated his firmness of spirit by burning all his Greek textbooks despite the pleas of his friend and fellow undergraduate at Queens', Arthur Wright, who was studying classics, to pass them on to him (Wright, 1912).

In his specialist subject of mathematics Osborne Reynolds also viewed his experience at Cambridge with some disappointment. Years later (13 October 1876) at a General Meeting of the Manchester Mechanical & Physical Society he declared:

> The mathematical education given at Cambridge, however much it might develop the power of mind of its possessors, [was] hardly calculated to forward the study which was its immediate object. Those who did attempt such a course found that they had spent several years in learning that which they had to lay aside on commencing their new work. Mathematics and the theory of mechanics, it is true, were then as now, the educational base most wanted; but these taught with a view to their application to the simpler problems of astronomy . . . were about as much use as the Latin grammar . . . for learning French.

Nevertheless, Reynolds graduated with a BA in 1867 as 7th Wrangler (Venn, 1954). However, this level of distinction did not come easily to him, as a quotation from one of his tutors will show. Thereafter, Osborne Reynolds took up employment with the well-known firm of civil engineering consultants, Lawson & Mansergh (Allen, 1970) in London. Within a few months, news reached him that Owens College, Manchester, had advertised the creation of a Chair in Civil and Mechanical Engineering. On January 18th, 1868 he duly wrote a letter of application (University of Manchester Archive) that began:

> Gentlemen, I beg leave to offer myself as a candidate for the Professorship of Engineering at Owen's [sic] College. I am in my twenty-sixth year. From my earliest recollection I have had an irresistible liking for mechanics; and the studies to which I have especially devoted my time are mechanics and the physical laws on which mechanics as a science are based. In my boyhood I had the constant guidance of my father, also a lover of mechanics and a man of no mean achievement in mathematics and their application to physics.

The hand-written opening of this letter is reproduced in Figure 1.1.

1.1.4 Chair application at Owens College

The Chair, which was initially advertised at a salary of £250 per annum (University of Manchester Archive), attracted 16 applications including those of Osborne Reynolds and a William Cawthorne Unwin who will also appear later in this account. Much has been made of Reynolds' youth and (thus) his

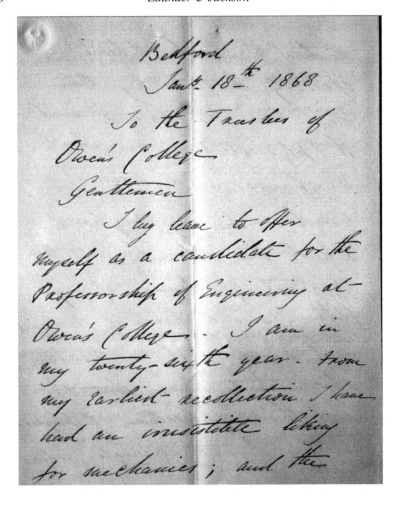

Figure 1.1 The opening page of Osborne Reynolds' letter of application for the Chair in Engineering. Reproduced with permission of the University of Manchester.

audacity in applying for the Chair and, equally, the wisdom of the appointing committee in eventually choosing him. However, Reynolds was by no means the youngest candidate: seven of the 16 were in their twenties, four of whom were younger than Reynolds. In the 1860s, a sound knowledge of mechanics and a vision of where it might lead in engineering applications must have been qualities predominantly possessed by younger candidates, much as knowledge and competency in certain aspects of software engineering are today.

W.C. Unwin must have felt that he had a good chance, having been assured of very strong support from his former employer and mentor, William Fairbairn, FRS (Allen, 1970), who was chairman of the committee of industrialists in Manchester which had raised the money to fund the Chair. The College set up an appointing committee made up of trustees of Owens College and a selection of professors. Fairbairn was not on that committee but, evidently, he must have been influential behind the scenes in guiding it with a view to ensuring that a suitable appointment was made. The procedure was for candidates to submit supporting testimonials and, in Reynolds' case, at least, the handwritten versions (of which there were 14) were complemented by a printed version of the same and of his letter of application itself (University of Manchester Archive). The testimonials included one from Mr Hayes from which the quotation above was taken, another from Archibald Sandeman, then Professor of Mathematics at Owens College but who had formerly been Reynolds' tutor at Queens' College, plus four others from Cambridge staff including one from James Clerk Maxwell, FRS, confirming Reynolds' standing in the graduation list and ending with the important observation that

> I had to examine Mr. Reynolds' papers for the Mathematical Tripos, including his solutions of many questions in mechanics and general physics; and found that he had knowledge of sound principles which will enable him in the study and teaching of engineering to exemplify the practical use of sound theoretical principles, and to show that all his practical rules are founded on general laws established by experiment.

Another referee, the Rev. W.M. Campion, BD, Fellow and Tutor of Queens' College, Cambridge, wrote:

> Mr. Reynolds is an accomplished Mathematician. But he is not a mere theorist. He possesses a considerable acquaintance with practical mechanics and engineering. For more than a year before he came to the University he studied the practice of the profession under Mr. Hayes of Stoney Stratford; and since taking his degree he has been occupied in like manner with Mr. Lawson of London. It would be difficult to find a Mathematician who combines such practical experience with theoretical knowledge.

Fulsome communications were also received from J.C. Challis, FRS, Professor of Astronomy, and the mathematics tutor, John Dunn, who commented that while on entry Reynolds had lacked knowledge in mathematics, "by innate talent and undeviating perseverance Mr Reynolds made the most rapid progress".

Perhaps most surprisingly to a 21st-century reader, his father, the Reverend Osborne Reynolds, also provided a testimonial (at the suggestion of another referee), a task which, in his words, had surprised and embarrassed him. He nevertheless praised his son's qualities and concluded: "The only point I can conceive against him is his youth – he is only in his 26th year. But this is compensated for by his early devotion to Science and the practice of his profession".

Despite the considerable number of applications, the minutes of a meeting of the Owens Committee of Trustees on 30 January 1868 reported reservations on the part of the appointing committee about the response to the advertisement. Accordingly, Mr Charles F. Beyer, a German who had come to Manchester as an impecunious young man to make his fortune (and had certainly done so!), offered to provide sufficient further funds to enable the post to be re-advertised with "an additional £250 p.a. for the first five years in the hope that the increased remuneration would enable the Trustees to obtain applications from gentlemen of higher scientific attainments and greater professional experience than could be expected under the moderate inducements held out in the earlier advertisement" (University of Manchester Archive). It is more than likely that this decision was influenced by the following sarcastic article which had appeared in the professional journal, *Engineering*, earlier that month (10 January 1868):

Technical Education

For all those who are interested in that subject of paramount national importance, upon which the future greatness of this country and its position in the civilised world are now recognised to depend – for all those who are speaking, and writing, and working for the spread of technical education in this country – we have gratifying news. The trustees of Owens College, in Manchester, are advertising for an able-bodied man-servant to act as performing professor of engineering for the rising generation in the metropolis of Manchester, at the liberal rate of wages of thirteen shillings and eight pence per day.

What a stir this grand opening will create in the scientific world! The greatest men of Great George street will close their offices and compete with each other; M. Flachat, Professor Conche, Baron Burg, and Professor Ruhlmann will leave their respective countries and professors' chairs; men like Rankine, Scott Russell, and Clausius will gather in long processions in the streets of Manchester, and vie with each other to answer the call in the newspapers. Thirteen and eight pence and a proportion of the fees paid by students (and perhaps the free loan of a sewing machine for the professor's wife to earn a little extra) are worth applying for in a country where the income of the head master at Eton is estimated at £6000, and that of an assistant master at the same school ranges from £1500 to £3500 a year.

In any event, a further 11 applications were received in response to the re-advertisement with its upgraded salary, and there had also clearly been discreet contact with Professor William Maquorn Rankine, FRS, at the University of Glasgow who, after initially showing some interest in the post, chose not to pursue the matter. Thus the Trustees decided to interview "Mr George Fuller, C.E., Associate of the Institution of Civil Engineers and Mr Osborne Reynolds, B.A., Fellow of Queens' College, Cambridge whom they believe to be the most eligible" (University of Manchester Archive). Both interviewees were drawn from the original list of applicants. Thus, with Rankine having eventually declined to become a candidate, the increased offer had served nothing other than to double the salary of the successful applicant. As the world of fluid mechanics gives thanks, the chosen candidate was Osborne Reynolds, a decision which has been described by Smith (1997) as "an inspired choice and one of the most successful gambles ever made by an appointing committee". A photograph of part of the formal terms of appointment is reproduced in Figure 1.2. As for W.C. Unwin, as soon as the Trustees' decision had been reached, his former employer wrote to him (Walker, 1938):

> My dear Unwin,
>
> I am very sorry I cannot forward to you the agreeable intelligence that you are elected to the position of professor. I so earnestly wished for you to occupy that position. It would have exactly suited your tastes, and I had every reason to believe you would have been an active and excellent professor.... In wishing you better luck in your next undertaking, I am,
>
> Yours,
>
> Wm. Fairbairn.

The commissioned biography of Unwin (Walker, 1938), as works of that kind inevitably are, was staunchly supportive of its subject. Ignoring the fact that Unwin was not one of those invited for interview, it nevertheless chose to present the chair appointment as a contest between Unwin and Reynolds that involved at least misjudgement by the interviewing committee and perhaps political intrigue to boot:

> Bearing in mind the researches on materials and on bridge design which he at that time had recently completed ... it is certainly remarkable that no better reason could be adduced by the College authorities for passing over Unwin in favour of one whose experience of civil engineering was less, and whose fame rests upon his work as a physicist rather than as an engineer. Owen's [sic] College at that date was, to a very great extent, a municipal undertaking and one cannot help thinking that, in the lively atmosphere that surrounded its early development, considerations other than academic may have played some part in the deliberations of the Senate.

It is noted for the record that none of the documents seen in the University of Manchester's archives lends any support to Walker's insinuation that "considerations other than academic may have played some part" in the decision. The

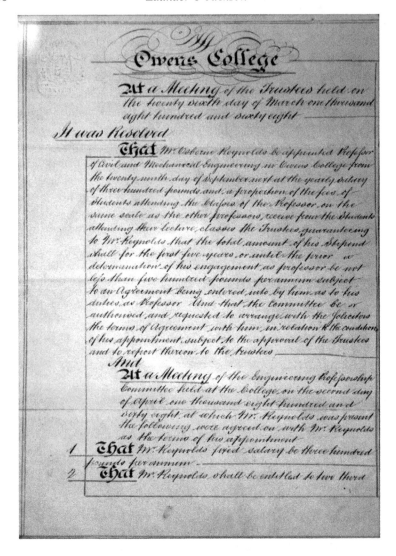

Figure 1.2 Opening portion of the terms of Reynolds' appointment as Professor of Civil & Mechanical Engineering, © The University of Manchester. Reproduced by courtesy of the University Librarian and Director, The John Rylands University Library, The University of Manchester.

creation of the Chair was the outcome of leading industrialists from Manchester recognizing the need to underpin the region's industrial strengths with a skilled and knowledgeable professional workforce. Moreover, few if any would agree with Walker's suggestion that Reynolds' subsequent "fame" (which in any event the appointing committee could hardly be expected to foresee) was due to his contributions as a physicist rather than as an engineer. As Reynolds' pioneering contributions have abundantly made clear, however, engineering breakthroughs habitually require a mind possessing deep physical insight!

1.2 Professorial career

1.2.1 Reynolds' inaugural lecture at Owens College

Osborne Reynolds gave the introductory address for the opening of the academic session 1868–69 at the start of his career in Manchester. In this lecture, entitled *The progress of engineering with respect to the social conditions of this country*, he firmly rejected any notion of engineering as an 'ivory tower' abstraction divorced from human context:

> The results, however, of the labour and invention of this century are not to be found in a network of railways, in superb bridges, in enormous guns, or in instantaneous communication. We must compare the social state of the inhabitants of the country with what it was. The change is apparent enough. The population is double what it was a century back; the people are better fed and better housed, and comforts and even luxuries that were only within the reach of the wealthy can now be obtained by all classes alike ... But with these advantages there are some drawbacks. These have in many cases assumed national importance, and it has become the province of the engineer to provide a remedy.

These remarks made at the outset of his career show the youthful Reynolds, whose contemporary portrait photograph appears in Figure 1.3, clear in his mind as to what needed to be done. Here was a figure charged with a sense of mission, full of ideas and ready to face the challenges ahead of him in serving the needs of society as Professor of Engineering at Owens College, Manchester.

1.2.2 The first decade

The year 1868 must have been one that brought Osborne Reynolds immense pleasure for, shortly after his success at the Chair interview, he married Charlotte Jemima, the eldest daughter of an eminent medical doctor in Leeds, Dr Charles Chadwick. However, the associated joy was short-lived for, after delivering their first-born child (also named Osborne) on 15 July 1869, Charlotte died 12 days later (*The Times*, 29 July 1869), leaving Reynolds with the

Figure 1.3 Osborne Reynolds, from the time of his appointment to the Owens College Chair, © University of Manchester. Reproduced by courtesy of the University Librarian and Director, The John Rylands University Library, The University of Manchester.

grief and isolation of being a widower accompanied by the responsibilities of bringing up his new-born son. The cause of death, a peritonal infection, was noted on 31 July 1869 in an article in *The Lancet* on Dr Chadwick and his bereavement. Clearly, these personal events posed a difficult beginning for Reynolds, even knowing that his father had managed to cope with a similarly challenging start to his own career. Fortunately, Reynolds' salary from Owens College was sufficient for him to be able to employ live-in domestic help to assist in running the household and caring for his son (1871 Census of Great Britain).

In November 1869, Osborne Reynolds became a member of the Manchester Literary and Philosophical Society. He read his first paper to the society in March 1870 on *The stability of a ball above a jet of water* (an interesting though rather academic fluid-dynamics problem). This marked the beginning of a close involvement on Reynolds' part with the 'Lit. & Phil.', in the course of which he contributed papers regularly, a total of 26 in all. These were mainly on scientific topics of general interest and broad appeal.

In a bid to further extend both his own and the College's contacts with the technical and industrial community of the area, Reynolds also actively involved himself with two other local societies, the Manchester Association of Employers, Foremen and Draughtsmen (a group consisting of men with technical expertise and experience, first formed in 1856) and the Manchester Scientific and Mechanical Society (formed in 1870 by William Fairbairn with the intention of linking academics with local industrialists). Between 1871 and 1874 Reynolds addressed the first of these bodies on a number of directly practical topics: *Elasticity and fracture*; *The use of high pressure steam*; and *Some properties of steel as a material for construction*. In contrast, his lectures to the Scientific and Mechanical Society, which he twice served as President, were of a more general nature, as indicated by titles such as *Future progress*, *Engineers as a profession* and *Mechanical advances*. In these ways, Osborne Reynolds set out to address the specific needs of the rather diverse sectors of his 'local social and technical constituency'.

At the time of Reynolds' appointment Owens College (which had been founded in 1851 following a generous bequest by John Owens) occupied a building on Quay Street, an early photograph of which appears as Figure 1.4, the former home of Richard Cobden, the distinguished MP for nearby Stockport. The building, now restored, is today used as chambers for barristers. In 1868 the newly formed Engineering Department was accommodated in what had been the stables at the rear of the building. In his recollections, Thomson (1936) (who had grown up in Cheetham Hill, Manchester, and enrolled as an engineering student in 1870 at the age of 14) noted that the stable itself was converted into a lecture room and the hayloft above it into a drawing office. Little was available in the way of facilities and equipment for experimental work and Reynolds had to rely on using other science laboratories in the College or performing simple experiments at home. This situation explains why his very early papers were concerned largely with explaining natural phenomena, what Thomson later termed "out-of-door physics". The work falling under this heading has been summarized in Jackson (1995) while the papers themselves all appear in Volume I of Reynolds' *Collected Works* (Reynolds, 1900). The tails of comets, the solar corona and the aurora, followed by the inductive role of the Sun on terrestrial magnetism, the electrical properties of clouds and the phenomenon of thunderstorms form some of the subjects of the early papers in this group. A further paper concerned the bursting of trees struck by lightning, the cause of which Reynolds was able to link to the rapid vaporization of moisture within a tree trunk by the sudden discharge of electricity through it. There were other papers on the destruction of sound by fog and the refraction of sound by the atmosphere. Thereafter, he tackled topics such as the

Figure 1.4 Early Photograph of Owens College. Reproduced with permission of the University of Manchester.

calming of seas (both by raindrops and by an oil film on the surface) and the formation of hailstones and snowflakes. For some of these studies he contrived simple but effective small-scale experiments using relatively unsophisticated apparatus. Even after the removal of the College in 1873 to a new purpose-built campus on a site south of the city centre (today the core of Manchester University) there was initially only limited scope for experimental work. A fuller summary of this phase of Reynolds' research appears in Jackson (1995) together with reproductions of the diagrams of the apparatus used in each case.

Lamb (1913) records in his obituary notice that for some time after Reynolds' arrival, while he was concentrating on these *out-of-door physics* problems, some shade of disappointment was felt by the eminent practical engineers and other friends of Owens College, who had worked for the creation of the professorship. This happened despite the fact, noted above, that Reynolds was at that time actively integrating with the societies that served local manufacturing and industrial management needs.

Of course, some delay in getting started on engineering research was inevitable but, as soon as it was practicable, Reynolds did begin to address a whole range of problems of a more practical nature, beginning with two experimental papers on heat transfer, one considering the effect of the presence of air in steam on the condensation rate at a cooled surface (Reynolds, 1873) and the other on the extent and action of the heating surface of steam boilers (Reynolds, 1874). In the latter he proposed and demonstrated links between pressure drop and the rate of heat transfer, establishing an analogy between the diffusion by means of turbulence of momentum and thermal energy. These studies accompanied others concerned with the efficiency of belt drives as communicators of work and the phenomenon of rolling friction.

At about that time Reynolds also initiated some work on the 'racing' of ships' screw propellers and also on the multi-staging of turbines and centrifugal pumps, the latter leading in 1875 to a detailed patent specification. Regarding this work, Gibson (1946) has pointed out that it anticipated the multi-stage Parsons turbine. Later, Reynolds went on to produce a series of important reports for the British Association for the Advancement of Science concerning the effect of propellers on the steering qualities of ships. Again, all the work referred to above is summarised in Jackson (1995) and the papers can be found in Reynolds (1900).

In the second half of the 1870s Reynolds' research in fluid mechanics underwent a shift towards more general phenomena such as the progression of dispersive surface waves in deep water and the motion of vortices, the latter made visible by coloured dye traces in water. He also made fundamental contributions in physical science concerned with forces caused by the communication of heat between a surface and a gas, explaining the operation of the Crookes light mill and constructing the experiment by means of which his ideas on this were validated. These provided valuable confirmation of the kinetic-theory model of gaseous fluids. He went on later to apply such theory to the phenomenon of thermal transpiration of gas through a porous medium (Reynolds, 1879), accompanying this with experimental work to validate his ideas. A fuller discussion of these aspects of Reynolds' work can again be found in Jackson (1995) (see also Allen, 1970 and Jackson, 2010).

In recognition of the wide-ranging contributions to both engineering and science which Osborne Reynolds had made within the decade following his appointment, he was admitted in 1877 to Fellowship of the Royal Society of London.

Throughout this very busy and productive period of his career at Owens College, Reynolds remained a devoted and involved father, as is well illustrated by a letter (Reynolds Society Archive Ref. RS RR.8.191) which he wrote on

22 July 1879 to George Stokes (at the time editor of *Phil. Trans. Roy. Soc.*) from St Leonards-on-Sea. The letter mainly concerned referees' comments on the manuscript of his paper 'On certain dimensional properties of matter in the gaseous state' (Jackson, 2010), but he also commented:

> I am here nursing my only child who is very ill and I do not like leaving him even for a day or I would try to see you and save you some of this writing. I expect to be here all the Summer.

In fact, despite this love and attention, Reynolds' son died on 27 September 1879 while they were still at St Leonards-on-Sea. (*Yorkshire Post*, 30 September 1879). This personal tragedy that again, curiously, paralleled events in his own father's life, may be said to mark the end of the first phase of Osborne Reynolds' professorial career.

1.2.3 New beginnings

Two years later, on 29 December 1881, Osborne Reynolds married Annie Charlotte, daughter of the Rev. Henry Wilkinson, Rector of Otley in Suffolk. No account is available of how the couple met but it is not all surprising that they did because Otley was the parish next to Debach-with-Boulge where, as already noted, his father had continued to live until 1880 after relinquishing the headmastership of Dedham Grammar School. The wedding took place in Otley church with the bride being given away by her father while the marriage service was conducted by the bridegroom's father, the Rev. Osborne Reynolds, then recently appointed as rector of Rockland St Mary, Norfolk (*The Ipswich Journal* Saturday, 31 December 1881). Reynolds' colleague at Owens College and close friend, Professor Arthur Schuster, FRS, acted as his best man.

Annie was born in December 1859 so there was an age difference of 17 years between them. The couple lived in the large red-brick semi-detached house, No. 23 Ladybarn Road, Fallowfield, which Reynolds had occupied since first taking up his chair at Manchester and which was to be to be their home for most of the remainder of their time in Manchester.

Their marriage appears to have been a successful one with three sons and a daughter resulting from the union. Reynolds' habitual style of working has been lovingly recounted by his daughter, Margaret, in a letter to a relative (University of Manchester Archive). Although it relates to a time in the late 1890s (for she was born only in 1885) it seems unlikely that these habits would have greatly changed with time:

> My father was a great worker. He breakfasted at 8 am, left for the College at 8:30 by tram, worked in the laboratory until 1pm, lunched at a cafe that catered for

the professors until 2 then lectured to students in the afternoon and walked the 2 miles home at 4pm. I sometimes came out of town by tram and joined him. He carried his stick or umbrella at the slope over a shoulder and he *never* knew I was there. He was far away, and I had to make myself known to take a short cut when we were nearly home. Then he worked again all the evening, after reading the [news]papers before dinner, until 2 or 3am. Every day alike, except Saturday afternoon and Sunday, when he went for long walks with his friends.

1.2.4 The turbulent flow papers

With his personal life having undergone such a satisfying transformation and, at a practical level, with Annie available to take charge of household management, assisted by a number of servants (1891 Census of Great Britain), Reynolds directed his efforts to a variety of further research preoccupations. In view of the subject of this book, the present account (which draws extensively from an earlier paper by the authors: Jackson and Launder, 2007) focuses particularly on his two major papers on turbulent flows (Reynolds, 1883; Reynolds, 1895), although this was by no means the only subject on which he was working at that time. While our principal attention will be focused on the later analytical study, Reynolds (1895), we first consider the experimental investigation, Reynolds (1883), as the discoveries in that paper shaped the later publication and, moreover, had made a significant impression on one referee who was called upon to review each of the papers.

The 1883 paper Most readers will be familiar with Reynolds' study of (to quote from the title of his *Phil. Trans. Roy. Soc.* paper) 'the circumstances which determine whether the motion of water shall be direct or sinuous'. However, as noted above, some nine years before that paper appeared, he made a presentation to the Manchester Literary and Philosophical Society, 'On the extent and action of the heating surfaces of steam boilers' (which is reprinted as Paper 14 in Vol. 1 of his *Collected Works*; Reynolds, 1900) in which he clearly signalled his awareness of the importance of turbulent eddies:

> The heat carried off by air, or any fluid, from a surface, apart from the effect of radiation, is proportional to the internal diffusion of the fluid at and near the surface. Now, the rate of this diffusion has been shown to depend on two things:– The natural internal [molecular] diffusion of the fluid [and] eddies caused by visible motion which mixes the fluid up and continually brings fresh particles into contact with the surface. The first of these [mechanisms], molecular diffusion, is independent of the velocity of the fluid and may be said to depend only on the nature of the fluid. The second, the effect of eddies, arises entirely from the motion of the fluid, and is proportional both to the density of the fluid and the velocity with which it flows past the surface.

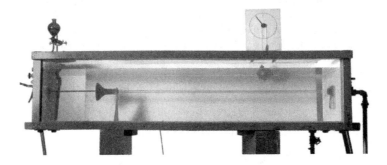

Figure 1.5 The Osborne Reynolds tank for visualizing distinctions between laminar and turbulent flow. © The University of Manchester.

In a paper read to the British Association in 1880, 'On the effect of oil in destroying waves on the surface of water', Reynolds associated that phenomenon with the production of turbulent eddies in the water below the surface generated as a result of the movement of the oil film under the action of the wind which caused a shear flow to be produced in the water (see Paper 38 of Vol. I of his *Collected Works*; Reynolds, 1900). As he related much later in a letter to his colleague, Horace Lamb (University of Manchester Archive; also reproduced in Allen, 1970) on conducting an experiment "on a windy day at Mr. Grundy's pond in Fallowfield", he observed that by throwing a small quantity of oil onto the surface, instead of waves being formed, eddies were produced in the water beneath the oil film ("which took on the appearance of plate glass"). This acute observation evidently provided him with his early insight into turbulence production in wall shear flows.

Returning now to the 1883 paper, the original print of the so-called Reynolds tank experiment has been reproduced in numerous articles and text books so, instead, Figure 1.5 shows a photograph of the apparatus as it is today at the University of Manchester. The glass tube with a flared entry which is itself housed within a tank filled with water is still used to provide students with a very clear indication of the starkly contrasting states of motion, whether 'direct' or 'sinuous' (or, in today's terminology, laminar or turbulent). In Reynolds' own words:

The internal motion of water assumes one or other of two broadly distinguishable forms – either the elements of the fluid follow one another along lines of motion which lead in the most direct manner to their destination, or they eddy about in sinuous paths the most indirect possible.

The existence of these two modes of fluid flow was, of course, already widely known. The first description of the transition between laminar and turbulent flow in a pipe had been provided by Hagen (1854) who used sawdust as a means of flow visualization (later, he recommended the use of filings of dark amber; Hagen, 1869). Reynolds' dye-streak studies and other data were, however, the first to show that, for a range of flow velocities, pipe diameters and viscosities, transition from the former mode to the latter occurred for roughly the same value of the unifying dimensionless parameter which today bears his name, the Reynolds number.

The first step in Reynolds' discovery of this parameter appears to have been his observation that 'the tendency of water to eddy becomes much greater as the temperature rises'. It occurred to him that this might be related to the fact that the viscosity of water decreased as the temperature rose. By examining the governing equations of motion he concluded that the forces involved were of two distinct types, inertial and viscous, and, further, that the ratio of these terms was related to the product of the mean velocity of the flow and the tube diameter divided by the kinematic viscosity. In his paper he states:

> This is a definite relation of the exact kind for which I was in search. Of course without integration the equations only gave the relation without showing at all in what way the motion might depend upon it. It seemed, however, to be certain, if the eddies were due to one particular cause, that integration would show the birth of eddies to depend on some definite value of [that group of variables].

Reynolds' earlier 1879 paper 'On certain dimensional properties of matter in the gaseous state' (one of his longest and most original) had a direct impact on his discovery of the Reynolds number, as he acknowledged in his 1883 paper where he observed that: "no idea of dimensional properties, as indicated by the dependence of the character of the motion on the size of the tube and the velocity of the fluid, occurred to me until after the completion of my investigation on the transpiration of gases".

In his 1883 paper Reynolds noted that kinematic viscosity "is a quantity of the nature of the product of a distance and a velocity" which, from his earlier work on the kinetic theory of gases, he was able to associate with molecular velocity and mean free path. Thus he saw the dimensionless grouping Ud/ν (where U is the average velocity, d the pipe diameter and ν the kinematic viscosity) as the ratio of characteristic velocity times characteristic length on the macroscopic and microscopic scales or the ratio of turbulent and molecular diffusivities of momentum. This provided him with interesting alternative physical interpretations of that unifying parameter.

Of course Reynolds recognized that the critical value arrived at in his experiments with the tank (sometimes called the 'higher critical number') was not unique and could be affected strongly by the level of background disturbances present. In a second series of experiments, with different apparatus, he thus set about determining the value of Reynolds number below which highly turbulent motion created at entry to the pipe would decay with distance, with the flow eventually becoming laminar. In this case, he made pressure drop measurements along a pipe for a range of velocities to delineate the mode of flow. Although in that paper Reynolds never cited the actual values, Allen (1970) concluded from the figures that he did quote that, for the two lead pipes used in this second set of experiments, the 'lower critical number' was 2010 and 2060 while in his later paper Reynolds (1895) put the critical value between 1900 and 2000.[1]

The two referees of the manuscript that Reynolds submitted to the Royal Society in 1883 were the considerable figures of Sir George Stokes and Lord Rayleigh, each of whom was broadly supportive of publication. Stokes was a pioneer in the use of the typewriter though it appears that the machine he used for his review only had available upper-case letters (see Figure 1.6), and that he found the process of typing sufficiently laborious that, rather than re-typing a final version, he chose to insert by hand his subsequent embellishments and corrections (though he failed to notice CHASS midway through).

Lord Rayleigh's review dated 30 March 1883 (Royal Society Archives Ref. 183) was spread over three pages but amounted to only 70 words. The first sentences gave his lofty, rather patronizing observation and verdict: "This paper records some well contrived experiments on a subject which has long needed investigation – the transition between the laws of flow in capillary tubes and in tubes of large diameter as employed in Engineering. I am of opinion that the results are important, and that the paper should be published in the *Phil. Trans*." It then concluded: "In several passages the Author refers to theoretical investigation whose nature is not sufficiently indicated. Rayleigh" The paper was duly published and, in the years that followed, each of the referees publicly signalled the exceptional importance of Reynolds' findings. First, Lord Rayleigh, in his 1884 Presidential Address to the British Association in Montreal, paid the following tribute:

---

[1] More than a century after Reynolds' experiments there has been considerable renewed interest in the mechanism of transition in pipe flow (Eckhardt, 2009; Fitzgerald, 2004) with the aim, *inter alia*, of acquiring a more complete understanding of the development of non-linear, finite instabilities leading to transition (for pipe flow is known to be stable to *infinitesimal* disturbances at all Reynolds numbers). Manchester is again making important contributions to this research through the work of Mullin and colleagues (Hof et al., 2003; Mullin, 2011) who have built an apparatus where the pipe length is 785 times the 20mm diameter and in which, importantly, the flow rate through it can be set as precisely constant in any experiment (rather than where, as in Reynolds' and later experiments, the head loss or pressure drop is fixed) thus keeping the Reynolds number truly constant.

REPORT ON PROF. O. REYNOLDS'S PAPER.

I CONSIDER PROFESSOR REYNOLDS'S PAPER A VALUABLE ONE, WHICH I RECOMMEND SHOULD BE PRINTED IN THE PHIL. TRANS. HE SHOWS FOR THE FIRST TIME THAT THE DISTINCTION BETWEEN REGULAR AND EDDYING MOTION DEPENDS ON A RELATION BETWEEN THE DIMENSIONS OF SPACE AND VELOCITY, OR WHAT COMES TO THE SAME OF SPACE AND TIME , INVOLVED IN THE EXPERIMENTS; A ~~DISTINCT~~ *relat*ION POINTED OUT BY THE KNOWN EQUATIONS OF MOTION OF A VISCOUS FLUID. HE SHOWS ALSO THAT THE ONE CLASS OF MOTIONS PASSES INTO THE OTHER WITH AN UNEXPECTED SUDDENNESS.

IN ONE PART THE LANGUAGE SEEMS TO IMPLY *(which was not perhaps intended)* THAT HE HAD DISCO-VERED NEW DIMENSIONAL PROPERTIES OF FLUIDS, AND MIGHT EVEN LEAD TO THE SUPPOSITION THAT HE SUPPOSED THAT HE HAD SHOWN *that* ANOTHER CONSTANT BEYOND THOSE RECOGNISED WAS NECESSARY IN ORDER TO DEFINE A FLUID MECHANICALLY. THIS CERTAINLY IS NOT THE CASE; THE DIMENSIONAL PROPERTIES ARE ALREADY *obviously* INVOLVED IN THE EQUATIONS OF MOTION; AND THERE IS ABSOLUTELY NOTHING TO PROVE THAT HE HAS DISCOVERED THE NECESSITY OF AN ADDITIONAL CONSTANT TO DEFINE A FLUID.

G. G. Stokes

19 April 1883

Figure 1.6 Sir George Stokes' review of Reynolds' 1883 paper © The Royal Society, reproduced with permission.

Professor Reynolds has traced with much success the passage from one state of things to the other, and has proved the applicability under these complicated conditions of the general laws of dynamic similarity as adapted to viscous fluids by Professor Stokes. In spite of the difficulties which beset both the theoretical and experimental treatment, we may hope to attain before long to a better understanding of a subject which is certainly second to none in scientific as well as practical interest.

Sir George Stokes served as President of the Royal Society from 1885 to 1890 and in this capacity, in November 1888, he presented the Society's Royal Medal to Osborne Reynolds "for his investigations in mathematical and experimental physics, and on the application of scientific theory to engineering". More than half of Stokes' citation was devoted to a summary of the 1883 paper.

Although the physical significance of the dimensionless parameter we know as the Reynolds number had thus quickly become widely recognized in Britain, it was only some years after Reynolds' retirement that his own name became attached to it through the publications of various German workers. Rott (1990) cites Sommerfeld (1908) as being the first to link Reynolds' name with the parameter, an attribution followed shortly thereafter by Prandtl (1910) in his early paper on the Reynolds analogy, while later, in an encyclopaedia entry on fluid motion, Prandtl (1913) unequivocally announces "The forementioned quantity, a dimensionless number, is named after the discoverer of this similarity, Osborne Reynolds, [and is called] the Reynolds number".

A passing fancy Besides the immediate acclaim accorded Reynolds' 1883 paper, the recent agreeable developments in his personal life (his marriage to Annie, and the safe delivery of Henry Osborne, the first of their four children), not to mention his admission in 1882 as an Honorary Fellow of Queens' College, Cambridge, would, one might have supposed, have had the effect of suppressing the desire, on his part, for bringing about any major upheaval in his life.

However, the Livery Companies of the City of London and the City Corporation, concerned at the limited provision in the capital of facilities in engineering, formed the City & Guilds of London Institute which secured a site in South Kensington where the Central Institution of the Institute was built (known from 1910 as the City & Guilds College, one of the constituent colleges of Imperial College). When the building was nearing completion, at the beginning of 1884, steps were taken to appoint key staff. Reynolds decided to apply for the advertised Chair in Civil & Mechanical Engineering and, not surprisingly, made the short-list along with W.C. Unwin, who following his disappointment in Manchester had, in 1872, been appointed to a chair at Cooper's Hill College, and A.B. Kennedy, who had been appointed Professor of Engineering at University College, London, in 1874 (Walker, 1938). On this occasion, however, it was Unwin who proved to be the successful candidate, the reverse of what had happened in the case of the Owens College appointment.

It is worthwhile pausing to reflect on the likely consequences for fluid mechanics if, instead, Reynolds had been appointed to the Chair. The building was new but unoccupied and presumably unequipped (since the professors would have been responsible for choosing the equipment for their laboratories). The first students were admitted in February 1885 from which time Unwin was appointed Dean of the Central Institution, with all the associated administrative responsibilities, on top of the task of teaching in his own department without, initially, any demonstrators or assistants (Walker, 1938). Thus, it seems at least questionable whether, had Reynolds been chosen for that position, his major remaining works on fluid mechanics would ever have been written, at least in the form we know them. The papers that would have been placed in jeopardy included not only his follow-up to the 1883 paper to which we shall shortly turn but also his very important paper on film-lubrication (Reynolds, 1886). Of that Lord Rayleigh, 32 years after its publication, felt able to remark "it includes most of what is now known on the subject" and in celebration of which a centennial international conference was held in 1986 (Dowson et al., 1987).

Reynolds' disappointment at failing to secure the chair in London must have been assuaged that summer, by the conferment on him of an honorary degree by the University of Glasgow, where W.J.M. Rankine had formerly been a professor and where the Thomson brothers (James, Rankine's successor, and Sir William [later Lord Kelvin]) then served. Whether this last distinction had any bearing on Reynolds' subsequent action is unknown but, later in 1884, he applied for the vacant Cavendish Professorship of Experimental Physics at Cambridge. Despite Reynolds' numerous distinctions, however, the appointment went to his former student, J.J. Thomson (then a young man of 27 working at Cambridge, later to become Sir Joseph Thomson, OM, PRS, Nobel laureate and discoverer of the electron). Although it has already been quoted (Gibson, 1946; Allen, 1970), it is worthwhile repeating part of Reynolds' generous letter of congratulations sent on Boxing Day, 1884:

> My dear Thomson,
> I do not like to let the occasion pass without offering you my congratulations, which are none the less sincere that we could not both hold the chair. Your election is in itself a matter of great pleasure and pride for me ... and I have no doubt but every hope that you will amply justify the wisdom of the election.
> Believe me yours sincerely
> Osborne Reynolds

Thus, Osborne Reynolds remained at Manchester. But what had provoked this desire to leave? Could his wife have applied pressure for them to move to a more attractive urban environment? This seems unlikely given that she had become settled in Manchester and, as a Victorian woman barely in her mid-twenties, would surely have deferred to the wishes of her husband on all things relating to his professional life. It seems more likely that the decisions were Reynolds' alone, perhaps feeling frustrated that, after 16 years in post, he

Figure 1.7 Osborne Reynolds *c*. 1895. © The University of Manchester. Reproduced by courtesy of the University Librarian and Director, The John Rylands University Library, The University of Manchester.

still did not have at his disposal what he considered to be adequate laboratory facilities. Indeed, Thompson (1886; as reported by Allen, 1970) notes that in that year (1884) Reynolds drew the attention of the University's Council to the urgent need for an engineering laboratory. It seems that, finally, this complaint may well have led in 1887 to the overdue provision of state-of-the-art laboratories (Gibson, 1946).

The 1895 paper

As the preceding section has indicated, in the years following publication of the 1883 paper, the unresolved questions stimulated by the work reported therein by no means fully occupied Reynolds' mind. Indeed, as the list of contents of Volume 2 of his *Collected Works* will confirm, he was actively exploring a number of other research interests. Perhaps for that reason, only in 1894 did he feel ready to respond to Lord Rayleigh's expressions of hope for progress on the theory. On 24 May he reported orally the results of his extensive analysis to the Royal Society and thereafter submitted a written version of this work that he had had printed at his own expense to be reviewed for publication in the *Phil. Trans.* By then Reynolds, whose contemporaneous photograph appears in Figure 1.7, had held his Chair for more than 25 years, had been a Fellow of the Royal Society for more than 15 and, as noted above, had received major awards. He was then unquestionably the leading engineering fluid mechanicist in England and quite possibly more widely than that.

Lord Rayleigh had meanwhile become Editor of the *Philosophical Transactions of the Royal Society*. Perhaps inevitably, on receiving this second manuscript on turbulent flow from Reynolds, he sent it for review by Sir George Stokes. This time, however, the referee's response was rather unsatisfactory. After a long period of silence, on 31 October 1894 Sir George, now equipped with a typewriter with both upper- and lower-case letters, sent his reply (Figure 1.8), effectively acknowledging that he did not understand the work. The letter is a copy-book example of the 'on-the-one-hand ... yet-on-the-other' style of review: Reynolds hadn't made his case – yet, he was an able man and the 1883 paper was sound; moreover the author had paid to have the present paper printed so obviously *he* thought it was important. However, the reviewer couldn't confirm that view ... but neither would he assert that it was wrong!

Stokes' concluding sentence seems to imply that he had finished with the matter, but Lord Rayleigh evidently had other ideas. Although the exchanges are incomplete it seems that Rayleigh pressed Stokes to go further and, when Stokes pleaded that he had mislaid the copy of the paper, he arranged for him to be sent another copy. (Since the paper had been printed, Reynolds had evidently submitted several copies.) On 5 December 1894, Sir George sent this second copy back indicating that he had now found the copy originally sent to him. He added his regrets that he was "not yet able to go beyond the rough indication contained in a letter sent to Lord Rayleigh some time ago" (Royal Society Archive Ref. 209 from Sir G.G. Stokes to Mr Rix).

Meanwhile, Lord Rayleigh had sent the paper to a second referee, Horace Lamb, Professor of Mathematics at Manchester, who a decade earlier had been elected a Fellow of the Royal Society. One can only speculate why Rayleigh approached the only other senior fluid mechanicist in Manchester to review his own colleague's work. Nevertheless, on 21 November 1894 Lamb sent his longhand assessment which began with the brisk summarizing statement:

> I think the paper should be published in the *Transactions* as containing the views of its author on a subject which he has to a great extent created, although much of it is obscure and there are some fundamental points which are not clearly established.

There followed three pages of detailed criticism including complaints at the inadequate definition of Reynolds' term 'mean–mean motion' and a misprint in the manuscript (Royal Society Archive Ref. 208).

Lensfield Cottage, Cambridge, 31 Oct. 1894.

Dear Lord Rayleigh,

   I must plead guilty to not having digested Professor Osborne
Reynolds's paper, though much time has passed since it was referred to me.

   I find it very difficult to make out what the author's notions
are. As far as I can conjecture his meaning, I must say I do
not think he has made out his point. He is however an able man,
and in his former paper did very good work in showing that the
conditions of dynamical similarity which follow from the dimensions of the hydrodynamical equations when viscosity is taken into
account are not confined to what I may call regular motions, but
continue to apply (in relation to mean effects) even when the motio
n is of that irregular kind which constituted eddies, and which
at first sight appears to defy mathematical treatment. The fact
that the author has gone to the expense of printing the paper
shows that he himself considers it as of much importance. I
confess I am not prepared to endorse that opinion myself, but
neither can I say that it may not be true.

   I do not know whether these remarks will be of any use in
assisting the Council to come to a decision.

     Yours very truly,

       G. G. Stokes

Figure 1.8 Sir George Stokes' initial review of the 1895 paper, © The Royal
Society, reproduced with permission.

   The Royal Society holds three further communications from the referees of
which only one is dated. There is thus some doubt as to the actual sequencing
though the most probable seems to be the following. At some point Sir George
Stokes does send his review to Lord Rayleigh, a two-page typed assessment
raising some of the problems with the paper he and, indeed, Lamb had aired
earlier. Thereafter (or, possibly, even before that communication), the referees had made contact with one another, presumably through the intervention
of Lord Rayleigh, which led Lamb to prepare a joint report that Sir George
attached to his letter of 30 January 1895 (Royal Society Archive Ref. 210):

Dear Lord Rayleigh,

I enclose what Lamb meant for a draft of remarks to be submitted to the author. I think we are both disposed to say let the paper be printed, but first let some remarks be submitted to the author. There was very good work in the former paper, and there *may* be something of importance in this, but the paper is very obscure. In its present state it would hardly be understood.

Yous very truly,

G.G. Stokes

The 'draft of remarks', in Lamb's handwriting followed:

Prof Reynolds' Paper

The referees have found great difficulty in following the argument of this paper; partly in consequence of the fact that such terms as 'mean-mean motion' and 'relative mean motion' are used without any precise definition. There is a well-known distinction between molecular and molar motion; but it is not clear in the case of molar motion how any physical distinction is to be drawn between what is 'mean' and what is 'relative'.

The introduction might be greatly shortened, as a good deal of it can only be understood after reading the rest of the paper. The purport of §5(a) p. 3 is not evident. The author's view does not appear to be different from that generally held, but it is insisted upon as something new.

The statement, in §5(b), that the ordinary equations of a viscous fluid are true only when the motion is approximately steady, is questionable. It is perhaps based on the investigation on p. 9; but this is purely mathematical; and there is besides a difficulty in seeing the connection between equations (7) and (8A). It would seem as if there had been a slip in writing u for $\overline{u}$; but at any rate there is need of explanation. It is to be noted that the argument, if valid, would show that there are *geometrical* difficulties in the way of applying the idea of mean velocity to cases other than steady homogeneous motion.

The essence of the paper lies in the equations on pp. 15,16[2]. If these are clearly established a great point would be secured, but its reasoning is somewhat obscure, and needs much amplification. The conception of 'mean-mean-motion' is a very delicate one and it is not made evident in what sense $\overline{u}$, $\overline{v}$, $\overline{w}$ are continuous functions, or on what conditions the derivatives, etc. are supposed to be formed. The whole argument turns on questions of this kind, and it is just here that explanations are wanting.

A margin instruction pencilled on the review in Rayleigh's hand, indicated that the report was to be copied (meaning that a clerk was to transcribe the review) presumably for onward transmission to Osborne Reynolds. Thus, the review phase of the 1895 manuscript was brought to completion. Figure 1.9 provides a photograph of each of the participants in this process.

[2] Taking account of the four-page insert made by Reynolds in the published version, the reference here is to Equations (13)–(19) of the published work.

Figure 1.9 Contemporary images of the distinguished fluid mechanicists involved with the review of Reynolds' 1895 manuscript. From left: Lord Rayleigh, Sir George Stokes, Horace Lamb.

On receiving the referees' assessment, Reynolds evidently reflected on the criticisms and on 19 February 1895 sent the following reply:

Dear Lord Rayleigh,

From the copy of the remarks on my paper on the criterion, which you sent me, it is clear that the referees have found great difficulty in understanding the drift of the main argument; namely that which relates to the geometrical separation of the components $\underline{u}$, $\underline{v}$, $\underline{w}$ at each point of a system into mean-components $\overline{u}$, $\overline{v}$, $\overline{w}$, and relative components $\underline{u}'$, $\underline{v}'$, $\underline{w}'$ and as to the conditions of distribution of $\overline{u}$, $\overline{v}$, $\overline{w}$ under which such separation is possible.

I am very glad to know of these difficulties and of the opportunity it afforded me of improving the paper in this particular. As it is by such separation of the simultaneous component of velocity at each point, introduced into the equations of viscous fluid, that the evidence of a geometrical limit to the criterion appears independently of all physical considerations, any want of clearness on this point, no doubt, confuses the whole argument.

That I should have scamped the preliminary explanation of this part of the argument and diffused it over the whole paper I can only explain as a consequence of its definite character having blinded me to the difficulties which would thereby result in distinguishing what was new from what was already accepted, and of my desire to set forth the proof of the actual maintenance of the geometrical conditions under which such separation is possible afforded by experiment, as well as to indicate the general character of the mechanical-actions, expressed in the equations of motion, on which such maintenance depends.

This head-reeling sentence, 100 words in length, is also remarkable for its naturalness; its innocent admission of the paper's weaknesses accompanied by its ready self-forgiveness. The letter then continues:

I now enclose you in M.S.S. a full preliminary description of this part of the argument which by permission I shall be glad to substitute for the first two lines of §5 p. 3. It contains, what I hope will be found, a clear definition of the terms mean-mean motion and relative-mean motion as well as of mean-motion and heat-motions and of the geometrical distinctions between these motions. And although no physical-distinction between mean-molar and relative-molar is draw[n] other than what is implied by the geometrical distinction that the integrals of $\rho\bar{u}$, etc, taken over the space determined by the scale or period-in-space of the relative mean motion $\rho u'$, etc, are the components of momentum of the molar motion of the mechanical system within S while the integrals of $\rho\underline{u}$, etc, taken over the same space are zero, it is shown that such physical distinction has no place in the argument any further than it is suppressed by the terms in the equations of motion.

This passage, like those cited earlier, brings out Reynolds' infatuation with long rambling sentences that stand starkly in contrast to Lamb's crisply stated criticisms. He finally acknowledges:

> With reference to the difficulties in logic of §8 p. 9, equations 7 and 8a, this is intirely removed by replacing the bar ($\bar{u}$) which has dropped from the $\underline{u}$ in the left of equation 4, p. 8.
>
> There are, I am sorry to say, certain other misprints in the paper which must have increased the inherent difficulties of the subject.
>
> Very truly yours,
>
> Osborne Reynolds

Apparently, no further exchanges between author and editor remain in existence and, since there is no copy of the original manuscript, it is not certain how extensive were the changes actually made. One clear indication of a change in the published version of the paper is that four pages of §5 of the Introduction are placed, entirely without explanation, within square parentheses and end with the date: Feb 18, 1895 (that is, the day before Reynolds sent his response). Thus, this passage clearly seems to be what Reynolds referred to in his reply to Rayleigh as "the full preliminary description of this part of the argument which by permission I shall be glad to substitute for the first two lines of §5. p. 3." Since this was the only significant change referred to by Reynolds it appears likely that all other changes were minor, mainly consisting of corrections to typographical errors in the original.

As a footnote to the above exchanges, within a few weeks of Reynolds responding to the (anonymous) Lamb–Stokes review, Horace Lamb was directly in touch with Reynolds seeking to check that he had correctly interpreted material in the 1883 paper when referring to it in his book *Hydrodynamics* which was about to be published. The enquiry elicited a distinctly cool response from Reynolds in a handwritten letter to Lamb dated 5 April 1895 (Allen, 1970):

"You have not noticed that just above the critical velocity the resistance dp/dz varies nearly as the cube of the velocity until dp/dz is about double what it is at the critical velocity. Of course these dimensional facts . . . are the definite clues to the physics and mechanics of the problem and the gist of the later research. Nor is it polite or true to speak of the 'empirical formulation adopted by Engineers' since it is Engineers who have done the scientific investigations which alone have given us accurate data." Readers may form their own opinion as to whether Reynolds' acerbic response was provoked by a perception that Lamb had been one of the referees of his 1895 paper[3].

Despite its rather lukewarm reception by the two eminent referees, the 1895 paper is seen today as a mighty beacon in the literature of fluid mechanics. First and foremost was the decomposition of the flow into mean and fluctuating parts leading to the averaged momentum equations (now known as the *Reynolds equations*) in which the *Reynolds stresses* appear as unknowns. In fact, throughout the analysis Reynolds treated the averaging in a form akin to what is now known as mass-weighted averaging, 60 years earlier than the source that is usually quoted for introducing that strategy. It was surely just that his experiments had used water as the fluid medium that has led to this feature being ignored. The paper's other major analytical result was the turbulent kinetic energy equation in which he observed that the terms comprising products of Reynolds stress and mean velocity gradient represented a transfer of kinetic energy from the mean flow to turbulence. As an indicator of just how far this discovery was ahead of its time, we note that the corresponding, albeit simpler, equation for the mean square temperature fluctuations was not published until the 1950s (Corrsin, 1952).

Reynolds' purpose in examining the turbulent kinetic energy equation was to provide an explanation of why the changeover from laminar to turbulent motion should occur at a particular value of the Reynolds number. Indeed, that was the driving rationale for the whole paper. For this purpose he considered fully developed laminar flow between parallel planes on which a small analytical disturbance was superimposed which permitted him to obtain expressions for the turbulence energy generation and viscous dissipation rates integrated over the channel. The relative magnitude of these two processes varied with Reynolds number and the lower critical Reynolds number he identified as being that where the overall turbulence energy generation rate had grown to balance the viscous dissipation rate. That his estimates were inaccurate is now seen as irrelevant since the paper contained more than enough novelty for the world of fluid mechanics to absorb over the ensuing decades.

[3] In response to Reynolds' criticism, Lamb replaced 'empirical formula adopted by Engineers' in the published book by 'practical formula adopted by writers on hydraulics'.

1.3 End piece

1.3.1 The final years

Publication of the second of his major works on turbulent flow did not mark the end of Reynolds' creative outpourings. In 1897 he gave the Bakerian Lecture to the Royal Society (Reynolds and Moorby, 1897) reporting measurements of the mechanical equivalent of heat. Of this huge experimental programme in which he obtained the equivalence within 0.2% of modern determinations, Lamb (1913) has written "The whole investigation is a model of scientific method and may claim to rank among the classical determinations of physical constants", an assessment that Gibson (1946) felt able to repeat more than 30 years later.

Osborne Reynolds' final years in Manchester were marked by his intense efforts to provide a mechanical theory of matter and the ether stemming from ideas contained in two earlier papers. This culminated in his work *The Sub-Mechanics of the Universe* being reported orally as the Royal Society's Reid Lecture in 1902 and published as Volume 3 of his *Collected Works* (Reynolds, 1903).

However, in his obituary notice Lamb (1913) remarks, in what must be seen as a kind understatement, "unfortunately illness had begun gravely to impair his powers of expression and the memoir as it stands is affected with omissions and discontinuities which make it unusually difficult to follow". Gibson (1946) has noted that 1903 was the last year in which Reynolds was able to take an active role in the department, his declining state (a condition that today might have been diagnosed as Alzheimer's disease) leading to his retirement from the University at the age of 63 in 1905. A photograph showing part of the fine retirement portrait of Reynolds by the distinguished portraitist, John Collier, OBE, appears in Figure 1.10.

In 1908, Osborne Reynolds left Manchester with his wife and daughter to live at the vicarage in St Decuman's, a hamlet on the hill above Watchet, a small though not insignificant historical port in north-west Somerset. The church and the vicarage are shown in Figure 1.11 in a photograph from *c.* 1900 (Wedlake, 1984). Why Reynolds or, perhaps more accurately, given the prevailing circumstances, his wife should have chosen Watchet as their retirement base is unknown though the fact that both their fathers had been clergymen probably provided the essential contacts for them to have been able to rent the vicarage.

The 1911 Census of Great Britain discloses that the return for the Reynolds household was completed by Annie on behalf of Osborne and that, numbered among the residents, in addition to their daughter, Margaret Charlotte, and two domestic staff, was a live-in sick nurse. Evidently, his final years were

Figure 1.10 From the retirement portrait in oil of Osborne Reynolds by John Collier, OBE. © The University of Manchester.

Figure 1.11 St Decuman's Church and Vicarage *c.* 1900.

Figure 1.12 Osborne Reynolds' gravestone, St Decuman's churchyard.

difficult ones both for him and his family. There he remained until his death from influenza on 21 February 1912.

His funeral in St Decuman's church was attended by Horace Lamb (The West Somerset Free Press, 2 March 1912) and Reynolds is buried in the church-yard, his gravestone being an elegant art nouveau cross with his name and the dates of his arrival and departure beautifully engraved thereon (Figure 1.12). His wife who lived until 1942 is interred with him while two grandsons (the sons of Henry Osborne Reynolds, one named Osborne Reynolds), both of whom were killed in action during the Second World War, are memorialized on the gravestone.

1.3.2 Recollections and reflections by contemporaries

In his recollections, Thomson (1936) wrote:

> The Professor I had most to do with in my first three years at Owens was Osborne Reynolds, the Professor of Engineering. My personal relations with him when I was a student are a very pleasant recollection; he was always very kind to me, had a winning way with him and a charming smile.
>
> He was one of the most original and independent of men. When he took up a problem, he did not begin by making a bibliography and reading the literature

about the subject, but thought it out for himself from the beginning before reading what others had written about it.

In his lectures Reynolds was often carried away by his subject and got into difficulties. Some humorous incidents are related with regard to the manner in which he got out of them. He was once explaining the slide rule to his class; holding one in his hand, he expounded in detail the steps necessary to perform a multiplication. "We take as a simple example three times four", he said, and after appropriate explanations he continued, "Now we arrive at the result; three times four is 11.8". The class smiles. "That is near enough for our purpose", says Reynolds.

Professor Arthur Schuster, an academic colleague and close friend of Reynolds for many years, wrote, in an article in *Biographical Byways* (Schuster, 1925), the following observations:

> In his writings, as in his speech, Reynolds was difficult to understand. His brain seemed to work along lines different from those of the majority of us. He looked upon all things in an original manner.
>
> In his later years Reynolds had difficulty in finding the right word, using sometimes one that had the opposite meaning to that required. This failing ultimately developed into a regular aphasia.

This is corroborated in a letter written by Reynolds' daughter, Margaret, to a relative (University of Manchester Archive), describing the decline in her father's condition at the beginning of the 20th century: "Father lost the power of speech when we were still at school and though we knew and loved him when we were children we completely lost touch with him when we grew older."

Gibson, who was in turn a student of Reynolds and a member of his teaching staff during the latter years of his active life, wrote (Gibson, 1946):

> My colleagues and myself looked on the 'old man' with veneration, not untinged with awe. His personal kindliness to us was limited only by his physical disability, and it was a sad day when we realised that he was to leave us.

The final photograph of Reynolds with students and his staff (in which Gibson is standing in the row behind Reynolds, immediately to his left) appears in Figure 1.13.

It is appropriate that the final word here should go to Horace Lamb, who in his obituary notice wrote (Lamb, 1913):

> The character of Reynolds was, like his writings, strongly individual. He was conscious of the value of his work, but was content to leave it to the mature judgement of the scientific world. For advertisement he had no taste; and undue pretensions on the part of others only elicited a tolerant smile. To his pupils he was most generous in the opportunities for valuable work which he put in

Figure 1.13 Osborne Reynolds with staff and students of the Engineering Department, 1903. © The University of Manchester. Reproduced by courtesy of the University Librarian and Director, The John Rylands University Library, The University of Manchester.

their way, and in the share of credit which he consigned to them in cases of co-operation. Somewhat reserved in serious or personal matters and occasionally combative and tenacious in debate, he was in the ordinary relations of life the most kindly and genial of companions.

1.3.3 Closure

In closing this chapter it is appropriate to ask why it was that, in his lifetime, Osborne Reynolds was never awarded any national honour. The obituary notice that appeared in *Nature* just a week after his death, began: "In Professor Osborne Reynolds... Great Britain has lost its most distinguished scientific engineer". Towards the end of the piece, after noting his admission to the Royal Society, the award of the Society's Royal Medal and his honorary doctorate from Glasgow University, it concluded by remarking that "this was the only public recognition he ever received". The tone and positioning of this last observation clearly leave the impression that the writer at least felt there was a measure of injustice in Reynolds not receiving other honours; why it was that he did not end his days as Sir Osborne Reynolds (as, in fact, many of his web entries do, erroneously, refer to him). One may remark that among the well-known fluid mechanicists of his time, George Stokes, Horace Lamb and Thomas Stanton[4] were all knighted while William Thomson was, as noted

---

[4] After whom the dimensionless heat-transfer coefficient, the Stanton number, is named.

earlier, first knighted and later admitted to the peerage as Lord Kelvin of Largs. If we exclude the last named who made notable contributions in several other walks of life, many would argue that none contributed as much to the advancement of fluid mechanics and thermodynamics in all its varied aspects as did Osborne Reynolds (not just in the particular studies of turbulent flow on which the present article has focused).

Was he, possibly, offered such an honour and declined it? This seems highly unlikely, first because, while he would have been at pains to dissociate himself from the formal trappings and snobbery of such a title, he would probably have been delighted if somewhat bemused by the award. Secondly, if such an offer had been made and declined, this fact (while kept secret during his lifetime) would surely have been disclosed following his death in one or more of the obituaries written by his colleagues.

Thus, there remains the question as to why he was not knighted, in response to which the authors offer three possible contributory factors. First, his public demeanour may have been perceived as lacking sufficient gravitas. As his daughter's letter (quoted in §1.2.3) implies, Reynolds was someone with his head in the clouds while he was also seen by some as rather eccentric. There are anecdotes of him setting puzzles for his listeners. According to Lamb (1913), for example: "He had a keen sense of humour, and delighted in startling paradoxes, which he would maintain half seriously and half playfully, with astonishing ingenuity and resource." Collectively, these foibles could well have made him seem unsuited for holding high office in public institutions which is so often a precursor to (if not quite a prerequisite for) a knighthood. By way of contrast, Stokes (like Kelvin) served as President of the Royal Society and his advancement was assured. Stanton, a student of Reynolds, after holding the Chair of Engineering at the University of Bristol, was appointed Superintendent of the Engineering Department at the National Physical Laboratory, a post he filled from 1901 to his retirement in 1930. Lamb served twice as Vice-President of the Royal Society and as President of the London Mathematical Society. His ability to cut through tricky problems, which must have served him very well throughout his career, is clearly illustrated by his review of Reynolds' 1895 paper cited earlier. Moreover, Lamb also possessed a further attribute that Reynolds unfortunately lacked: longevity! He was knighted only at the age of 82 in 1931. Thus, the relatively young age at which Reynolds dropped from professional visibility was quite possibly a second contributory factor in his case being overlooked.

Thirdly, it is incontrovertible that the importance of Reynolds' major works was simply not widely recognized until after his death. As his obituary in *Nature* observed: "Well in advance of his time, in many cases years elapsed before the practical bearing of his researches was fully appreciated; even now the sphere of his influence on engineering progress is still widening." This was arguably the major contributor to his failure to receive the accolade of a knighthood. Indeed, we may note, wryly, the correctness of this assertion in *Nature*'s obituary for, while summarizing many of his important research contributions, it made no reference at all to the turbulent flow papers which are central to the present appreciation of his work. We should be indulgent of that lapse, however, for, when, in 1895, his strategy for the analysis of turbulent flows was published in *Phil. Trans. Roy. Soc.*, could anyone, even Osborne Reynolds, have foreseen that it was destined to shape the direction of research in engineering fluid mechanics for the next century?

Acknowledgements We wish to acknowledge the efforts of those who have earlier researched the life and work of Osborne Reynolds and helped to build up the collection of material to which we have had access in the course of preparing this account. Particular mention is made of the contribution of Professor Jack Allen whose detailed article on the academic life and work of Osborne Reynolds has been an invaluable source of material. In the present article we have quoted extensively from that account and from other sources that he identified. Biographical material researched by Ron Hayward in the 1970s and Ian Fishwick in the 1990s has enabled many further details to be included in the present article and their contributions are also gratefully acknowledged. Support from The Leverhulme Trust in the form of an Emeritus Fellowship during the period August 2003 to January 2006 helped JDJ to continue his research on Osborne Reynolds.

Sections 1.3.1 and 1.3.3 of the present chapter are edited versions of the account previously published by the present authors in Jackson and Launder (2007) while other aspects have been aired in preliminary form in Launder (2009). Our sincere thanks go to Dr James Peters, the University of Manchester's archivist, who assisted in locating and facilitating the photographing of archive papers.

References

Allen, J., 1970. The life and work of Osborne Reynolds. In *Osborne Reynolds and Engineering Science Today*, edited by D.M. McDowell and J.D. Jackson, 1–82, Manchester University Press.

Corrsin, S.C., 1952. Heat transfer in isotropic turbulence, *J. Appl. Phys.*, **23**, 113–118.

Crisp, F.A., 1911. The Reynolds of Suffolk, *Visitation of England and Wales*, **17**, 172–176. Privately printed by Grove Park Press, Essex.

Dowson, D., Taylor, C.M., Godet, M. and Berthe, D. (Editors), 1987. *Fluid-film Lubrication: Osborne Reynolds Centenary*, Elsevier, Amsterdam.

Eckhardt, B. (Editor), 2009. Turbulent transition in pipe flow – 125th anniversary of Reynolds' paper, *Phil. Trans. Roy. Soc.*, **367A**, 448–559.

Fitzgerald, R., 2004. New experiments set the scale for the onset of turbulence in pipe flow, *Physics Today*, **57**(2), 21–23.

Gibson, A.H., 1946. *Osborne Reynolds and his Work in Hydraulics and Hydrodynamics*, Longmans Green (commissioned by the British Council).

Hagen, G., 1854. Über den Einfluss der Temperatur auf die Bewegung des Wassers in Röhren, *Math. Abh. Akad. Wiss. Berlin*, 17–98.

Hagen, G., 1869. Über die Bewegung des Wassers in cylindrischen, nahe horizontalen Leitungen. *Math. Abh. Akad. Wiss. Berlin*, 1–29.

Hof, B., Juel, A. and Mullin, T., 2003. Scaling of the turbulence transition threshold in a pipe, *Phys. Rev. Lett.*, **91**, 244502.

Jackson, J.D., 1995. Osborne Reynolds: scientist, engineer and pioneer, *Proc. Roy. Soc.*, **451A**, 41–86.

Jackson, J.D., 2010. Osborne Reynolds – Victorian scientist, engineer and pioneer, *Memoirs of the Manchester Literary & Philosophical Society for the Session 2009–2010*, Vol. 147.

Jackson, J.D. and Launder, B.E., 2007. Osborne Reynolds and the publication of his papers on turbulent flow, *Ann. Rev. Fluid Mech.*, **39**, 18–35.

Jones, C.A., 1907. *History of Dedham*, Reprinted from *Dedham Parish Magazine* with Additions, Wilks and Son, Colchester.

Lamb, H., 1913. Osborne Reynolds, 1842–1912, Obituary notices, *Proc. Roy. Soc.*, **88**, xv–xxi.

Launder, B.E., 2009. Osborne Reynolds: the turbulent years. Invited keynote paper, in *Proc. Conf. Modelling Fluid Flow (CMFF '09)*, edited by J. Vad, Budapest University of Technology & Economics, Budapest, 28–40.

Mullin, T., 2011. Experimental studies of transition to turbulence in a pipe, *Ann. Rev. Fluid Mech.*, **43**, 1–24.

Prandtl, L., 1910. Eine Beziehung zwischen Warmeaustausch und Strömungswiderstand der Flüssigkeiten. *Phys. Z.*, **11**, 1072–1078.

Prandtl, L., 1913. Flüssigkeitsbewegung. In *Handwörterbuch der Naturwissenschaften*, **4**, 101–140.

Ramsey, R.J., 1949. Early Essex Patents, *Essex Review*, **LVI**, C17.

Reynolds, O., 1873. On the condensation of a mixture of air and steam upon cold surfaces, *Proc. Roy. Soc.*, **21**, 274–281.

Reynolds, O., 1874. On the extent and action of heating surfaces of steam boilers, *Manchester Lit. & Phil.*, Vol. 14, Session 1884–5.

Reynolds, O., 1879. On certain dimensional properties of matter in the gaseous state, *Phil. Trans. Roy. Soc.*, **170**, 727–845.

Reynolds, O., 1883. An experimental investigation of the circumstances which determine whether the motion of water shall be direct or sinuous and the law of resistance in parallel channels, *Phil. Trans. Roy. Soc.*, **174**, 935–982.

Reynolds, O., 1886. On the theory of lubrication and its application to Mr Beauchamp-Tower's experiments, *Phil. Trans. Roy. Soc.*, **187**, 157–234.

Reynolds, O., 1895. On the dynamical theory of incompressible viscous fluids and the determination of the criterion, *Phil. Trans. Roy. Soc.*, **186A**, 123–164.

Reynolds, O., 1900. *Papers on Mechanical and Physical Subjects, 1870–1880*, Collected Works, Volume I, Cambridge University Press.

Reynolds, O., 1901. *Papers on Mechanical and Physical Subjects 1881–1900.* Collected Works, Volume II, Cambridge University Press.

Reynolds, O., 1903. *Papers on Mechanical and Physical Subjects – the Sub-Mechanics of the Universe.* Collected Works, Volume III, Cambridge University Press.

Reynolds, O. and Moorby, W.H., 1897. On the mechanical equivalent of heat, *Phil. Trans. Roy. Soc.*, **190A**, 301–422.

Rott, N., 1990. Note on the history of the Reynolds number, *Ann. Rev. Fluid Mech.*, **22**, 1–12.

Schuster, A., 1925. *Biographical Fragments*, Section VII, 228–233, MacMillan and Co. Ltd., London.

Smith, R.A., 1997. Osborne Reynolds – the most distinguished engineering professor and an inspired choice, *Queens College Record*, 14–15.

Sommerfeld, A., 1908. Ein Beitrag zur hydrodynamischen Erklarung der turbulenten Flüssigkeits-bewegung. In *Proc. 4th Int. Congr. Math.*, Rome, 3: 116–124.

Thompson, J., 1886. *The Owens College: Its Foundation and Growth; and its Connection with the Victoria University of Manchester*, Cornish, Manchester.

Thomson, J.J., 1936. *Recollections and Reflections*, G. Bell & Sons Ltd.

Venn, J.A., 1954. *Alumni Cantabrigienses*, Part 11, V, 279 and 281, Cambridge University Press.

Walker, E.G., 1938. *The life and work of William Cawthorne Unwin*, 54–58, and 75, Unwin Memorial Committee, London.

Wedlake, A.L., 1984. *Old Watchet, Williton and Around*, Exmoor Press.

White, W., 1844. *History Gazetteer and Directory of Suffolk.* Reissued 1970, David & Charles.

Wright, A., 1912. Old Queens' Men, *The Dial*, **3**, No. 13, 42.

2

Prandtl and the Göttingen school

Eberhard Bodenschatz and Michael Eckert

2.1 Introduction

In the early decades of the 20th century Göttingen was the center for mathematics. The foundations were laid by Carl Friedrich Gauss (1777–1855) who from 1808 was head of the observatory and professor for astronomy at the Georg August University (founded in 1737). At the turn of the 20th century, the well-known mathematician Felix Klein (1849–1925), who joined the University in 1886, established a research center and brought leading scientists to Göttingen. In 1895 David Hilbert (1862–1943) became Chair of Mathematics and in 1902 Hermann Minkowski (1864–1909) joined the mathematics department. At that time, pure and applied mathematics pursued diverging paths, and mathematicians at Technical Universities were met with distrust from their engineering colleagues with regard to their ability to satisfy their practical needs (Hensel, 1989). Klein was particularly eager to demonstrate the power of mathematics in applied fields (Prandtl, 1926b; Manegold, 1970). In 1905 he established an Institute for Applied Mathematics and Mechanics in Göttingen by bringing the young Ludwig Prandtl (1875–1953) and the more senior Carl Runge (1856–1927), both from the nearby Hanover. A picture of Prandtl at his water tunnel around 1935 is shown in Figure 2.1.

Prandtl had studied mechanical engineering at the Technische Hochschule (TH, Technical University) in Munich in the late 1890s. In his studies he was deeply influenced by August Föppl (1854–1924), whose textbooks on technical mechanics became legendary. After finishing his studies as mechanical engineer in 1898, Prandtl became Föppl's assistant and remained closely related to him throughout his life, intellectually by his devotion to technical mechanics and privately as Föppl's son-in-law (Vogel-Prandtl, 1993). Under Föppl's supervision Prandtl wrote his doctoral dissertation on a problem of

40

Figure 2.1 Ludwig Prandtl at his water tunnel in the mid to late 1930s. Reproduction from the original photograph DLR: FS-258.

technical mechanics (*Kipp-Erscheinungen, ein Fall von instabilem elastischem Gleichgewicht*: *On Tilting Phenomena, an Example of Unstable Elastic Equilibrium*). Technische Hochschulen were not then authorized to grant doctoral degrees, so that Prandtl had to perform the required academic rituals at the Philosophical Faculty of the neighboring Ludwig Maximilian University of Munich on 29 January 1900. At the time Technische Hochschulen were fighting a bitter struggle until they were granted equal rights with the Universities. The institutional schism affected in particular the academic disciplines at the interface of science and engineering, such as applied mathematics and technical mechanics (Oswatitsch and Wieghardt, 1987). On 1 January 1900, before receiving his doctorate, Prandtl had taken up an engineering position at the Maschinenbaugesellschaft in Nuremberg, which had just merged with Maschinenfabriken Augsburg to become MAN (Maschinenfabrik Augsburg Nürnberg: Machine Works of Augsburg and Nuremberg). At MAN he was first introduced to problems in fluid dynamics through designing a blower. Very shortly thereafter, he received an offer for the Chair of Mechanics at the Technische Hochschule Hanover. He left Nuremberg on 30 September 1901 to become at age 26 the youngest professor in Prussia (Vogel-Prandtl, 1993). In 1904 Felix Klein was able to convince Prandtl to take a non-full professor position at Göttingen University to become Head of the Department of Technical Physics at the Institute of Physics with the prospect of a co-directorship of a new Institute for Applied Mathematics and Mechanics. In the same vein, Klein had arranged Runge's call to Göttingen. In autumn 1905, Klein's institutional plans materialized. Göttingen University opened a new Institute for Applied Mathematics and Mechanics under the joint directorship of Runge and Prandtl. Klein also involved Prandtl as Director in the planning of an extramural Aerodynamic Research Institute, the Motorluftschiffmodell-Versuchsanstalt, which started its operation with the first Göttingen design windtunnnel in 1907 (Rotta, 1990; Oswatitsch and Wieghardt, 1987). Klein regarded Prandtl's "strong power of intuition and great originality of thought with the expertise of the engineer and the mastery of the mathematical apparatus" (Manegold, 1970, p. 232) ideal qualities for what he had planned to establish at Göttingen.

With these institutional measures, the stage was set for Prandtl's unique career between science and technology – and for the foundation of an academic school with a strong focus on basic fluid dynamics and their applications. Prandtl directed the Institut für Angewandte Mechanik of Göttingen University, the Aerodynamische Versuchsanstalt (AVA), as the rapidly expanding Motorluftschiffmodell-Versuchsanstalt (airship model test facility) was renamed after the First World War, and, after 1925, the associated Kaiser-Wilhelm-Institut (KWI) für Strömungsforschung. His ambitions and the history leading to the establishment of the KWI are well summarized in his opening speech at his institute, which has been translated into English (Prandtl, 1925E).

During the half century of Prandtl's Göttingen period, from 1904 until his death, his school extended Göttingen's fame from mathematics to applied mechanics, a specialty which acquired in this period the status of a self-contained discipline. Prandtl had more than eighty doctoral students, among them Heinrich Blasius, Theodore von Kármán, Max Munk, Johann Nikuradse, Walter Tollmien, Hermann Schlichting, Karl Wieghardt, and others who, like Prandtl, perceived fluid mechanics in general, and turbulence in particular, as a paramount challenge to bridge the gulf between theory and practice. Like Prandtl's institutional affiliations, his approach towards turbulence reflects a broad spectrum of 'pure' and 'applied' research (if such dichotomies make sense in turbulence research). We have to consider the circumstances and occasions in these settings in order to better characterize the approach of the Göttingen school on turbulence.

2.2 The boundary layer concept, 1904–1914

When Prandtl arrived in Göttingen in autumn 1904, he came with an asset: the boundary layer concept (Eckert, 2006, chapter 2; Meier, 2006). He was led to this concept during his short industrial occupation when he tried to account for the phenomenon of flow separation in diverging ducts. Prandtl presented the concept together with photographs of flow around obstacles in a water trough at the Third International Congress of Mathematicians in Heidelberg in August 1904 (Prandtl, 1905). In a summary, prepared at the request of the American Mathematical Society, he declared[1] that the "most important result" of this concept was that it offered an "explanation for the formation of discontinuity surfaces (vortex sheets) along continuously curved boundaries". In his Heidelberg presentation he expressed the same message in these words: "A fluid layer set in rotational motion by the friction at the wall moves into the free fluid and, exerting a complete change of motion, plays there a similar role as Helmholtz' discontinuity sheets" (Prandtl, 1905, p. 578). (For more on the emergence of Helmholtz's concept of discontinuity surfaces, see Darrigol, 2005, chapter 4.3).

According to the recollection of one participant at the Heidelberg congress, Klein recognized the momentousness of Prandtl's method immediately (Sommerfeld, 1935). However, if this recollection from many years later may be trusted, Klein's reaction was exceptional. The boundary layer concept required elaboration before its potential was more widely recognized (Dryden, 1955; Goldstein, 1969; Tani, 1977; Grossmann et al., 2004). Its modern understanding in terms of singular perturbation theory (O'Malley Jr., 2010) emerged only decades later. The first tangible evidence that Prandtl's concept provided more than qualitative ideas was offered by Blasius, who derived in his doctoral dissertation the coefficient for (laminar) skin friction from the boundary layer equations for the flow along a flat plate ('Blasius flow': Blasius, 1908; Hager, 2003). However, this achievement added little understanding to what Prandtl had considered the most important result of his concept, namely how vortical motion, to say nothing of turbulence, is created at the boundary.

Even before Prandtl arrived in Göttingen, the riddle of turbulence was a recurrent theme in Klein's lectures and seminars. In a seminar on hydraulics in the winter semester 1903/04 Klein called it a "true need of our time to bridge the gap between separate developments". The notorious gulf between hydraulics and hydrodynamics served to illustrate this need with many examples. The seminar presentations were expected to focus on the comparison between theory and experiment in a number of specific problems with the flow of water, such as the outflow through an orifice, the flow over a weir, pipe

---

[1] Undated draft in response to a request from 13 August 1904, Blatt 43, Cod. Ms. L. Prandtl 14, Acc. Mss. 1999.2, SUB.

flow, waves, the water jump ('hydraulic jump'), or the natural water flow in rivers.[2]

In the winter semester 1907/08 Klein dedicated another seminar to fluid mechanics, this time with the focus on 'Hydrodynamics, with particular emphasis of the hydrodynamics of ships'. With Prandtl and Runge as co-organizers, the seminar again involved a broad spectrum of problems from fluid mechanics that Klein and his colleagues regarded as suitable for mathematical approaches.[3] Theodore von Kármán, who made then his first steps towards an outstanding career in Prandtl's institute, presented a talk on unsteady potential motion. Blasius, who was finishing his dissertation on the laminar boundary layer in 1907, reviewed in two sessions contemporary research on turbulent flows. Other students and collaborators of Prandtl dealt with vortical motion (Karl Hiemenz) and boundary layers and the detachment of vortices (Georg Fuhrmann). Although little was published about these themes at the time, Klein's seminar served as a testing ground for debates on the notorious problems of fluid mechanics like the creation of vorticity in ideal fluids ('Klein's Kaffeelöffelexperiment': see Klein, 1910; Saffman, 1992, chapter 6).

With regard to turbulence, the records of Blasius' presentation from this seminar illustrate what Prandtl and his collaborators must have regarded as the main problems at that time. After reviewing the empirical laws, such as Chezy's law for channel flow and Reynolds' findings about the transition to turbulence in pipe flow (for these and other pioneering 19th-century efforts, see Darrigol, 2005, chapter 6), Blasius concluded that the problems "addressed to hydrodynamics" should be sorted into two categories: I. Explanation of instability; and II. Description of turbulent motion. These had to address the dichotomy of hydraulic description versus rational hydrodynamic explanations. Concerning the first category, the onset of turbulence, Blasius reviewed Hendrik Antoon Lorentz's recent approach where a criterion for the instability of laminar flow was derived from a consideration of the energy added to the flow by a superposed fluctuation (Lorentz, 1897, 1907). With regard to the second category, fully developed turbulence, Blasius referred mainly to Boussinesq's pioneering work (Boussinesq, 1897) where the effect of turbulence was described as an additional viscous term in the Navier–Stokes equation. In contrast to the normal viscosity, this additional 'turbulent' viscosity term was due to the exchange of momentum by the eddying motion in turbulent flow.

---

[2] Klein, handwritten notes. SUB Cod. Ms. Klein 19 E (Hydraulik, 1903/04), and the seminar protocol book, no. 20. Göttingen, Lesezimmer des Mathematischen Instituts. Available online at librarieswithoutwalls.org/klein.html.

[3] Klein's seminar protocol book, no. 27. Göttingen, Lesezimmer des Mathematischen Instituts. Available online at librarieswithoutwalls.org/klein.html.

Boussinesq's concept had already been the subject of the preceding seminar in 1903/04, where the astronomer Karl Schwarzschild and the mathematicians Hans Hahn and Gustav Herglotz reviewed the state of turbulence (Hahn et al., 1904). However, the efforts in the seminar to determine the (unknown) eddy viscosity of Boussinesq's approach proved futile. "Agreement between this theory and empirical observations is not achieved," Blasius concluded in his presentation.[4]

In spite of the emphasis on the riddles of turbulence in these seminars, it is commonly reported that Prandtl ignored turbulence as a research theme until many years later. For example, the editors of his *Collected Papers* dated his first publication in the category *Turbulence and Vortex Formation* to the year 1921 (see below). The preserved archival sources, however, belie this impression. Prandtl started to articulate his ideas on turbulence much earlier. "Turbulence I: Vortices within laminar motion", he wrote on an envelope with dozens of loose manuscript pages. The first of these pages is dated by himself as 3 October 1910, with the heading *Origin of turbulence*. Prandtl considered there "a vortex line in the boundary layer close to a wall" and argued that such a vortical motion "fetches (by frictional action) something out of the boundary layer which, because of the initial rotation, becomes rolled up to another vortex which enhances the initial vortex." Thus he imagined how flows become vortical due to processes that originate in the initially laminar boundary layer.[5] In the same year he published a paper on *A relation between heat exchange and flow resistance in fluids* (Prandtl, 1910) which extended the boundary layer concept to heat conduction. Although it did not explicitly address turbulence – the article is more renowned because Prandtl introduced here what was later called the 'Prandtl number' – it reveals Prandtl's awareness for the differences of laminar and turbulent flow with regard to heat exchange and illustrates from a different perspective how turbulence entered Prandtl's research agenda (Rotta, 2000).

Another opportunity to think about turbulence from the perspective of the boundary layer concept came in 1912 when wind tunnel measurements about the drag of spheres displayed discrepant results. When Otto Föppl (1885–1963), Prandtl's brother-in-law and collaborator at the airship model test facility, compared the data from his own measurements in the Göttingen wind tunnel with those from the laboratory of Gustave Eiffel (1832–1921) in Paris, he found a blatant discrepancy and supposed that Eiffel or his collaborator had omitted a factor of 2 in the final evaluation of their data (Föppl, 1912). Provoked by this claim, Eiffel performed a new test series and found that

---

[4] Klein's seminar protocol book, no. 27, p. 80. Göttingen, Lesezimmer des Mathematischen Instituts. Available online at http://www.librarieswithoutwalls.org/felixKlein.html.

[5] Cod. Ms. L. Prandtl 18 (*Turbulenz I: Wirbel in Laminarbewegung*), Acc. Mss. 1999.2, SUB.

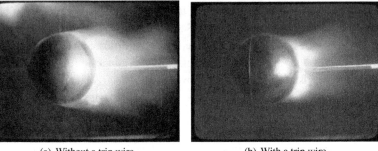

(a) Without a trip wire.   (b) With a trip wire.

Figure 2.2 Turbulence behind a sphere made visible with smoke. Reproduction from the original 1914 photograph. GOAR: GK-0116 and GK-0118.

the discrepancy was not the result of an erroneous data evaluation but a new phenomenon which could be observed only at higher air speeds than those attained in the Göttingen wind tunnel (Eiffel, 1912). After inserting a nozzle into their wind tunnel, Prandtl and his collaborators were able to reproduce Eiffel's discovery: at a critical air speed the drag coefficient suddenly dropped to a much lower value. Prandtl also offered an explanation of the new phenomenon. He assumed that the initially laminar boundary layer around the sphere becomes turbulent beyond a critical air speed. On the assumption that the transition from laminar to turbulent flow in the boundary layer is analogous to Reynolds' case of pipe flow, Prandtl displayed the sphere drag coefficient as a function of the Reynolds number, UD/v (flow velocity U, sphere diameter D, kinematic viscosity v), rather than, as did Eiffel, of the velocity; thus he demonstrated that the effect occurred at roughly the same Reynolds number even if the individual quantities differed widely (the diameters of the spheres ranged from 7 to 28 cm; the speed in the wind tunnel was varied between 5 and 23 m/s). Prandtl further argued that the turbulent boundary layer flow entrains fluid from the wake so that the boundary layer stays attached to the surface of the sphere longer than in the laminar case. In other words, the onset of turbulence in the boundary layer reduces the wake behind the sphere and thus also its drag. But the argument that turbulence decreases the drag seemed so paradoxical that Prandtl conceived an experimental test: when the transition to turbulence in the boundary layer was induced otherwise, e.g. with a thin 'trip wire' around the sphere or a rough surface, the same phenomenon occured. When smoke was added to the air stream, the reduction of drag became visible by the reduced extension of the wake behind the sphere (Wieselsberger, 1914; Prandtl, 1914) (see Figure 2.2).

2.3 A working program for a theory of turbulence

During the First World War turbulence became pertinent in many guises. Arnold Sommerfeld (1868–1951), theoretical physicist at the University of Munich, once forwarded to Prandtl a request "concerning the fall of bombs in water and air". Sommerfeld was involved at that time in other war-related research (about wireless telegraphy) and had heard about this problem from a Major whom he had met during a visit in Berlin. "It deals with the drag of a sphere (radius a) moving uniformly through water (density ρ) at a velocity V. At Re (Reynolds number) > 1000 the drag is $W = \psi\rho a^2 V^2$." By similarity, "ψ should be universal and also independent of the fluid", Sommerfeld alluded to Prandtl's recent study about the drag of spheres in air; but according to older measurements of the friction coefficient ψ for water this was not the case.[6] Prandtl suspected an error with the assessment of the experimental measurements. Furthermore, the impact of a falling bomb on the water surface involved additional effects so that a comparison was difficult.[7] A few months later, the aerodynamics of bomb shapes became officially part of Prandtl's war work.[8]

The turbulence effect as observed with the drag of spheres became also pertinent for the design of airplanes. The struts and wires which connected the wings of bi- and triplanes were subject to the same sudden changes of drag. For this reason, Prandtl's institute was charged with a systematic wind tunnel investigation of struts and wires. The goal was to find out how the sudden change of drag could be avoided by choosing appropriate strut and wire shapes. "The critical range [of Reynolds numbers] is considered as the interval within which there are two fundamentally different modes of flow", Prandtl's collaborator, Max Munk, explained in a technical war report on measurements of the drag of struts. The report also mentioned how this phenomenon occurred in practice. In particular, a reduction of the speed, for example, when the plane changes from horizontal flight into a climb, results in a sudden increase of the drag coefficient, and often of a considerable increase of the drag itself. It was therefore not sufficient to minimize the drag by streamlining the profile of a strut, but also to give it a shape that did not experience the sudden change of drag when the airplane passed through the critical speed range (Munk, 1917).

---

[6] Sommerfeld to Prandtl, 9 May 1915. GOAR 2666.

[7] Prandtl to Sommerfeld, 14 May 1915. GOAR 2666.

[8] He received, for example, contracts from the Bombenabteilung der Prüfanstalt u. Werft der Fliegertruppen, dated 23 December 1915, concerning 'Fliegerbombe, M 237', and on 'Carbonit-Bomben, Kugelform', dated 1 September 1916. GOAR 2704B.

In view of such practical relevance, Prandtl sketched[9] in March 1916 a *Working program about the theory of turbulence*. Like Blasius in his presentation in Klein's seminar, Prandtl discriminated between the onset of turbulence, i.e. the transition from laminar to turbulent flow, and what he called accomplished turbulence, i.e. fully developed turbulence, as the two pillars of this research program. The onset of turbulence was generally perceived as the consequence of a hydrodynamic instability, a problem with a long history of frustrated efforts (Darrigol, 2005); although it had been revived during the preceding decade by William McFadden Orr, Sommerfeld, Ludwig Hopf, Fritz Noether and others, a solution seemed out of sight (Eckert, 2010). Prandtl sketched plane flows with different piece-wise linear velocity profiles. The stability of such flow configurations had been extensively studied since the 1880s by Lord Rayleigh for the inviscid case. Profiles with an inflection were unstable according to Rayleigh's (1887) analysis. Prandtl's strategy seemed clear: he approached the stability analysis from the limiting case of infinite Reynolds numbers, i.e. the inviscid case treated by Rayleigh, in order to derive from this limit approximations for flows of low viscosity. Like his boundary layer concept, this approach would be restricted to high Reynolds numbers (unlike the Orr–Sommerfeld approach which applied to the full range of Reynolds numbers). According to his sketches and somewhat cryptic descriptions, Prandtl attempted to study the behavior of "a sinusoidal discontinuity" in a "stripe flow". Prandtl's 'stripes,' i.e. piece-wise linear flow profiles, indicate that he aimed at a theory for the onset of turbulence in the plane flow bounded by two fixed walls and a flow bounded by a single wall. The latter configuration obviously was perceived as an approximation of the 'Blasius flow', i.e. the velocity profile of the laminar boundary layer flow along a flat plate. According to Rayleigh's inflection theorem, both flows were stable in the inviscid case because the curvature of the velocity profile did not change direction. The focus was on the boundary layer motion with Rayleigh oscillation, as Prandtl concluded this part of his turbulence program.

With regard to fully developed turbulence, the other part of his working program, Prandtl apparently had no particular study in mind as a starting point. "Statistical equilibrium of a set of vortices in an ideal fluid in the vicinity of a wall", he noted as one topic for future research. For the goal of a "complete approach for very small friction" he started from the assumption that vortices from the wall ("in the boundary layer") are swept into the fluid by "disordered motion". He envisioned a balance between the vortex creation at the wall and the vortices destroyed in the fluid as a result of friction. For a closer analysis

---

[9] Page 15 (dated 6 March 1916) in Cod. Ms. L. Prandtl, 18, Acc. Mss. 1999.2, SUB.

of the involved vortex interaction he introduced what he called the rough assumption that the vortex remains unchanged for a certain time $\sim r^2/\nu$ and then suddenly disappears, whereby it communicates its angular momentum to the mean flow.[10]

During the war Prandtl had more urgent items on his agenda (Rotta, 1990, pp. 115–193). But the riddle of turbulence as a paramount challenge did not disappear from his mind. Nor from that of his former student, the prodigy Theodore von Kármán, who returned after the War to the Technische Hochschule Aachen as Director of the then fledgling Aerodynamic Institute. Both the Aachen and the Göttingen fluid dynamicists pursued the quest for a theory of turbulence in a fierce rivalry. "The competition was gentlemanly, of course. But it was first-class rivalry nonetheless," Kármán later recalled, "a kind of Olympic Games, between Prandtl and me, and beyond that between Göttingen and Aachen" (von Kármán, 1967, p. 135). Since they had nothing published on turbulence, both Prandtl and Kármán pondered how to ascertain their priority in this quest. In summer 1920, Prandtl supposed that von Kármán used a forthcoming science meeting in Bad Nauheim to present a paper on turbulence at this occasion. "I do not yet know whether I can come", he wrote[11] to his rival, "but I wish to be oriented about your plans. As the case may be I will announce something on turbulence (experimental) as well. I have now visualized turbulence with lycopodium in a 6 cm wide channel." The Aachen–Göttingen rivalry had not yet surfaced publicly at this occasion. By correspondence, however, it was further developing. The range of topics encompassed Prandtl's entire working program. Early in 1921 Prandtl learned that von Kármán was busy elaborating a theory of fully developed turbulence in the boundary layer along a flat wall – with "fabulous agreement with observations". Ludwig Hopf and another collaborator of the Aachen group had by this time started with hot-wire experiments. Hopf revealed[12] that in Aachen they planned to measure in a water channel the mean square fluctuation and the spectral distribution of the fluctuations.

Little seems to have resulted from these experiments, neither in Aachen by means of the hot-wire technique nor in Prandtl's laboratory by visualizing turbulence with lycopodium. Von Kármán's theoretical effort, however, appeared promising. "Dear Master, colleague, and former boss", Kármán addressed Prandtl in a five-page letter with ideas for a turbulent boundary layer

---

[10] Page 16 in Cod. Ms. L. Prandtl, 18, Acc. Mss. 1999.2, SUB. Apparently r and ν are the radius of the vortex and the kinematic viscosity of the fluid, respectively. Prandtl did not define the quantities involved here. His remarks are rather sketchy and do not lend themselves for a precise determination of the beginnings of his future mixing length approach.

[11] Prandtl to Kármán, 11 August 1920. GOAR 1364.

[12] Hopf to Prandtl, 3 February 1921. MPGA, Abt. III, Rep. 61, Nr. 704.

theory (see below) and about the onset of turbulence.[13] The latter was regarded as *the* turbulence problem. The difficulty in explaining the transition from laminar to turbulent flow had been rated as a paramount riddle since the late 19th century. In his dissertation performed under Sommerfeld in 1909, Ludwig Hopf had titled the introductory section *The turbulence problem*, because neither the energy considerations of Reynolds and Lorentz nor the stability approaches of Lord Kelvin and Lord Rayleigh were successful. Hopf was confronted with the problem in the wake of Sommerfeld's own stability approach to viscous flows, but "the consequent analysis of the problem according to the method of small oscillations by Sommerfeld is not yet accomplished" (Hopf, 1910, pp. 6–7). In the decade that followed the problem was vigorously attacked by this technique (later labeled as the Orr–Sommerfeld method) – with the discrepant result that plane Couette flow seemed stable for all Reynolds numbers (Eckert, 2010).

In comparison with these efforts, Prandtl's approach as sketched in his working program appeared like a return to the futile attempts of the 19th century: "At large Reynolds number the difference between viscous and inviscid fluids is certainly imperceptible," Hopf commented[14] on Prandtl's idea to start from the inviscid limit, but at the same time he regarded it "questionable whether one is able to arrive at a useful approximation that leads down to the critical number from this end". In response to such doubts Prandtl began to execute his working program about the onset of turbulence in plane flows with piecewise linear flow profiles. "Calculation according to Rayleigh's papers III, p. 17ff," he noted on a piece of paper in January 1921, followed by several pages of mathematical calculations.[15] Despite their initial reservations, the Aachen rivals were excited about Prandtl's approach. Von Kármán immediately rushed his collaborators to undertake a stability analysis for certain piecewise linear flow profiles, Hopf confided to Prandtl.[16] Prandtl had by this time already asked a doctoral student to perform a similar study. "Because it deals with a doctoral work, I would be sorry if the Aachener would publish away part of his dissertation," he asked Hopf, so as not to interfere in this effort. Von Kármán responded that the Aachen stability study was aiming at quite different goals, namely the formation of vortices in the wake of an obstacle (labeled later as the 'Kármán vortex street' after von Kármán's earlier theory about this phenomenon; Eckert, 2006, chapter 2). The new study was motivated by "the hope to determine perhaps the constants that have been left indetermined in my old theory," wrote Kármán, attempting to calm Prandtl's worry. Why not arrange

[13] Kármán to Prandtl, 12 February 1921. GOAR 3684.
[14] Hopf to Prandtl, 27 October 1919. GOAR 3684.
[15] Pages 22–26 in Cod. Ms. L. Prandtl, 18, Acc. Mss. 1999.2, SUB.
[16] Hopf to Prandtl, 3 February 1921. MPGA, III, Rep. 61, Nr. 704.

a division of labor between Göttingen and Aachen, he further suggested[17], so that his group would deal with these wake phenomena and Prandtl's doctoral student with boundary layer instability. Prandtl agreed and suggested that von Kármán should not feel hemmed in by his plans. He explained[18] once more that the focus at Göttingen was to study the onset of turbulence in the boundary layer. "We have now a method for approximately taking into account friction."

A few months later Prandtl reported that the calculations of his doctoral student were terribly complicated and yielded "a peculiar and unpleasant result". If the corresponding flow is unstable according to Rayleigh's inviscid theory, the instability was not reduced by taking viscosity into account – as they had expected – but *increased*. The calculation was done in first-order approximation, but its extension to the second-order seemed almost hopeless, Prandtl wrote[19] in frustration, "and so, once more, we do not obtain a critical Reynolds number. There seems to be a very nasty devil in turbulence so that all mathematical efforts are doomed to failure."

At this stage Prandtl published his and his doctoral student's, Oskar Tietjens (1893–1971), effort. In addition to the profiles which were unstable in the inviscid case, the study was extended to those that were stable in the inviscid case (i.e. ones without an inflection) – with the surprising result that these profiles also became unstable if viscosity was included. Contrary to the stability deadlock of the earlier studies concerning plane Couette flow, Prandtl's approach left the theory in an instability deadlock. "We did not want to believe in this result and have performed the calculation three times independently in different ways. There was always the same sign which indicated instability" (Prandtl, 1921a, p. 434).

Prandtl's paper appeared in a new journal edited by the applied mathematician Richard von Mises, the *Zeitschrift für Angewandte Mathematik und Mechanik* (*ZAMM*) where the turbulence problem was presented as a major challenge. In his editorial von Mises described the the present state of the theory as completely open. He regarded it as undecided whether the viscous flow approach is able to explain turbulence at sufficient mathematical depth (von Mises, 1921, p. 12). Fritz Noether, like Hopf, a Sommerfeld disciple who had struggled with this matter for years, introduced the subject with a review article titled *The Turbulence Problem* (Noether, 1921). He summarized the series of futile attempts of the preceding decades and presented the problem in a generic manner. (Noether presented the 'stability equation' or 'perturbation differential equation' – to quote the contemporary designations – in the form in which it became familiar later as the Orr–Sommerfeld equation. His paper

---

[17] Kármán to Prandtl, 12 February 1921. GOAR 3684.
[18] Prandtl to Kármán, 16 February 1921. MPGA, III, Rep. 61, Nr. 792.
[19] Prandtl to Kármán, 14 June 1921. MPGA, III, Rep. 61, Nr. 792.

became the door-opener for many subsequent studies of the Orr–Sommerfeld approach.) Noether was also well informed about the Göttingen effort, as is evident from his correspondence with Prandtl. In one of his letters[20] he expressed some doubts about Prandtl's approach, but he belittled his dissent and regarded it merely as a difference of mindset and expression. Another contributor to the turbulence problem in this first volume of *ZAMM* was Ludwig Schiller, a physicist working temporarily in Prandtl's laboratory; Schiller (1921) surveyed the experimental efforts at measuring the onset of turbulence.

The turbulence problem was also discussed extensively in September 1921 at a conference in Jena, where the Deutsche Physikalische Gesellschaft, the Deutsche Gesellschaft für Technische Physik and the Deutsche Mathematiker-Vereinigung convened their annual meetings of that year in a common event. At this occasion, Prandtl's *Remarks about the Onset of Turbulence* caused quite a stir. "With regard to the theoretical results which have always yielded stability it should be noted that these referred to the so-called Couette case," said Prandtl, explaining the difference between his result and the earlier studies. But Sommerfeld found it "very strange and at first glance unlikely" that all flows are unstable except Couette flow. "What causes the special position of Couette flow?" asked Sommerfeld. Kármán pointed to kinks at arbitrary positions of the piecewise linear profiles as a source of arbitrariness. Hopf criticized Prandtl's approximation $Re \rightarrow \infty$ (Prandtl, 1922, pp. 22–24).

The Jena conference and the articles on the turbulence problem in *ZAMM* from the year 1921 marked the beginning of a new period of research on the onset of turbulence. From then on Prandtl did not participate with his own contributions to this research. But he continued to supervise doctoral dissertations about this part of his working program. Tietjens (1925) paved the way along which Walter Tollmien (1900–1968), in another Göttingen doctoral dissertation, achieved the first complete solution of the Orr–Sommerfeld equation for a special flow (Tollmien, 1929). A few years later, another disciple of Prandtl, Hermann Schlichting (1907–1982), further extended this theory (Schlichting, 1933), so that the process of instability could be analysed in more detail. But the 'Tollmien–Schlichting' approach remained disputed until it was experimentally corroborated in World War II (Eckert, 2008).

2.4 Skin friction and turbulence I: the 1/7th law

Originally, Prandtl's boundary layer concept had focused on laminar flow. Ten years later, with the interpretation of Eiffel's drag phenomenon as a turbulence

---

[20] Noether to Prandtl, 29 June 1921. GOAR 3684.

effect, boundary layer flow could also be imagined as fully turbulent. From a practical perspective, the latter appeared much more important than the former. Data on fluid resistance in pipes, as measured for decades in hydraulic laboratories, offered plenty of problems for testing theories about turbulent friction. Blasius, who had moved in 1911 to the Berlin Testing Establishment for Hydraulics and Ship Building (Versuchsanstalt für Wasserbau und Schiffbau), published in 1913 a survey of pipe flow data: when displayed as a function of the Reynolds number Re, the coefficient for 'hydraulic' (i.e. turbulent) friction varied in proportion to $Re^{-1/4}$ (in contrast to laminar friction at low Reynolds numbers, where it is proportional to Re^{-1}) (Blasius, 1913).

No theory could explain this empirical 'Blasius law' for turbulent pipe flow. But it could be used to derive other semi-empirical laws, such as the distribution of velocity in the turbulent boundary layer along a plane smooth wall. When Kármán challenged Prandtl in 1921 with the outline of such a theory, he recalled that Prandtl had told him earlier how one could extrapolate from pipe flow to the flow along a plate, and that Prandtl already knew that the velocity distribution was proportional to $y^{1/7}$, where y was the distance from the wall. Prandtl responded that he had known this "already for a pretty long time, say since 1913". He claimed that he had already in earlier times attempted to calculate boundary layers in which he had assumed a viscosity enhanced by turbulence, which he chose for simplicity as proportional to the distance from the wall and proportional to the velocity in the free flow. But he admitted that Kármán had advanced further with regard to a full-fledged turbulent boundary layer theory. "I have planned something like this only for the future and have not yet begun with the elaboration." Because he was busy with other work he suggested[21] that Kármán should proceed with the publication of this theory: "I will see afterwards how I can gain recognition with my different derivation, and I can get over it if the priority of publishing has gone over to friendly territory."

Kármán published his derivation without further delay in the first volume of *ZAMM* (von Kármán, 1921) with the acknowledgement that it resulted from a suggestion "by Mr. Prandtl in an oral communication in Autumn 1920". Prandtl's derivation appeared in print only in 1927 – with the remark that "the preceding treatment dates back to Autumn 1920" (Prandtl, 1927a). Johann Nikuradse (1894–1979), whom Prandtl assigned by that time an experimental study about the velocity distribution in turbulent flows as subject of a doctoral work, dated Prandtl's derivation more precisely to a discourse in Göttingen on 5 November, during the winter semester of 1920 (Nikuradse,

---

[21] Prandtl to Kármán, 16 February 1921. MPGA, Abt. III, Rep. 61, Nr. 792.

1926, p. 15). Indeed, Prandtl outlined this derivation in notices dated (by himself) to 28 November 1920.[22] Further evidence is contained in the first volume of the *Ergebnisse der Aerodynamischen Versuchsanstalt zu Göttingen*, accomplished at "Christmas 1920" (according to the preface), where Prandtl offered a formula for the friction coefficient proportional to $Re^{-1/5}$, with the Reynolds number Re related to the length of the plate (Prandtl, 1921b, p. 136). Although Prandtl did not present the derivation, he could not have arrived at this friction coefficient without the 1/7th law for the velocity distribution. (The derivation was based on the assumption that the shear stress at the wall inside the tube only depends on the flow in the immediate vicinity of the wall; hence it should not depend on the radius of the tube. Under the additional assumption that the velocity grows according to a power law with increasing distance from the wall, the derivation was straightforward.)

Kármán presented his theory on turbulent skin friction again in 1922 at a conference in Innsbruck (von Kármán, 1924). He perceived it only as a first step on the way towards a more fundamental understanding of turbulent friction. The solution, he speculated at the end of his Innsbruck talk, would probably come from a statistical consideration. But in order to pursue such an investigation "a fortunate idea" was necessary, "which so far has not yet been found" (von Kármán, 1924, p. 167). (For more on the quest for a statistical theory in the 1920s, see Battimelli, 1984.) Prandtl, too, raised little hope for a more fundamental theory of turbulence from which empirical laws, such as that of Blasius, could be derived from first principles: "You ask for the theoretical derivation of Blasius' law for pipe friction," Prandtl responded[23] to the question of a colleague in 1923. "The one who will find it will thereby become a famous man!"

2.5 The mixing length approach

Prandtl's ideas concerning fully developed turbulence remained the subject of informal conversations and private correspondence for several more years after 1921. "I myself have brought nothing to paper concerning the 1/7-law," Prandtl wrote[24] to Kármán in continuation of their correspondence about the turbulent boundary layer theory in summer 1921. A few years later, the velocity distribution in the turbulent boundary layer of a smooth plate in a wind tunnel was measured directly in Johannes M. Burgers' (1895–1981) laboratory in Delft

[22] Prandtl, notices, MPGA, Abt. III, Rep. 61, Nr. 2296, page 65.
[23] Prandtl to Birnbaum, 7 June 1923. MPGA, Abt. III, Rep. 61, Nr. 137.
[24] Prandtl to Kármán, 14 June 1921. MPGA, Abt. III, Rep. 61, Nr. 792.

by the new method of hot wire anemometry (Burgers, 1925). Prandtl had hesitated in 1921 to publish his derivation of the 1/7th law because, as he revealed in another letter[25] to his rival at Aachen, he aimed at a theory in which the experimental evidence would play a crucial role. Four years later, with the data from Burgers' laboratory, from the dissertation of Nikuradse (1926), and from other investigations about the resistance of water flow in smooth pipes (Jakob and Erk, 1924), this evidence was available. The experiments confirmed the 1/7th law within the range of Reynolds numbers for which the Blasius 1/4th law was valid. But they raised doubts whether it was valid for higher Reynolds numbers. Prandtl, therefore, attempted to generalize his theoretical approach so that he could derive from any empirical resistance law a formula for the velocity distribution. He wrote[26] to von Kármán in October 1924 thus:

> I myself have occupied myself recently with the task to set up a differential equation for the mean motion in turbulent flow, which is derived from rather simple assumptions and seems appropriate for very different cases. ... The empirical is condensed in a length which is entirely adjusted to the boundary conditions and which plays the role of a free path length.

Thus he alluded to the 'mixing length' approach, as it would be labeled later. He published this approach together with the derivation of the 1/7th law. Prandtl's basic idea was to replace the unknown eddy viscosity ϵ in Boussinesq's formula for the turbulent shear stress, $\tau = \rho\epsilon\frac{dU}{dy}$, by an expression which could be tested by experiments. The dimension of ϵ is $\frac{m^2}{s}$, i.e. the product of a length and a velocity. Prandtl made the Ansatz

$$\epsilon = l \cdot l \left|\frac{dU}{dy}\right|,$$

with $l\left|\frac{dU}{dy}\right|$ as the mean fluctuating velocity with which a 'Flüssigkeitsballen' ('fluid bale' or 'fluid eddy') caused a lateral exchange of momentum. Formally, the approach was analogous to the kinetic theory of gases, where a particle could travel a mean free path length before it exchanged momentum with other particles. In turbulent flow, however, the exchange process was less obvious. Prandtl visualized l first as a braking distance (Prandtl, 1925, p. 716, or, in English, Prandtl, 1949E) then as a mixing length (Prandtl, 1926a, p. 726). He made this approach also the subject of his presentation at the Second International Congress of Applied Mechanics, held in Zürich during 12–17 September 1926 (Prandtl, 1927b).

[25] Prandtl to Kármán, 16 February 1921. MPGA, Abt. III, Rep. 61, Nr. 792.
[26] Prandtl to Kármán, 10 October 1924. MPGA, Abt. III, Rep. 61, Nr. 792.

The historic papers on turbulent stress and eddy viscosity by Reynolds and Boussinesq were of course familiar to Prandtl since Klein's seminars in 1904 and 1907, but without further assumptions these approaches could not be turned into practical theories. At a first look, Prandtl had just exchanged one unknown quantity (ϵ) with another (l). However, in contrast to the eddy viscosity ϵ, the mixing length l was a quantity which, as Prandtl had written[27] to Kármán, "is entirely adjusted to the boundary conditions" of the problem under consideration. The problem of turbulent wall friction, however, required rather sophisticated assumptions about the mixing length. Without wall interactions, the mixing length could be adjusted less arbitrarily. Prandtl resorted to other phenomena for illustrating the mixing length approach, such as the mixing of a turbulent jet ejected from a nozzle into an ambient fluid at rest. In this case the assumption that the mixing length is proportional to the width of the jet in each cross-section gave rise to a differential equation from which the broadening of the jet behind the nozzle could be calculated. The theoretical distribution of mean flow velocities obtained by this approach was in excellent agreement with experimental measurements (Prandtl, 1927b; Tollmien, 1926).

For the turbulent shear flow along a wall, however, the assumption of proportionality between the mixing length l and the distance y from the wall did not yield the 1/7th law as Prandtl had hoped. Instead, when he attempted to derive the distribution of velocity for plane channel flow, he arrived at a logarithmic law – which he dismissed because of unpleasant behavior at the centerline of the channel (see Figure 2.3).[28] From his notes in summer 1924 it is obvious that he struggled hard to derive an appropriate distribution of velocity from one or another plausible assumption for the mixing length – and appropriate meant to him that the mean flow $U(y) \propto y^{1/7}$, not some logarithmic law.

Three years later (Prandtl, 1930, p. 794) in a lecture in Tokyo in 1929, he dismissed the logarithmic velocity distribution again. He argued that "l proportional y does not lead to the desired result because it leads to U prop. $\log y$, which would yield $U = -\infty$ for $y = 0$."

This provided an opportunity for Kármán to win the next round in their 'gentlemanly' competition.

2.6 Skin friction and turbulence II: the logarithmic law and beyond

In June 1928, Walter Fritsch, a student of Kármán, published the results of an experimental study of turbulent channel flow with different wall surfaces

---

[27] Prandtl to Kármán, 10 October 1924. MPGA, Abt. III, Rep. 61, Nr. 792.
[28] Prandtl, notices, MPGA, Abt. III, Rep. 61, Nr. 2276, page 12.

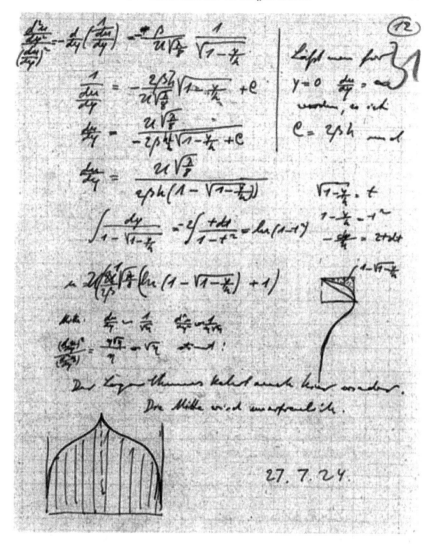

Figure 2.3 Excerpt of Prandtl's 'back of the envelope' calculations from 1924.

(Fritsch, 1928). He found that the velocity profiles line up with each other in the middle parts if they are shifted parallel. This suggested that the velocity distribution in the fluid depends only on the shear stress transferred to the wall and not on the particular wall surface structure. Kármán derived from this empirical observation a similarity approach. In a letter to Burgers he praised this approach for its simplicity: "The only important constant thereby is the proportionality factor in the vicinity of the wall." As a result, he was led to

logarithmic laws both for the velocity distribution in the turbulent boundary
layer and for the turbulent skin friction coefficient. "The resistance law fits
very well with measurements in all known regions," he concluded, with a hint
to recent measurements.[29]

The recent measurements to which Kármán alluded where those of Fritsch in
Aachen and Nikuradse in Göttingen. The latter, in particular, showed a marked
deviation from Blasius' law, and hence from the 1/7th law for the distribution of velocity, at higher Reynolds numbers. Nikuradse had presented some
of his results in June 1929 at a conference in Aachen (Nikuradse, 1930); the
comprehensive study appeared only in 1932 (Nikuradse, 1932). By introducing a dimensionless wall distance $\eta = v_* y / v$ and velocity $\varphi = u/v_*$, where
$v_* = \sqrt{\tau_0/\rho}$ is the friction velocity, τ_0 the shear stress at the wall and ρ the
density, Nikuradse's data suggested a logarithmic velocity distribution of the
form $\varphi = a + b \log \eta$.

Backed by these results from Prandtl's laboratory, Kármán submitted a paper entitled *Mechanical Similarity and Turbulence* to the Göttingen Academy
of Science. Unlike Prandtl, he introduced the mixing length as a characteristic
scale of the fluctuating velocities determined by $l = kU'/U''$, where k is a dimensionless constant (later called the 'Kármán constant') and U', U'' are the
first and second derivatives of the mean velocity of a plane parallel flow in the
x-direction with respect to the perpendicular coordinate y. He derived this formula from the hypothesis that the velocity fluctuations are similar anywhere
and anytime in fully developed turbulent flow at some distance from a wall.
He had plane channel flow in mind, because he chose his coordinate system so
that the x-axis coincided with the centerline between the walls at $y = \pm h$. The
approach would fail both at the center line and at the walls, but was supposed
to yield reasonable results in between. (For more detail on Kármán's approach,
see Chapter 3 by Leonard and Peters.) Whereas Prandtl's approach required a
further assumption about the mixing length, Kármán's l was an explicit function of y at any point in the cross-section of the flow. Kármán obtained a logarithmic velocity distribution and a logarithmic formula for the turbulent friction
coefficient (von Kármán, 1930a).

A few months later, Kármán presented his theory at the Third International
Congress of Applied Mechanics, held in Stockholm during 24–29 August 1930.
For this occasion he also derived the resistance formula for the turbulent skin
friction of a smooth plate. "The resistance law is no power law," hinting at
the earlier efforts of Prandtl and himself. "I am convinced that the form of the
resistance law as derived here is irrevocable." He presented a diagram about

---

[29] Kármán to Burgers, 12 December 1929. TKC 4.22.

the plate skin friction where he compared the 'Prandtl v. Kármán 1921' theory with the 'new' one, and with recent measurements from the Hamburgische Schiffbau–Versuchsanstalt. "It appears to me that for smooth plates the last mismatch between theory and experiment has disappeared," he concluded his Stockholm presentation (von Kármán, 1930b).

Prandtl was by this time preparing a new edition of the *Ergebnisse der Aerodynamischen Versuchsanstalt zu Göttingen* and eager to include the most recent results.[30] The practical relevance of Kármán's theory was obvious. In May 1932, the Hamburgische Schiffbau–Versuchsanstalt convened a conference where the recent theories and experiments about turbulent friction were reviewed. Kármán was invited for a talk on the theory of the fluid resistance, but he could not attend so that he contributed only in the form of a paper which was read by another attendee (von Kármán, 1932). Franz Eisner, a scientist from the Preussische Versuchsanstalt für Wasserbau und Schiffbau in Berlin, addressed the same theme from a broader perspective, and Günther Kempf from the Hamburg Schiffbau–Versuchsanstalt presented recent results about friction on smooth and rough plates (Eisner, 1932b; Kempf, 1932). Prandtl and others were invited to present commentaries and additions (Prandtl et al., 1932). By and large, this conference served to acquaint practitioners, particularly engineers in shipbuilding, with the recent advances achieved in the research laboratories in Göttingen, Aachen and elsewhere.

Two months after this conference, the Schiffbautechnische Gesellschaft published short versions of these presentations in its journal *Werft, Reederei, Hafen*. From Eisner's presentation a diagram about plate resistance was shown which characterized the logarithmic law "after Prandtl (Ergebnisse AVA Göttingen, IV. Lieferung 1932" as the best fit of the experimental values. According to this presentation, the "interregnum of power laws" had lasted until 1931, when Prandtl formulated the correct logarithmic law (Eisner, 1932a). When Kármán saw this article he was upset. He felt that his breakthrough for the correct plate formula in 1930 as he had presented it in Stockholm was ignored. He complained in a letter to Prandtl[31] that from the article about the Hamburg conference "it looks as if I have given up working on this problem after 1921, and that everything has been done in 1931/32 in Göttingen". He asked Prandtl to correct this erroneous view in the Göttingen *Ergebnisse*, which he regarded as the standard reference work for all future reviews. "I write so frankly how I think in this matter because I know you as the role model of a just man," appealing to Prandtl's fairness. But he had little sympathy for "your lieutenants who

---

[30] Prandtl to Kármán's colleagues at Aachen, 30 October 1930. TKC 23.43; Prandtl to Kármán, 29 November 1930; Kármán to Prandtl, 16 December 1930. MPGA, Abt. III, Rep. 61, Nr. 792.

[31] Kármán to Prandtl, 26 September 1932. MPGA, Abt. III, Rep. 61, Nr. 793.

understandably do not know other gods beside you. They wish to claim every-thing for Göttingen." He was so worried that he also sent Prandtl a telegram[32] with the essence of his complaint.

Prandtl responded immediately. He claimed[33] that he had no influence on the publications in *Werft, Reederei, Hafen*. With regard to the *Ergebnisse der Aero-dynamischen Versuchsanstalt zu Göttingen* he calmed Kármán's worries saying that in the publication they would of course refer to the latter's papers. As in the preceeding volumes of the *Ergebnisse*, the emphasis was on experimental results. The news about the logarithmic laws were presented in a rather short theoretical part (12 out of 148 pages) entitled *On turbulent flow in pipes and along plates*. By and large, Prandtl arrived at the same results as Kármán. He duly acknowledged Kármán's publications from the year 1930, but he claimed that he had arrived at the same results at a time when Kármán's papers had not yet been known, so that once more, like ten years before with the same prob-lem, the thoughts in Aachen and Göttingen followed parallel paths (Prandtl, 1932, p. 637). For the Hamburg conference proceedings, Prandtl and Eisner (1932) formulated a short appendix where they declared "that the priority for the formal [formelmässige] solution for the resistance of the smooth plate un-doubtedly is due to Mr. v. Kármán who talked about it in August 1930 at the Stockholm Mechanics Congress."

When Kármán was finally aware of these publications, he felt embarrassed: "I hope that there will not remain an aftertaste from this debate," he wrote to Prandtl[34]. Prandtl admitted that he had "perhaps not without guilt" contributed to Kármán's misgivings. But he insisted[35] that his own version of the theory of plate resistance was better suited for practical use. Although the final results of Prandtl's and Kármán's approaches agreed with each other, there were differ-ences with regard to the underlying assumptions and the ensuing derivations. Unlike Kármán, Prandtl did not start from a similarity hypothesis. There was no 'Kármán's constant' in Prandtl's version. Instead, when Prandtl accepted the logarithmic law as empirically given, he used the same dimensional con-siderations from which he had derived the 1/7th law from Blasius' empirical law. In retrospect, with the hindsight of Prandtl's notes[36], it is obvious that he came close to Kármán's reasoning – but the problem of how to account for the viscous range close to the wall (which Kármán bypassed by using the center-line of the channel as his vantage point) prevented a solution. In his textbook

---

[32] Kármán to Prandtl, 28 September 1932. MPGA, Abt. III, Rep. 61, Nr. 793.
[33] Prandtl to Kármán, 29 September 1932. MPGA, Abt. III, Rep. 61, Nr. 793.
[34] Kármán to Prandtl, 9 December 1932. MPGA, Abt. III, Rep. 61, Nr. 793.
[35] Prandtl to Karman, 19 December 1932. MPGA, Abt. III, Rep. 61, Nr. 793.
[36] Prandtl, notes, MPGA, Abt. III, Rep. 61, Nr. 2276, 2278.

Figure 2.4 Picture of the Rauhigkeitskanal at the Max Planck Institute for Dynamics and Self-Organization. It was built in 1935 and reconstituted by Helmut Eckelmann and James Wallace in the 1970s.

presentations (Prandtl, 1931, 1942a), he avoided the impression of a rivalry about the 'universal wall law' and duly acknowledged Kármán's priority.

But the rivalry between Prandtl and Kármán did not end with the conciliatory exchange of letters in December 1932. Kármán, who had moved in 1933 permanently to the USA, presented his own version of *Turbulence and Skin Friction* in the first issue of the new *Journal of the Aeronautical Sciences* (von Kármán, 1934). The Göttingen viewpoint was presented in textbooks such as Schlichting's *Boundary Layer Theory*, which emerged from wartime lectures that were translated after the war and first published as Technical Memoranda of NACA (Schlichting, 1949). The Göttingen school was also most active in elaborating the theory for practical applications which involved the consideration of pressure gradients (Gruschwitz, 1931) and roughness (Nikuradse, 1933; Prandtl, 1933E; Prandtl and Schlichting, 1934; Prandtl, 1934). After these basic studies, the turbulent boundary continued to be a major concern at Göttingen. By 1937–1938 the engineer Fritz Schultz-Grunow had joined the Institute and built (see Figure 2.4) a special 'Rauhigkeitskanal' – roughness channel – for the study of airplane surfaces (Schultz-Grunow, 1940), which was used subsequently for a variety of war-related turbulence research (Wieghardt, 1947; Prandtl, 1948a). This wind tunnel, which became the workhorse at KWI for measurements during 1939–45 (Wieghardt, 1942, 1943, 1944;

Wieghardt and Tillmann, 1944, 1951E) is the only tunnel that survived the dismantling at the end of the war as it was part of the KWI and not the AVA[37].

2.7 Fully developed turbulence I: 1932 to 1937

Prandtl's interest in fully developed turbulence[38] – beyond the quest for a 'universal wall law' – started in the early 1930s and is best captured following the regular correspondence he had with G.I. Taylor. Starting in 1923, Prandtl was regularly communicating with Taylor on topics in turbulence and instabilities. In 1923, after reading the seminal paper by Taylor (1923) on Taylor–Couette flow, Prandtl in his reply sent him a package of iron-glance powder (hematite) for flow visualization. It was the same material Prandtl had used for his visualization studies that led him to his 1904 discovery. He proposed[39] to Taylor to use it in his experiments, which Taylor immediately and successfully did. This initial contact led later to a very close relationship between the two giants of fluid mechanics. Although their relationship broke 15 years later on a disagreement about the politics of the Third Reich, their close relationship and the openness with which they communicated is impressive. Prandtl and Taylor were sometimes exchanging letters weekly. Prandtl visited Cambridge three times. The first time in 1927 was on the invitation by Taylor, the second time in 1934 for the Fourth International Congress of Applied Mechanics and the last time in 1936 to receive an honorary doctorate from Cambridge University. Prandtl would usually write[40] in typewritten German (in which Taylor's wife was fluent) while Taylor would reply in handwriting in English.

A letter from 1932 from Prandtl to Taylor marks a new stage with regard to turbulence. Prandtl refers to Taylor's recent work following the measurements of Fage and Falkner (see Chapter 4 by Sreenivasan):

> Your new theory of the wake behind a body and the experimental statement by Mr. Fage and Mr. Falkner on this matter reveals a very important new fact

[37] Private communication Helmut Eckelmann, and, as commented upon in the British Intelligence Objectives Sub-Committee report 760 that summarizes a visit at the KWI 26–30 April 1946, "Much of the equipment of the AVA has been or is in process of being shipped to the UK under MAP direction, but the present proposals for the future of the KWI Göttingen, appear to be that it shall be reconstituted as an institute for fundamental research in Germany under Allied control, in all branches of physics, not solely in fluid motion as hitherto. Scientific celebrities now at the KWI include Profs. Planck, Heisenberg, Hahn and Prandtl among others. In the view of this policy, it is only with difficulty that equipment can be removed from the KWI. The KWI records and library have already been reconstituted." Five months later the Max Planck Society was founded.

[38] It is important to note that all the research on turbulence in Göttingen was conducted at Prandtl's Kaiser Willhem Institute and not at the more technically oriented AVA.

[39] Prandtl to Taylor, 25 April 1923. MPGA, Abt. III, Rep. 61, Nr. 1653.

[40] Prandtl to Taylor, 5 June 1934. MPGA, Abt. III, Rep. 61, Nr. 1653.

concerning turbulence. It demonstrates, that there are two different forms of turbulence, one belonging to the fluid motions along walls and the other belonging to mixture of free jets. In the first the principal axis of vorticity is parallel to the direction of the main flow, in the other this direction is perpendicular to the flow.

He continued that they found agreement with Taylor's theory at Göttingen by measuring the flow of a cold jet of air through a warm room. He closed the letter[41] with the following footnote:

> In the last weeks I studied your old papers from 1915 and 1922 with the greatest interest. I think, that if I had known these papers. I would have found the way to turbulence earlier.

Thus Prandtl concluded that there are two kinds of turbulence, one being wall-turbulence and the other jet-turbulence (Prandtl, 1933). A third kind of fully developed turbulence, the turbulence in a wind tunnel, had appeared on Prandtl's agenda as early as in 1921, when Hugh Dryden from the National Bureau of Standards in Washington, DC, had asked[42] him about "a proper method of defining numerically the turbulence of tunnels and your idea as to the physical conception of the turbulence". Already then Prandtl was considered the pioneer in wind tunnel design as is reflected in his instructions that he wrote in 1932 in the *Handbook of Experimental Physics* and that were translated shortly thereafter into English (Prandtl, 1933E2).

In his response[43] Prandtl had pointed to the vortices in the turbulent air stream that

> are carried with the flow and are in time consumed by the viscosity of the air. In a turbulent flow the velocity of the flow is changing in space and time. Characteristic quantities are the average angular velocity of the vortex and the diameter, whereby one has to think of a statistical distribution, in which vortices of different sizes and intensity coexist next to each other.

However, without the appropriate means for measuring these quantities, the problem disappeared again from his agenda – until the 1930s, when the isotropic turbulence behind a grid in a wind tunnel was measured with sophisticated new techniques. Fage and Townend (1932) (see also Collar, 1978) had investigated the full three-dimensional mean flow and the associated three-dimensional average velocity fluctuations in turbulent channel and pipe flow using particle tracking streak images of micron size tracers with a microscope. In addition, Dryden and Kuethe (1929) (see also Kuethe, 1988) invented compensated hotwire measurements, which were to revolutionize the field of

---

[41] Prandtl to Taylor, 25 July 1932. MPGA, Abt. III, Rep. 61, Nr. 1653.
[42] Dryden to Prandtl, 6 March 1921. MPGA, Abt. III, Rep. 61, Nr. 362.
[43] Prandtl to Dryden, 20 April 1921. MPGA, Abt. III, Rep. 61, Nr. 362.

turbulence measurements. This set the stage for spectral measurements of turbulent velocity fluctuations.

From two letters between Taylor and Prandtl in August and December 1932, following the correspondence[44] discussed above, it is apparent that both had started to conduct hotwire experiments to investigate the turbulent velocity fluctuations: Taylor in collaboration with researchers at the National Physical Laboratory (NPL) – most likely Simmons and Salter (see Simmons et al., 1938) – and Prandtl with Reichardt (see Reichardt, 1933; Prandtl and Reichardt, 1934). Taylor responded[45] with suggestions for pressure correlation measurements and argued:

> The same kind of analysis can be applied to hotwire measurements and I am hoping to begin some work on those lines. In particular the 'spectrum of turbulence' has not received much attention.

Prandtl replied[46]:

> I do not believe that one can achieve a clear result with pressure measurements, as there is no instrument that can measure these small pressure fluctuations with sufficient speed. Instead hotwire measurements should lead to good results. We ourselves have conducted an experiment in which two hotwires are placed at larger or smaller distances from each other and are, with an amplifier, connected to a cathode ray tube such that the fluctuations of the one hotwire appear as horizontal paths, and those of the other as perpendicular paths on the fluorescent screen[47] ... To measure also the magnitude of the correlation my collaborator Dr. Reichardt built an electrodynamometer with which he can observe the mean of u'_1, u'_2 and $u'_1 u'_2$. In any case, I am as convinced as you that from the study of those correlations as well as between the direction and magnitude fluctuations, for which we have prepared a hotwire setup, very important insights into turbulent flows can be gained.

In the same letter Prandtl sketched three pictures of the deflections of the oscilloscope that are also published in Prandtl and Reichardt (1934). In this article the authors reported that the hotwire measurements leading to the figures had been finished in August 1932 (date of the letter of Taylor to Prandtl), and that in October 1933 a micro-pressure manometer had been developed to measure the very weak turbulent fluctuations.[48] It is very remarkable that it took less than a year for Prandtl and Reichardt to pick up the pressure measurement proposal by Taylor. It also shows the technical ingenuity and the excellent mechanics workshop at the Göttingen KWI. The micro-pressure

---

[44] Prandtl to Taylor, 25 July 1932. MPGA, Abt. III, Rep. 61, Nr. 1653.
[45] Taylor to Prandtl, 18 August 1932. MPGA, Abt. III, Rep. 61, Nr. 1653.
[46] Prandtl to Taylor, 23 December 1932. MPGA, Abt. III, Rep. 61, Nr. 1653.
[47] This way of showing correlations had been used at the KWI since 1930 (Reichardt, 1938b).
[48] See the paper Reichardt (1934) which was at that time in preparation.

gauge first described in Reichardt (1935, 1948E) is still a very useful design.

This exchange of letters marks the beginning of the correlation and spectral analysis of turbulent fluctuations that are at the foundation of turbulence research even today. Only three weeks later, Taylor replied[49] from a skiing vacation in Switzerland. He relied on the NPL with regard to wind tunnel measurements. They measured the spectrum of turbulence behind a screen of equally spaced rods and found it to settle down to a time error function for which he had no theoretical explanation. Prandtl suggested in his reply that the frequency spectrum behind a grid made of rods may be attributable to von Kàrmàn vortices. He added that "apart from this one needs to wait for the publication". Finally he asked[50] whether Taylor could have his letters rewritten by someone else in more legible writing as he had problems in deciphering Taylor's handwriting. This seems to have caused an interruption of their communication on turbulence for a while.

The next letter[51] in the MPG-Archive is from June 1934. Taylor invited Prandtl to stay in his house during the upcoming Fourth International Congress for Applied Mechanics, to be held in Cambridge during 3–9 July 1934. Prandtl answered[52] in a quite formal and apologetic manner: "In reply to your exceedingly friendly lines from 1.6.34 I may reply to you in German, as I know that your wife understands German without difficulty."

The discussion on turbulence came back to full swing after Prandtl's 60th birthday on 4 February 1935, with almost weekly correspondence. The year 1935 was the one in which Taylor published what many regard as his most important papers in turbulence (Taylor, 1935a,b,d,e). The correspondence between the two that year seems to have greatly influenced those papers. Taylor contributed as the only non-German scientist to the Festschrift published in *ZAMM* and handed to Prandtl at the occasion of his birthday. In his article Taylor compared his calculation of the development of turbulence in a contraction with independent measurements by Salter using a hot wire, as well as photographs by Townend of spots of air heated by a spark and by Fage using his ultramicroscope (Taylor, 1935a). In other words, the best English fluid-dynamicists contributed to this Festschrift.

Prandtl thanked Taylor immediately[53] asking him about details of the paper. The reply from Taylor convinced Prandtl of the correctness of Taylor's work. As a sideline, Taylor also mentioned that turbulence after a constriction

---

[49] Taylor to Prandtl, 14 January 1933. MPGA, Abt. III, Rep. 61, Nr. 1653.
[50] Prandtl to Taylor, 25 January 1933. MPGA, Abt. III, Rep. 61, Nr. 1653.
[51] Taylor to Prandtl, 1 June 1934. MPGA, Abt. III, Rep. 61, Nr. 1653.
[52] Prandtl to Taylor, 5 June 1934. MPGA, Abt. III, Rep. 61, Nr. 1653.
[53] Prandtl to Taylor, 28 February 1935. MPGA, Abt. III, Rep. 61, Nr. 1654.

readjusts itself into a condition where the turbulent velocities are much more nearly equally distributed in space.

(This was later investigated in detail in Comte-Bellot and Corrsin, 1966.) In the same letter[54] Taylor informed Prandtl that

> I lately have been doing a great deal of work on turbulence... In the course of my work I have brought out two formulae which seem to have practical interest. The first concerns the rate of decay of energy in a windstream... and I have compared them with some of Dryden's measurements behind a honeycomb – it seems to fit. It also fits Simmons' measurements with turbulence made on a very different scale ...

The second formula was concerned with the "theory of the critical Reynolds number of a sphere behind a turbulence-grid", as Prandtl replied in his letter[55] pointing him to his own experimental work from 1914. Prandtl also mentioned that measurement from Göttingen see a signature of the grid. Taylor interpreted this as the "shadow of a screen", which according to Dryden's experiments dies away after a point, where the turbulence is still fully developed.[56] It was this region where Taylor expected his theory to apply. Taylor submitted his results in four consecutive papers "On the statistics of turbulence" on 4 July 1935 (Taylor, 1935a, 1935b, 1935d, 1935e). Later Prandtl re-derived Taylor's decay law of turbulence (Wieghardt, 1941, 1942E; Prandtl and Wieghardt, 1945). A detailed discussion of the physics of the decay law of grid-generated turbulence can be found in Chapter 4 by Sreenivasan.

A month later Taylor[57] thanked Prandtl for sending him his paper with Reichardt (Prandtl and Reichardt, 1934) on measurements of the correlations of turbulent velocity fluctuations that Prandtl had already referred to in his letter[58] in 1932. Taylor needed these data for "comparison with my theory of energy dissipation".

Again the correspondence with Prandtl surely contributed to Taylor's understanding and finally led to the third paper in the 1935 sequence (Taylor, 1935d). After returning from the 5th Volta Congress in Rome (on high speeds in aviation), which both attended, Prandtl mentioned[59] to Taylor that "Mr. Reichardt conducts new correlation measurements this time correlations between u' and v'. The results we will send in the future". Again, just as in 1933[60],

[54] Taylor to Prandtl, 2 March 1935. MPGA, Abt. III, Rep. 61, Nr. 1654.
[55] Prandtl to Taylor, 12 March 1935. MPGA, Abt. III, Rep. 61, Nr. 1654.
[56] Taylor to Prandtl, 14 March 1935. MPGA, Abt. III, Rep. 61, Nr. 1654.
[57] Taylor to Prandtl, 21 April 1935. MPGA, Abt. III, Rep. 61, Nr. 1654.
[58] Prandtl to Taylor, 23 December 1932. MPGA, Abt. III, Rep. 61, Nr. 1653.
[59] Prandtl to Taylor, 12 November 1935. MPGA, Abt. III, Rep. 61, Nr. 1654.
[60] Prandtl to Taylor, 25 January 1933. MPGA, Abt. III, Rep. 61, Nr. 1653.

where Prandtl used a similar formulation, the correspondence does not return to the matter of turbulence until more than a year later.

They resumed the discussion again when Taylor sent Prandtl a copy of his paper on "Correlation measurements in a turbulent flow through a pipe" (Taylor, 1936). Prandtl responded[61] that "currently we are most interested in the measurement of correlations between locations in the pipe", suggesting that Taylor may consider measurements away from the center of the pipe and mentioned that "the measurements in Fig. 4 agree qualitatively well with our u' measurements. A better agreement is not to be expected as we measured in a rectangular channel and you in a round pipe." Taylor replied on 11 January 1937 and also on[62] 23 January 1937 when he sent Prandtl "our best measurements so that you may compare with your measurements in a flat pipe".

Thus by 1937 the stage was set at Göttingen and Cambridge for the most important measurements about the statistics of turbulent fluctuations. At the same time, Dryden and his co-workers at the National Bureau of Standards in Washington measured the decay of the longitudinal correlations behind grids of different mesh sizes M (Dryden et al., 1937) and calculated from it by integration of the correlation function the integral scale of the flow, what Taylor (1938b, p. 296) called the "the scale of turbulence". By using different grids they were able to show that the grid mesh size M determined the large-scale L of the flow, just as Taylor had assumed in 1935. They also found that the relative integral scale L/M increased with the relative distance x/M from the grid, independently of M. This was later analyzed in more detail by Taylor (1938b) with data from the National Physical Laboratory in Teddington. The paper by Dryden and collaborators (Dryden et al., 1937) was very important for the further development of turbulence research, as it was data from here that Kolmogorov used in 1941 for comparison with his theory (see Chapter 6 on the Russian school by Falkovich).

2.8 Fully developed turbulence II: 1938

After Taylor's paper on the "Spectrum of turbulence" appeared (Taylor, 1938a), Taylor answered a previous letter[63] by Prandtl. He first thanked Prandtl for sending him Reichardt's $u'v'$ correlation data in a channel flow (Reichardt, 1938a,b; 1951E), which he regarded as "certainly of the same type" as those

---

[61] Prandtl to Taylor, 9 January 1937. MPGA, Abt. III, Rep. 61, Nr. 1654.

[62] Taylor to Prandtl, 23 January 1937. MPGA, Abt. III, Rep. 61, Nr. 1654.

[63] Taylor to Prandtl, 18 March 1938. MPGA, Abt. III, Rep. 61, Nr. 1654. Prandtl's letter which prompted this response has not yet been found.

of Simmons for the round pipe. Then he answered a question of Prandtl about the recent paper on the spectrum of velocity fluctuations (Taylor, 1938a) and explained to him what we now know as 'Taylor's frozen flow hypothesis', i.e.

> that the formula depends only on the assumption that u is small compared to U so that the succession of events at a point fixed in the turbulent stream is <u>assumed</u> to be related directly to the Fourier analysis of the (u, x) curve obtained from simultaneous measurements of u and x along a line parallel to the direction of U.

It is interesting to note that Taylor had a clear concept of the self-similarity of grid-generated turbulence:

> The fact that increasing the speed of turbulent motion leaves the curve $\{UF(n), n/U\}$ unchanged except at the highest levels of n means that an increase in the 'Reynolds number of turbulence' leaves the turbulence pattern unchanged in all its features except in the components of the highest frequency.

Six months after this exchange, Prandtl and Taylor met in Cambridge, MA, for the Fifth International Congress for Applied Mechanics, held at Harvard University and the Massachusetts Institute of Technology during 12–16 September 1938. By now the foundations for seminal discoveries in turbulence research were set. For the next 60 years, experimental turbulence research was dominated by the Eulerian approach, i.e. spatial and equal time measurements of turbulent fluctuations as introduced in the period 1932–1938.

At this Fifth International Congress turbulence was the most important topic.

> In the view of the great interest in the problem of turbulence at the Fourth Congress and of the important changes in accepted views since 1934 it was decided to hold a Turbulence Symposium at the Fifth Congress. Professor Prandtl kindly consented to organize this Symposium and... The Organizing committee is grateful to Professor Prandtl and considers his Turbulence Symposium not only the principal feature of this Congress, but perhaps the Congress activity that will materially affect the orientation of future research

wrote Hunsaker and von Kármán in the 'Report of the Secretaries' in the conference proceedings. Prandtl had gathered the leading turbulence researchers of his time to this event. The most important talks, other than the one by Prandtl, were the overview lecture by Taylor on "Some recent developments in the study of turbulence" (Taylor, 1938b) and Dryden's presentation with his measurements of the energy spectrum (Dryden, 1938). It is interesting to note that, in concluding his paper, Dryden presented a single hotwire technique that he intended to use to measure the turbulent shearing stress $u'v'$. Prandtl's laboratory under Reichardt's leadership had already found a solution earlier[64]

---

[64] Taylor to Prandtl, 18 March 1938. MPGA, Abt. III, Rep. 61, Nr. 1654.

and Prandtl presented this data in his talk. Reichardt used hotwire anemome-try with a probe consisting of three parallel wires, where the center wire was mounted a few millimeters downstream and used as a temperature probe (Re-ichardt, 1938a,b; 1951E). The transverse component of velocity was sensed by the wake of one of the front wires. This probe was calibrated in oscillating lam-inar flow. During the discussion of Dryden's paper, Prandtl made the following important remark concerning the turbulent boundary layer: "One can assume that the boundary zone represents the true 'eddy factory' and the spread to-wards the middle would be more passive." In his comment, he also showed a copy of the fluctuation measurements conducted by Reichardt and Motzfeld (Reichardt, 1938b, Fig. 3) in a wind tunnel of 1 m width and 24 cm height.

Prandtl's 1938 paper at the turbulence symposium deserves a closer review because, on the one hand, it became the foundation for his further work on tur-bulence and, on the other, as an illustration of Prandtl's style. It reflects beau-tifully and exemplarily, what von Kármán, for example, admired as Prandtl's "ability to establish systems of simplified equations which expressed the essen-tial physical relations and dropped the nonessentials"; von Kármán regarded this ability "unique" and even compared Prandtl in this regard with "his great predecessors in the field of mechanics – men like Leonhard Euler (1707–1783) and d'Alembert (1717–1783)".[65]

In this paper, Prandtl distinguished four types of turbulence: wall turbulence, free turbulence, turbulence in stratified flows (see also Prandtl and Reichardt, 1934), and the decaying isotropic turbulence. He first considered the decay law of turbulence behind a grid using his mixing length approach by assuming that the fluctuating velocity is generated at a time t' and then decays:

$$u'^2 = \int_{-\infty}^{t} \frac{dt'}{T} \left[\left(l_1 \frac{dU}{dy} \right)_{t'} \times f\left(\frac{t-t'}{T} \right) \right]^2 ,$$

with $f(\frac{t-t'}{T}) \approx \frac{T}{T+t-t'}$ justified by Dryden's measurements. With the shear stress $\tau = \rho u' l_2 \frac{dU}{dy}$ and $l_2 = km$, where m is the grid spacing and $k \approx 0.103$, he derived

$$u' = \frac{\text{const}}{t+T} = \frac{cU_m}{x} ,$$

where x is the downstream distance from the grid, c is related to the thickness of the rods, and U_m is the mean flow. From the equation of motion to lowest order,

$$U_m \frac{\partial U}{\partial x} = \frac{1}{\rho} \frac{\partial \tau}{\partial y} ,$$

---

[65] See Kármán (1957); Anderson (2005); see also Prandtl's own enlightening contribution to this topic (Prandtl, 1948b).

and the Ansatz

$$U = U_m + Ax^{-n}\cos\left(\frac{2\pi y}{m}\right),$$

he obtained $n = 4\pi^2 k\frac{c}{m}$. By analyzing the largest frequency component of $U - U_m$ (from data provided by Dryden) he determined $n \approx 4.5$. This result led him to assume a transition from anisotropic flow near the grid to isotropic turbulence further downstream. How this transition occurs was left open.

As a next item, Prandtl considered the change of a wall-bounded, turbulent flow at the transition from a smooth to a rough wall and vice versa. He derived model equations and found reasonable agreement with the measurements of his student Willi Jacobs.

The third item of Prandtl's conference paper concerned an ingeniously simple experiment. By visualizing the flow with iron-glance flakes (the same as he had used in his 1904 experiments) he measured what he perceived as the 'Taylor scale' of turbulence in a grid-generated turbulent water flow (see Figure 2.5; in the region of large shear the flakes align and make visible the eddies in the turbulent flow). From the surface area per eddy as a function of mesh distances behind the grid (see Prandtl, 1938, Fig. 1) he found that these areas grew linearly starting from about 16 mesh distances downstream from the grid. From this observation Prandtl concluded that the Taylor scale λ (Taylor, 1935b) increases as $(x - x_0)^{0.5}$, where $x_0 \approx 10$ is the 'starting length'. This was in contradiction with Taylor's own result, but agreed with the prediction by Kármán and Howarth (1938) from eight months earlier. (It is not clear whether Prandtl knew about their work – it is not credited in his paper. Later the von Kármán and Howarth prediction was quantitatively measured with hot wires by Comte-Bellot and Corrsin, 1966.)

Finally, and in retrospect most importantly, Prandtl discussed Reichardt's and Motzfeld's measurements of wall-generated turbulence in channel flow (Reichardt, 1938a,b; 1951E; Motzfeld, 1938). In Figure 2.6(a) we reproduce his Fig. 2. It displays the mean fluctuating quantities $\sqrt{\overline{u'^2}}$, $\sqrt{\overline{v'^2}}$, $\overline{v'u'}$, and $\Psi = \overline{v'u'}/(\sqrt{\overline{u'^2}}\ \sqrt{\overline{v'^2}})$ as a function of distance from the wall to the middle of the tunnel at 12 cm. Here u is the streamwise, v the wall normal, and w wall parallel velocity.

The spectral analysis of the streamwise velocity fluctuation u as a function of frequency revealed that the frequency power spectra were indistinguishable for distances of 1 to 12 cm from the wall, although $\sqrt{\overline{u'^2}}$ decreased by more than a factor of 2 over the same range (see Figure 2.6(a), reproduced from Motzfeld, 1938). Prandtl showed in his Fig. 4 (here Figure 2.6(b)) the semi-log plot of $nf(n)$, which displays a maximum at the largest scales of

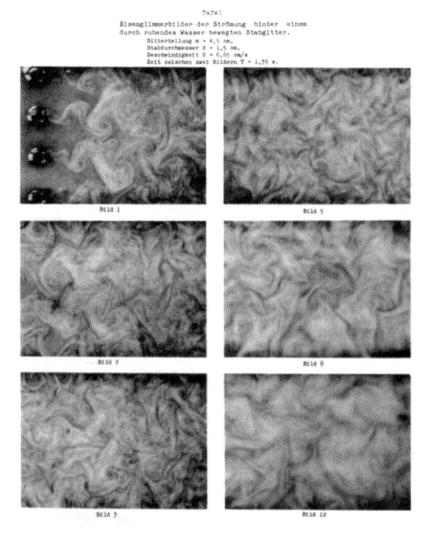

Figure 2.5 Prandtl's visualization of the development of grid-generated isotropic turbulence. Pictures were taken at relative grid spacings of 2, 4, 6, 10, 16 and 24.

the flow. Later (Lumley and Panofsky, 1964), the location of this maximum was proposed as a surrogate for the integral scale. As shown in Figure 2.6(c), Motzfeld also compared his data with the 1938 wind tunnel data in Simmons et al. (1938) by rescaling both datasets with the mean velocity U and the

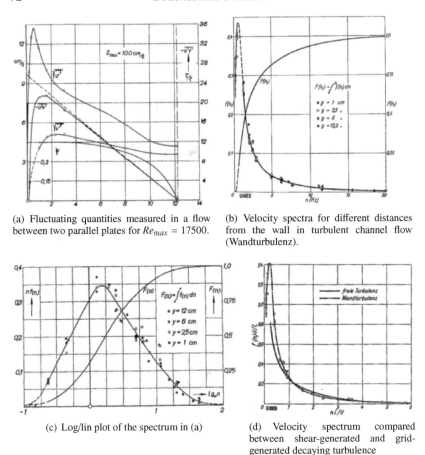

(a) Fluctuating quantities measured in a flow between two parallel plates for $Re_{max} = 17500$.

(b) Velocity spectra for different distances from the wall in turbulent channel flow (Wandturbulenz).

(c) Log/lin plot of the spectrum in (a)

(d) Velocity spectrum compared between shear-generated and grid-generated decaying turbulence

Figure 2.6 Turbulence spectra as measured by Motzfeld and Reichardt in 1938.

channel height L or, for the wind tunnel, with the grid spacing L. As we can see both datasets agree reasonably well. Prandtl remarked about the surprising collapse of the data in Figure 2.6(a) (Fig. 3 in Prandtl, 1938): "The most remarkable about these measurements is that *de facto* the same frequency distribution was found." From the perspective of experimental techniques, the electromechanical measurement technique employed by Motzfeld and Reichardt is also remarkable. As shown in Figure 2.7, they used the amplitude of an electromechanically driven and viscously damped torsion wire resonant oscillator to measure, by tuning resonance frequencies and damping, the frequency components of the hotwire signal.

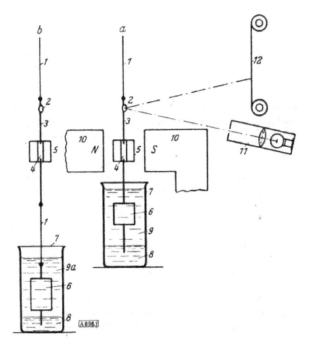

Figure 2.7 Schematic of the electromechanical spectral analysis system used by Motzfeld in 1937/38. Two different designs of a damped torsion pendulum were used: (a) below 20 Hz and (b) above 20 Hz. The design (a) consisted of a torsion wire (1), a thin metal rod (3) with mirror (2), an insulating glass rod (4) around which a coil was wound (5), a swinger consisting of a thin metal rod with a cylindrical body to add inertia (6). The swinger was placed in a beaker (7) that was filled on the top with a damping fluid (9) and on the bottom with mercury (8). Electric currents could flow from (1) into the coil and from there to (8). The electromagnet was placed in a permanent magnetic field. The deflections of the wire were recorded on photographic film that was transported with a motor. In the alternative design (b) for more than 20 Hz the swinger was replaced with a torsion wire (1) and a weight (6). The weight was placed into very viscous oils so that it did not move (9a). Otherwise the design was the same. Altogether ten swingers were used with six of kind (a) and four of kind (b). The resonance frequencies were between 0.2 Hz and 43 Hz.

As we will see, these results would lead (Prandtl and Wieghardt, 1945) not only to the derivation of what is now known as the 'one equation model' (Spalding, 1991), but also to the assumption of a universal energy cascade of turbulence cut off at the dissipation scale (Prandtl, 1945, 1948a). Thus, at age 70, Prandtl had finally found what he had been looking for all his life albeit at the worst time – when the Second World War ended and he was not allowed to conduct scientific research.[66]

Prandtl's sojourn in the USA in September 1938 was also remarkable in another respect – because it marked the beginning of his, and for that matter Germany's, alienation from the international community. When he tried to convince the conference committee to have him organize the next Congress in Germany, he encountered strong opposition based on political and humanitarian reasons. Against many of his foreign colleagues, Prandtl defended Hitler's politics and actions. Taylor attempted to cure Prandtl of his political views.[67] As discussed also in Chapter 4 by Sreenivasan, Taylor was concerned with the humanitarian situation of the Jewish population and the political situation in general. However, Taylor's candor (he called Hitler "a criminal lunatic") did not bode well with Prandtl, who responded again by defending German politics.[68] Only a few days before Prandtl wrote his letter (on 18 October 1938) 12,000 Polish-born Jews were expelled from Germany. On 11 November 1938, the atrocities of the 'Kristallnacht' started the genocide and Holocaust (Gilbert, 2006). Taylor replied on 16 November 1938 with a report about the very bad experiences in Germany of his own family members.[69] Nevertheless he ended his letter still quite amicably:

> You will see that we are not likely to agree on political matters so it would be best to say no more about them. Fortunately there is no reason why people who do not agree politically should not be best friends.

Then he continued to make a remark that he does not understand why Prandtl plotted $nf(n)$ instead of $f(n)$ (shown in Figure 5b) (Fig. 4 in Prandtl, 1938). As far as we know Prandtl never replied. After this correspondence Prandtl wrote one more letter to Mrs Taylor.[70] Only a month later World War II started and cut off their communication. Prandtl tried[71] to resume contact with Taylor after the war, but there is no evidence that Taylor ever responded to this effort.

2.9 Fully developed turbulence III: 1939 to 1945

With the beginning of WWII on 1 September 1939, German research in fluid dynamics became isolated from the rest of the world. This may explain why

---

[66] Prandtl to Taylor, 18 July 1945 and 11 October 1945. MPGA, Abt. III, Rep. 61, Nr. 1654
[67] Taylor to Prandtl, 27 September 1938. MPGA, Abt. III, Rep. 61, Nr. 1654.
[68] Prandtl to Taylor, 29 October 1938. MPGA, Abt. III, Rep. 61, Nr. 1654.
[69] Taylor to Prandtl, 16 November 1938. GOAR 3670-1
[70] Prandtl to Mrs Taylor, 5 August 1939. MPGA, Abt. III, Rep. 61, Nr. 1654.
[71] Prandtl to Taylor, 18 July 1945 and 11 October 1945. MPGA, Abt. III, Rep. 61, Nr. 1654

the very important discovery by Motzfeld and Reichardt was not recognized abroad. We have found no reference to Motzfeld's 1938 publication other than in the unpublished 1945 paper by Prandtl (see below). As described above their discovery showed that the spectrum of the streamwise velocity fluctuation in a channel flow did not depend on the location of the measurements in the channel and did agree qualitatively with those by Simmons and Salter for decaying isotropic turbulence. The 1938 Göttingen results show beautifully the universal behavior that Kolmogorov postulated in his revolutionary 1941 work (see Chapter 6 on the Russian school by Falkovich).

Prandtl and his co-workers were not aware of the developments in Russia and continued their program in turbulence at a slower pace. According to a British Intelligence report[72] after the war, based on an interrogation of Prandtl,

> due to more urgent practical problems little fundamental work, either experimental or theoretical, had been conducted during the war. No work had been done in Germany similar to that of G.I. Taylor or Kármán and Howarth on the statistical theory of turbulence. Experiments had been planned on the decay of turbulence behind grids in a wind tunnel analogous to those undertaken by Simmons at the National Physical Laboratories, but these were shelved at the outbreak of the war.

Indeed as far as fully developed turbulence was concerned the progress was mostly theoretical and mainly relying on measurements made before the war. In his response to the military interrogators, Prandtl was very modest. From late autumn of 1944 till the middle of 1945 he worked on the theory of fully developed turbulence almost daily (see Figure 2.8). This was his most active period in which he pulled together the threads outlined earlier.

We will now review briefly the development from 1939 to 1944 that led to this stage. The status of knowledge of turbulence in 1941 is well summarized in Wieghardt (1941; 1942E), and that between 1941 and 1944 in Prandtl's (1948a) FIAT article entitled *Turbulenz*. Prandtl reviewed in 23 tightly written pages the work at the KWI in chronological order and by these topics:

(i) Turbulence in the presence of walls

    (a) Pipeflow

    (b) Flat plates

    (c) Flow along walls with pressure increase and decrease

(ii) Free turbulence

    (a) General laws

---

[72] British Intelligence Objectives Sub-Committee report 760 that summarizes a visit to the KWI, 26–30 April 1946.

October 1944

Mo	Tu	We	Th	Fr	Sa	Su
						1
2	3	4	5	6	**7**	**8**
9	10	11	12	13	T1	T1
T1	17	T1	19	20	T1	**22**
23	24	25	26	27	**28**	**29**
30	"E

November 1944

Mo	Tu	We	Th	Fr	Sa	Su
		1	T2	3	T2	**5**
T2	7	8	9	10	**11**	**12**
13	14	15	16	17	**18**	**19**
20	21	22	23	24	**25**	T1
T1	T1	29	30

December 1944

Mo	Tu	We	Th	Fr	Sa	Su
				1	**2**	T1
4	T1	T1	T1	8	**9**	T1
11	12	T2	14	15	**16**	T2
T2	19	T2	T1	T1	**23**	**24**
25	T2	27	28	29	**30**	**31**

January 1945

Mo	Tu	We	Th	Fr	Sa	Su
T1	2	T1	S	5	6	T1
8	9	10	11	12	**13**	**14**
15	16	17	18	19	**20**	T2
22	23	24	25	A	**27**	28
K41	30	31

February 1945

Mo	Tu	We	Th	Fr	Sa	Su
			1	2	**3**	B
5	6	7	9	T2	**10**	C
12	13	14	T2	T2	**17**	**18**
19	20	21	22	23	T2	**25**
26	27	28

March 1945

Mo	Tu	We	Th	Fr	Sa	Su
			1	2	**3**	**4**
5	6	7	8	9	**10**	**11**
12	13	14	15	16	**17**	**18**
19	20	21	22	23	**24**	**25**
26	R2	T2	T2	30	**31**

April 1945

Mo	Tu	We	Th	Fr	Sa	Su
						1
2	3	4	5	6	**7**	O
9	10	11	12	13	**14**	**15**
16	T2	T2	19	20	**21**	**22**
23	24	25	26	25	**28**	**29**
30

May 1945

Mo	Tu	We	Th	Fr	Sa	Su
	1	2	3	4	5	6
7	8	9	10	11	**12**	13
14	15	16	17	18	19	20
21	22	23	24	25	**26**	27
28	29	30	31

June 1945

Mo	Tu	We	Th	Fr	Sa	Su
				1	**2**	**3**
4	5	6	7	8	**9**	T2
11	12	T2	14	15	T2	**17**
18	19	20	21	22	**23**	T2
25	26	27	28	29	**30**

July 1945

Mo	Tu	We	Th	Fr	Sa	Su
						1
2	3	*DP*	5	6	**7**	**8**
9	10	11	12	13	**14**	V
V	17	18	19	V	**21**	**22**
23	24	25	26	27	**28**	TS
TS	31

August 1945

Mo	Tu	We	Th	Fr	Sa	Su
		1	TS	3	**4**	**5**
6	7	8	9	10	**11**	TS

Figure 2.8 Prandtl worked continuously on the topic of fully developed turbulence. T1 marks Prandtl's work on the energy equation of turbulence, T2 his investigations on the effect of dissipation, V a derivation of the vorticity equation in a plane shear flow, and TS his attempts to develop a statistical theory of velocity fluctuations. The other letters mark important dates: on 31 October 1944 he formulated for the first time the 'one equation model' (E); on 4 January 1945 he presented it at a theory seminar (S) and on 26 January 1945 at a meeting of the Göttingen Academy of Science (A); 29 January 1945 marks his discovery of what is known as the Kolmogorov length scale (K41); on 4 February 1945 he had his 70th birthday (B); on 11 February 1945 he formulated for the first time his cascade model (C); on 27 March 1945 he is reworking the draft for the paper on dissipation (R2); on 8 April 1945 Göttingen was occupied by American forces; on 4 July 1945 Prandtl entered remarks on the already typewritten draft revisions of the dissipation paper. The period in May, where he had no access to the Institute as it was used by American forces, is light gray – the Institute reopened on 4 June 1945 to close again briefly thereafter.

(b) Special tasks

(c) Properties of jets in jet engines

(iii) Various investigations

(a) Turbulence measurement technologies

(b) Heat exchange

(c) Geophysical applications

(d) Fundamental questions

Prandtl identified as fundamental and important in particular the work by Schultz-Grunow (1940; 1941E) and Wieghardt (1944) on the measurements of the turbulent boundary layer. Even today, these very careful and now classical experiments provide the data for quantitative comparisons (Nagib et al., 2007).

Furthermore, Prandtl singled out the investigations on heat transfer in turbulent boundary layers by Reichardt (1944), who applied ideas from earlier papers on turbulent transport of momentum in a free jet (Reichardt, 1941, 1942). Reichardt had found experimentally for a planar jet that the PDF of transverse variations of the streamwise velocity profile in the middle of a jet was Gaussian. In the middle of such flows $\frac{\partial \bar{u}}{\partial y} = 0$, where the mixing length approach failed by design, as Prandtl (1925) had already noted, when he suggested another way around this problem. Based on the observation of the Gaussian distribution he conjectured inductively that the transfer of momentum was similar to that of heat. By neglecting viscosity he wrote down the two-dimensional planar momentum equation

$$\frac{\partial}{\partial x}(p/\rho + \overline{u^2}) + \frac{\partial \overline{(uv)}}{\partial y} = 0$$

and

$$\overline{uv} = -\lambda \frac{\partial \overline{u^2}}{\partial y},$$

with λ as *Übertragunsgrösse* (transfer quantity). Reichardt calculated some examples and showed that the new theory worked reasonably well. Prandtl (1942c) had already published about it in *ZAMM*. He showed that if the pressure term in lowest order is zero the two equations by Reichardt lead to

$$\frac{\partial}{\partial x}\overline{u^2} = -\lambda \frac{\partial^2 \overline{u^2}}{\partial y^2}.$$

In a subsequent paper Henry Görtler (1942) applied the theory to four cases: the plane mixing layer, the plane jet, the plane wake and the plane grid. He compared the first two cases with the measurements by Reichardt and found good agreement.

As another important result Prandtl highlighted improvements of the hot-wire measurement system by H. Schuh who had found a method for circumventing the otherwise very tedious calibration of each new hotwire probe in a calibration tunnel (Schuh, 1945, 1946). With regard to theoretical achievements, Prandtl reported on the work of the mathematician Georg Hamel who had proved von Kármán's 1930 similarity hypothesis for the two-dimensional flow in a channel as well as Prandtl's log law (Hamel, 1943; Prandtl, 1925).

At the end of the FIAT paper, Prandtl mentioned rather briefly what he had been so deeply engaged in from the autumn of 1944 to the summer of 1945. In only a little more than a page he summarized his energy model of turbulence (the 'one equation model'; Spalding, 1991) and his own derivation of the Kolmogorov length scales, for which he used a cascade model of energy transfer to the smallest scales. The latter he attributed to his unpublished manuscript from 1945 (see the discussion below). Then he reviewed the work by Weizsäcker (1948) and Heisenberg (1948; 1958E) that they had conducted while detained in England from July 1945 to January 1946.[73] Weizsäcker's work was similar, but Prandtl considered it to be mathematically more rigorous than his phenomenologically driven approach. In addition to the results that Prandtl had obtained, Weizsäcker also calculated from the energy transport the $k^{-5/3}$ scaling of the energy spectrum. Prandtl then reviewed the Fourier mode analysis of Heisenberg and commented on the good agreement with experiments. He closed with a hint at a paper in preparation by Weizsäcker concerning the influence of turbulence on cosmogony.

2.10 Prandtl's two manuscripts on turbulence, 1944–1945

When the American troops occupied Göttingen on 8 April 1945, Prandtl had already published the 'one equation model' (Prandtl and Wieghardt, 1945) and drafted a first typewritten manuscript of a paper entitled 'The role of viscosity in the mechanism of developed turbulence' that was last dated by him 4 July 1945 (Prandtl, 1945; see Figure 2.8). In this paper he derived from a cascade model the dissipation length scale, i.e. the 'Kolmogorov length'. Before we describe in more detail his discoveries, it is important to ask why he did not publish this work. Clearly this was an important discovery and would have retrospectively placed him next to Kolmogorov in the "remarkable series of coincidences" (Batchelor, 1946, p. 883) now known as the K41 theory.

---

[73] Their work had also been reviewed by Batchelor in December 1946 together with the work of Kolmogorov and Onsager (Batchelor, 1946). Of course Batchelor had no knowledge of the fact that Prandtl had derived the same results based on similar reasoning already in January 1945.

His drafting of the paper fell right into the end of WWII. By July 1945 the Institute was under British administration and had[74] "many British and American visitors". Prandtl was still allowed[75] "to work on some problems that were not finished during the war and from which also reports were expected. Starting any new work was forbidden." By 11 October 1945 the chances for publication were even worse because[76] "all research was shelved completely" and "any continuation of research was forbidden by the Director of Scientific Research in London". So it seems that as of August 1945 Prandtl followed orders and stopped writing the paper and stopped working on turbulence (see Figure 2.8). In addition, in that period the AVA was being disassembled and the parts were sent to England. Then in January 1946, Heisenberg and Weizsäcker returned to Göttingen from being interned in England and brought along their calculations that superseded Prandtl's work. So by January 1946 the window of opportunity for publication had passed. In addition, he was busily writing the FIAT report on turbulence – and that is where he at least mentioned his work.

Let us now discuss briefly Prandtl's last known work on turbulence. His very carefully written notes[77] cover the period from 14 October 1944 until 12 August 1945; they allow us to understand his achievement better. These notes comprise 65 numbered pages, 5 pages on his talk in a Theory Colloquium on 4 January 1945 where he presented his energy equation of turbulence, 7 pages on an attempt at understanding the distribution function of velocity from probability arguments, and 19 pages of sketches and calculations. Figure 2.8 summarizes the days he entered careful handwritten notes in his workbook. From these entries we see that he devoted a large part of his time to the study of turbulence. It seems remarkable how much time he was able to dedicate to this topic, considering that he also directed the research at his Institute and that he was engaged as an adviser to the Air Ministry concerning the direction of aeronautical research for the war. How much effort he dedicated to the latter activity is open to further historical inquiry.

In order to discern the subsequent stages of Prandtl's approach we proceed chronologically:

On 31 October 1944 we find the first complete formulation of the 'one equation model' for the evolution of turbulent kinetic energy per unit volume in terms of the square of the fluctuating velocity (see Figure 2.9). The same

---

[74] Prandtl to Taylor, 18 July 1945. MPGA, Abt. III, Rep. 61, Nr. 1654.

[75] Prandtl to Taylor, 18 July 1945. MPGA, Abt. III, Rep. 61, Nr. 1654. See also Prandtl to the President of the Royal Society, London, 11 October 1945. MPGA, Abt. III, Rep. 61, Nr. 1402.

[76] Prandtl to Taylor, 18 October 1945. MPGA, Abt. III, Rep. 61, Nr. 1654.

[77] GOAR 3727.

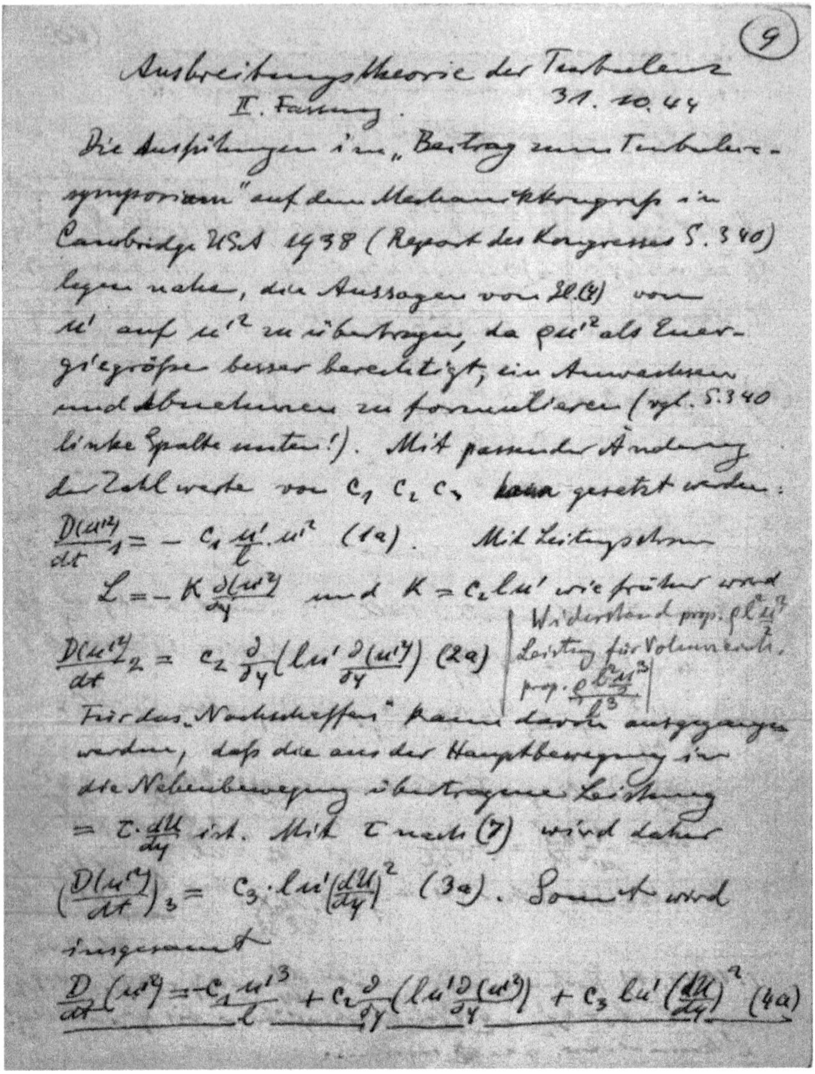

(9)

Ausbreitungstheorie der Turbulenz
II. Fassung. 31.10.44

Die Ausführungen im „Beitrag zum Turbulenz-symposium" auf dem Mechanikkongreß in Cambridge USA 1938 (Report des Kongresses S. 340) legen nahe, die Aussagen von Gl.(8) von u' auf u'^2 zu übertragen, da $\rho u'^2$ als Energiegröße besser berechtigt; ein Anwachsen und Abnehmen zu formulieren (vgl. S. 340 linke Spalte unten!). Mit passender Änderung der Zahlwerte von c_1 c_2 c_3 kann gesetzt werden:

$$\frac{D(u'^2)}{dt}_1 = -c_1 \frac{u'}{l} \cdot u'^2 \quad (1a).$$

Mit Leitungsstrom

$$L = -K\frac{\partial(u'^2)}{\partial y} \quad \text{und} \quad K = c_2 l u'$$

wie früher wird

$$\frac{D(u'^2)}{dt}_2 = c_2 \frac{\partial}{\partial y}\left(l u' \frac{\partial(u'^2)}{\partial y}\right) \quad (2a)$$

Widerstand prop. $\rho l u'^2$; Leistung für Volumeneinh. prop. $\rho \frac{l u'^3}{l^3}$

Für das „Nachschaffen" kann davon ausgegangen werden, daß die aus der Hauptbewegung in die Nebenbewegung übertragene Leistung $= \tau \cdot \frac{dU}{dy}$ ist. Mit τ nach (7) wird daher

$$\left(\frac{D(u'^2)}{dt}\right)_3 = c_3 \cdot l u' \left(\frac{dU}{dy}\right)^2 \quad (3a).$$

Somit wird insgesamt

$$\frac{D}{dt}(u'^2) = -c_1 \frac{u'^3}{l} + c_2 \frac{\partial}{\partial y}\left(l u' \frac{\partial(u'^2)}{\partial y}\right) + c_3 l u' \left(\frac{dU}{dy}\right)^2 \quad (4a)$$

Figure 2.9 Derivation of energy-balance equation without dissipation.

formula now written in terms of kinetic energy per unit volume was presented by him in his paper at the meeting of the Göttingen Academy of Science on 26 January 1945 (Prandtl and Wieghardt, 1945). There Wieghardt also presented his determination of the parameters from measurements in grid turbulence and channel flow and found good agreement with the theory. His differential equation marked as Eq. 4a in Figure 2.9 determined the change of energy per unit volume from three terms: the first term on the right-hand side gives the turbulent energy flux for a "bale of turbulence" (in German: Turbulenzballen) of size l (which he equated with a mixing length), the second term represents the diffusion of turbulence in the direction of the gradient of turbulent energy and the third term represents the source of turbulent energy from the mean shear. Here $cu'l$ is the eddy viscosity. Please note that this equation is what is now called a k-model. This equation was independently derived later by Howard Emmons in 1954 and by Peter Bradshaw in 1967 (Spalding, 1991). Prandtl and Wieghardt also pointed out the deficiencies of the model, namely the role of viscosity at the wall and for the inner structure of a bale of turbulence. For the latter, Prandtl argued that, as long as the Reynolds number of each bale of turbulence is large, a three-dimensional version of his equation should also be applicable to the inner dynamics of the bale of turbulence. He then introduced the idea of bales of turbulence within bales of turbulence, which we now know as the turbulent cascade. He called them steps (in German: Stufen) in the sequence that goes from large to small. He pointed out that by going from step to step the Reynolds number will decrease with increasing number of the step (decreasing size of the turbulent eddy) until viscosity is dominant and all energy is transformed to heat. Finally he conjectured that a general understanding of the process can be obtained.

And indeed he discovered it within a short time. On 29 January 1945, only three days after the presentation at the Academy, he entered in his notes the derivation of the Kolmogorov length scale that at this time he called in analogy to Taylor's smallest length scale λ (Figure 2.10). In Eq. (1) the decay rate of the kinetic energy per unit mass is equated with the dissipation at the smallest scales. Eq. (2) connects the final step of the cascade process with the Kolmogorov velocity. By putting (1) and (2) together Prandtl arrived at the Kolmogorov length scale given by Eq. (3). This seemed to him so remarkable that he commented on the side of the page "Checked it multiple times! But only equilibrium of turbulence."

At this stage he was almost done, but as we can tell from his typewritten manuscript (Prandtl, 1945) and from his notes he was not satisfied. He had to put this on more formal grounds. So only two weeks later, as shown in Figure 2.11 he used a cascade model for each step of turbulence, with β as the ratio in length scale from step to step. This allowed him to derive the Kolmogorov length more rigorously from a geometrical series.

This became the content of his draft paper from 1945 that we shall discuss in detail in a separate publication. It is clear that the spectral data from Motzfeld (1938) were instrumental for his progress (those which Prandtl had presented at the Cambridge Congress in 1938 and which are reproduced above

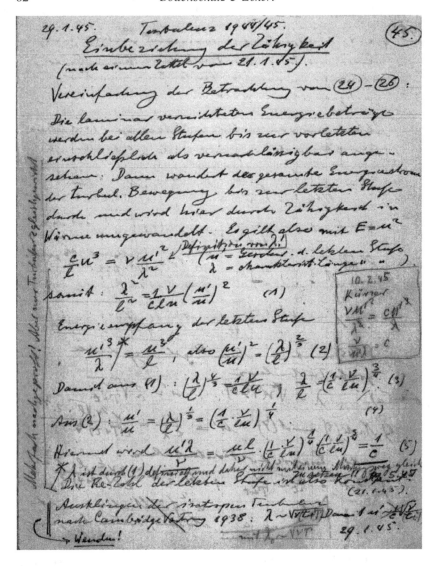

Figure 2.10 First known derivation of the Kolmogorov length scale (here called λ).

in Figure 2.6). Here we close our review with a quote from the introduction to his unpublished paper "The role of viscosity in the mechanism of developed turbulence" (Prandtl, 1945) which beautifully reflects his thinking and needs no further analysis. This is only a short excerpt from the introduction to the paper. A full translation is in preparation and will be published elsewhere. Also,

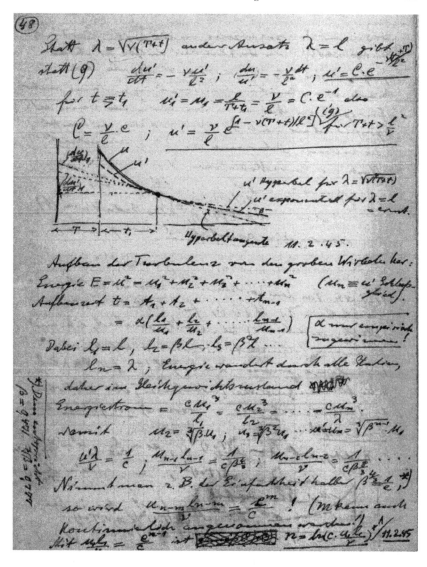

Figure 2.11 Prandtl's cascade model for the fluctuating velocities at different steps in the cascade.

our translation is very close to the original German text and therefore some sentences are a bit long.

The following analyses, which consider in detail the inner processes of a turbulent flow, will prove that the solution for λ by Taylor that he obtained from energy considerations does not yet give the smallest element of turbulence. The mechanism of turbulence generation is not resolved in all details. So much is however known [here Prandtl referred in a footnote to work by Tollmien published in *Göttinger Nachr*. Heft 1 (1935) p. 79] that flows with an inflection point in the velocity profile may become unstable at sufficiently large Reynolds numbers. Therefore one has to expect that at sufficiently high Reynolds number $\frac{u'l}{\nu}$ the motion of an individual bale of turbulence is by itself turbulent, and that for this secondary turbulence the same is true, and so on. Indeed one observes already at very modest Reynolds numbers a frequency spectrum that extends over many decades. That it is mostly the smallest eddies that are responsible for the conversion of the energy of main motion into heat can easily be understood, as for them, the deformation velocities $\frac{\partial u}{\partial y}$, etc. are the largest.

The earlier discussion is the simplest explanation of the fact that in turbulent motion always the smaller eddies are present next to the larger ones. G.I. Taylor, 1935, used a different explanation. He pointed out that according to general statistical relations the probability of two particles separating in time is larger than for them to come closer, and he applied this relationship to two particles on a vortex line. From the well-known Helmholtz theorem it would follow that – as long as the viscosity does not act in an opposing sense – the increase of the angular velocity of the vortex line is more probable than its decay. He shows this tendency with an example, whose series expansion clearly shows the evolution towards smaller eddies. However, it could not be continued, so the processes could only be followed for short time intervals. One can counter Taylor's deductions insofar that through the increase of the angular velocities, pressure fields develop, which oppose a further increase of the vortex lines. It thus cannot be expected that the extension would reach the expected strength. It seems, however, that the action in the sense of Taylor is surely present, if, though, with weaker magnitude than expected from a purely kinematic study.

For the development of smaller eddy diameters in the turbulence, one can also note that wall turbulence starts with thin boundary layers and that free turbulence has equally thin separating sheets. Therefore, in the beginning, only the smallest vortices are present and the larger ones appear one after another. Opposing this, however, is the result that in the fully developed channel flow, the frequency spectrum *de facto* does not depend on the distance from the wall (Motzfeld, 1938). One would not expect this if all of the fine turbulence originated at the wall. This strongly supports the validity of the conjecture for stationary turbulence presented here. Further support is given by investigations conducted later, which concerned isotropic, temporally decaying, turbulence and which have been quite satisfactorily justified by experiments. The two descriptions of the re-creation of the smaller eddies by turbulence of second and higher order, and the one that relates to the Helmholtz theorem, are, by the way, intricately related: they are both, so to say, descriptions that elucidate one and the same process only from different perspectives.

In the following, initially temporally stationary turbulence may be assumed, as is found, for example, in a stationary channel or pipe flow. Of the dissipated power D in a unit volume per unit time, a very small fraction $\mu(\frac{\partial U}{\partial y})^2$ will be dissipated immediately into heat (U is the velocity of the mean flow); the rest, which one may call D_1, increases the kinetic energy of the turbulent submotion [Nebenbewegung] and generates, according to Taylor, secondary turbulence . . .

[Here we leave out some equations.]

We now establish corresponding equations for turbulence of the second step (third step etc.). Instead of the velocity U, here a suitably smoothed velocity of turbulence of first step, second step etc. must be used. The instantaneous values of the velocity u, which is used as a representative for the triple (u, v, w) for simplicity, will thus be separated into a sum of partial velocities of which u_1 is the smoothest main part of u and represents the 'first step'; correspondingly, the smaller, but finer-structured part, u_2, the second step etc.; the nth order shall be the last one in the series that will no longer become turbulent [here Prandtl added in a footnote: "The separation into steps thereby creates difficulties, namely that the elements of the first step do not all have the same size and that in the following the differences may increase even further. As the purpose of the analysis is only a rough estimate one may conjecture that the elements in each step have the same well-defined size. A more detailed analysis by considering the statistical ensemble of turbulence elements is an aim for the future."]

Motivated by the way the u_i are introduced, it seems natural to assume that their effective values u_i^2 build a geometric series, at least with the exclusion of the final members of the series, for which viscosity is already noticeable. As a first approximation one may assume also that the final members of the series, other than the very last one, are members of the geometric series.

By this reasoning Prandtl ended the geometric series by closing it with a single last step at which all energy dissipation occurs. His final derivation of the Kolmogorov length scale is then quite similar to what he calculated on 29 January 1945 (see Figure 2.10).

This did not conclude Ludwig Prandtl's quest for an understanding of turbulence. In mid July 1945 he had realized that his 'one equation model' was missing a second equation that allowed him to determine the mixing length. Therefore he resorted to the vorticity equation that Kármán had investigated. As shown in Figures 2.12–2.15 he calculated with help of his vorticity equation B9 (see Figure 2.13) for the case of plane shear flow under the assumption of 'homologue' turbulence (for which the correlation coefficients of the velocity components are independent of space) the mixing length and $\frac{dU}{dy}$.

So from October 1944 to August 1945, Prandtl had returned to his lifelong quest to understand turbulence. On 17 September 1945 the Georg August University was re-opened as the first in post-war Germany and Prandtl taught again. By January 1946, Otto Hahn, Werner Heisenberg, Max von Laue and Carl Friedrich von Weizsäcker returned from England to Prandtl's Institute that was reopened on 1 August 1946. Max Planck became interim President of the Kaiser-Wilhelm-Society, which now had its headquarters in the buildings of Prandtl's Institute. On 11 September 1946 the Max Planck Society was founded in Bad Driburg as the successor of the Kaiser-Wilhelm-Society. On

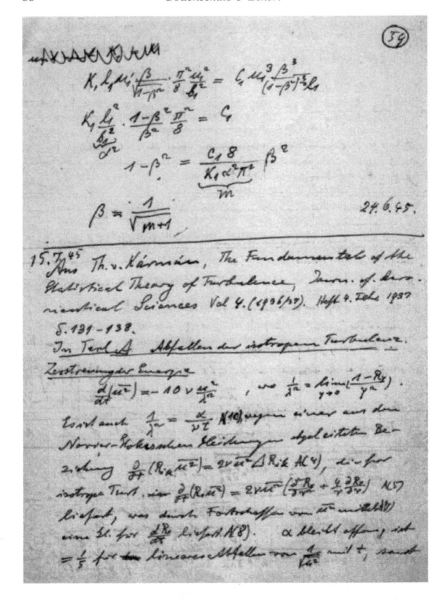

Figure 2.12 Calculation of the mixing length from the vorticity equation; 1 of 4.

26 February 1948 the Max Planck Society convened its constitutional meeting in the cafeteria of Prandtl's Institute.

Prandtl himself retired from the University and Institute's Directorships in the fall of 1946 and continued working on problems in meteorology until his death on 15 August 1953.

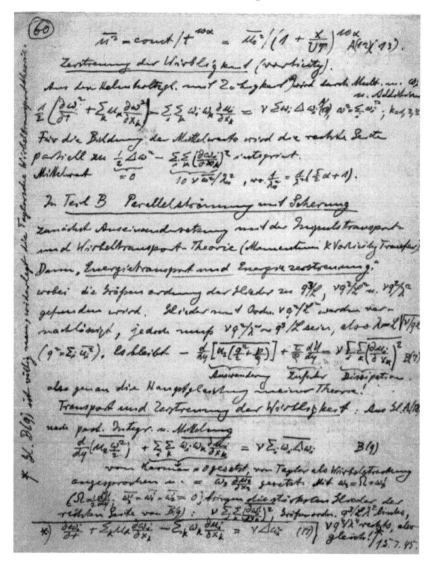

Figure 2.13 Calculation of the mixing length from the vorticity equation; 2 of 4.

2.11 Conclusion

Prandtl's achievements in fluid mechanics generally, and in turbulence in particular, are often characterized by the label 'theory'. However, it is important to note that he did not perceive himself as a theoretician. When the German Physical Society of the British Zone awarded him honorary membership two years

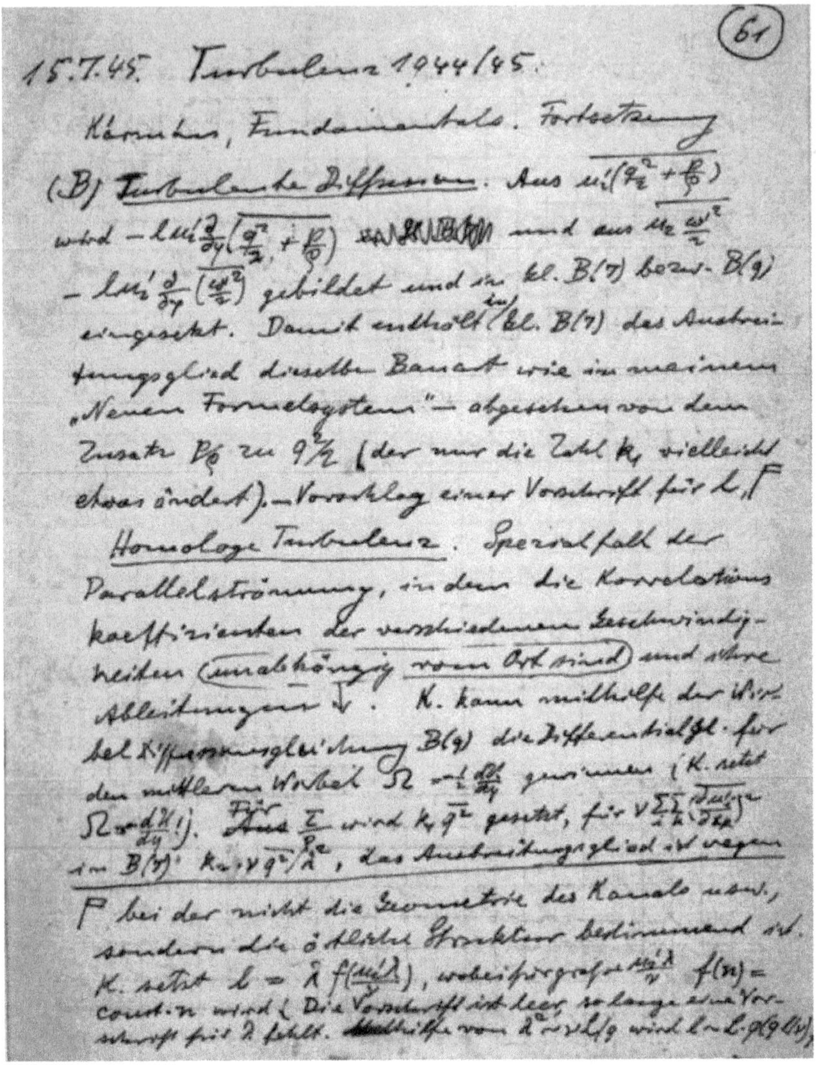
61

15.7.45. Turbulenz 1944/45.

Kármán, Fundamentals. Fortsetzung

(B) Turbulente Diffusion. Aus ... wird ... und aus ... gebildet und in Gl. B(7) bzw. B(9) eingesetzt. Damit enthält Gl. B(7) das Ausbreitungsglied dieselbe Bauart wie in meinem „Neuen Formelsystem" – abgesehen von dem Zusatz ... zu ... (der nur die Zahl k_1 vielleicht etwas ändert). – Vorschlag einer Vorschrift für l, Γ

Homologe Turbulenz. Spezialfall der Parallelströmung, in dem die Korrelationskoeffizienten der verschiedenen Geschwindigkeiten (unabhängig vom Ort sind) und ihre Ableitungen ... K. kann mithilfe der Wirbel Differenzengleichung B(9) die Differentialgl. für den mittleren Wirbel Ω gewinnen (K. setzt ...). Für ... wird ... gesetzt, für ... das Ausbreitungsglied ist wegen ...

Γ bei der nicht die Geometrie des Kanals usw., sondern die örtliche Struktur bestimmend ist. K. setzt ..., wobei für große ... $f(x)$ = const. wird (Die Vorschrift ist leer, so lange eine Vorschrift für λ fehlt. ...

Figure 2.14 Calculation of the mixing length from the vorticity equation; 3 of 4.

after the war, he used this occasion to clarify his research style in a lecture entitled "My approach towards hydrodynamical theories". With regard to boundary layer theory, for example, he argued that he was guided by a 'heuristic principle' of this kind: "If the whole problem appears mathematically hopeless, see what happens if an essential parameter of the problem approaches zero" (Prandtl, 1948b, p. 1606). His notes amply illustrate how he used one

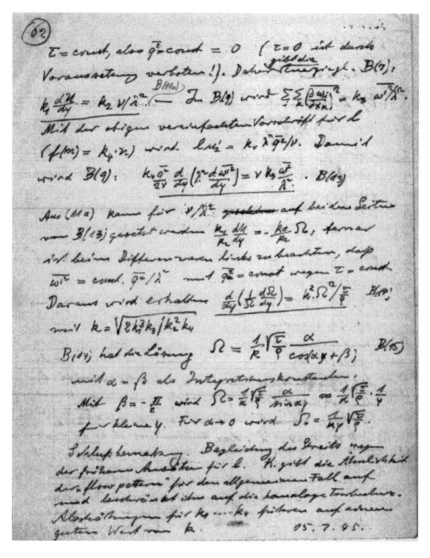

Figure 2.15 Calculation of the mixing length from the vorticity equation; 4 of 4.

or another assumption, often combined with clever dimensional arguments, in order to single out those features of a problem which he regarded as crucial. He always attempted to gain "a thorough visual impression" about the problems with which he was concerned. "The equations come later when I think that I have grasped the matter" (Prandtl, 1948b, p. 1604).

By the same token, Prandtl's approach to theory relied heavily on practice. For that matter, practice could be an observation of flow phenomena in a water channel, an *experimentum crucis* like the trip-wire test, or a challenge posed by practical applications such as skin friction. Prandtl's FIAT review on turbulence, in particular, illustrates how his theoretical research was motivated and guided by practice. As we have seen, Prandtl named explicitly, among others, Schultz-Grunow (1940), Wieghardt (1944), Reichardt (1944) and Schuh (1945) as important roots for the theoretical insight expressed in Prandtl and Wieghardt (1945). His closest collaborator for the fundamental studies on fully developed turbulence, Wieghardt, was by that time developing technical expertise for studying the skin friction of rubber with regard to a possible use for the hull of submarines (Prandtl, 1948a, p. 58). These and other war-related studies were based on experimental turbulence measurements in the same 'roughness tunnel' that provided the data for the more fundamental inquiries.

Prandtl's style as well as the closeness of theory and practice is also reflected in the third edition of his famous *Essentials of Fluid Mechanics* (Prandtl, 1948c). In a paragraph about the onset of turbulence, for example, Prandtl reported about the recent confirmation of the Tollmien–Schlichting theory by the experiments in Dryden's laboratory at the National Bureau of Standards in Washington. Turbulent jets and turbulent shear flow along walls were discussed in terms of the mixing length approach (Prandtl, 1948c, pp. 115–123). Isotropic turbulence was summarized rather cursorily, with a reference to his FIAT review and the recent work by Weizsäcker and Heisenberg (Prandtl, 1948c, p. 127). In general, he preferred textual and pictorial presentations supported by experiments over sophisticated mathematical derivations.

For a deeper understanding of Prandtl's and his Göttingen school's contributions to turbulence it would be necessary to account for the broader research conducted at the KWI and the AVA, which covered a host of fundamental and applied topics, from solid elasticity to gas dynamics and meteorology. Research on turbulence was never pursued as an isolated topic. But in view of its ultimate importance for engineering, turbulence always remained an important and challenging problem. Among the variety of research problems dealt with at Göttingen in the era of Prandtl, turbulence may be regarded as the one with the longest tradition – from Klein's seminar in 1907 to the climax of Prandtl's unpublished manuscripts in 1945.

A number of questions have been left unanswered. The timing of Prandtl's breakthrough during the last months of the Second World War, in particular, suggests further inquiries: to what extent was fundamental research on turbulence interrupted during the war by Prandtl's involvement as a Scientific Adviser to the Ministry of Aviation (Reichsluftfahrtsministerium) with regards to aeronautical war research?[78] Or was the renewed interest in the basic riddles of turbulence sparked by the wartime applications? Or, on the contrary, did Prandtl at the end of the war find the time to work on what he was really interested in?

Both Prandtl's advisory role as well as his local responsibilities for fluid dynamics research at Göttingen came to a sudden stop when the American and British troops occupied his Institute and prevented further research – a prohibition which Prandtl perceived as unwarranted. Not only did he write[79] to Taylor for help, but also he requested[80] help from the President of the Royal Society, of which he had been a Foreign Member since 1928. "The continuation of the research activity that had to be shelved during the War should not be hindered any more!" demanded Prandtl in this letter. His request remained unanswered.

This correspondence provokes further questions regarding Prandtl's political attitude. Biographical knowledge of Prandtl has been provided by his family (Vogel-Prandtl, 1993), by admiring disciples (Flügge-Lotz and Flügge, 1973; Oswatitsch and Wieghardt, 1987), and by reviews on German wartime aeronautical research (Trischler, 1994); a more complete view based on the rich sources preserved in the archives in Göttingen, Berlin and elsewhere seems expedient.[81] Recent historical studies on the war research at various Kaiser-Wilhelm-Institutes (see, for example, Maier, 2002; Schmaltz, 2005; Sachse and Walker, 2005; Maier, 2007; Heim et al., 2009; Gruss and Ruerup, 2011) call for further inquiries into Prandtl's motivations for research into turbulence. An important question of course is: What can we learn from the position of great men like Prandtl and others in the political web of Nazi Germany? What consequences arise for the responsibilities of scientists or engineers? Another lacuna which needs to be addressed in greater detail concerns the relationship of Prandtl with his colleagues abroad and in Germany, in particular with von Kármán, Taylor, Sommerfeld and Heisenberg. Last, but not least, one may ask about the fate of turbulence research at Göttingen under Prandtl's successors after the war. We leave these and many other questions for future studies.

Acknowledgement It is our pleasure to express our gratitude to a large number of people without whom this work would not have been possible. We are very thankful to the science historian Florian Schmaltz for sharing his insights and for giving us copies of documents from his collections. We express

---

[78] From 6 July 1942 Prandtl became the Chair of the Scientific Research Council of the Ministry of Aviation led by Göring and was pushing for fundamental research in war-related matters (Maier, 2007).

[79] Prandtl to Taylor, 18 October 1945. MPGA, Abt. III, Rep. 61, Nr. 1654.

[80] Prandtl to Royal Society, 18 October 1945. MPGA, Abt III, Rep. 61, Nr. 1402.

[81] An almost complete set of his correspondence is preserved.

our gratitude to the Max Planck Society Archives, especially Lorenz Beck, Bernd Hofmann, Susanne Übele and Simone Pelzer; and to the Archives of the German Aerospace Center (DLR), especially Jessika Wichner and Andrea Missling. We are also very grateful to K. Sreenivasan, N. Peters and G. Falkovich for sharing their insights on this topic and to the editors of this book for motivating us to write this review and helpful suggestions for improvements of the text. EB also thanks Haitao Xu and Zellman Warhaft for valuable comments, and Helmut Eckelmann for sharing his memories about the Institute. We are most grateful for the understanding and support from our families that were missing their husband and father for long evenings and weekends. This work was generously supported by the Max Planck Society and the Research Institute of the German Museum in Munich. Part of this work was written at the Kavli Institute for Theoretical Physics and was supported in part by the National Science Foundation under Grant No. NSF PHY05-51164.

Abbreviations

DMA: Deutsches Museum, Archiv, München.

GOAR: Göttinger Archiv des DLR, Göttingen.

LPGA: Ludwig Prandtls Gesammelte Abhandlungen, herausgegeben von Walter Tollmien, Hermann Schlichting und Henry Görtler. 3 Bände, Berlin u. a. 1961.

MPGA: Max-Planck-Gesellschaft, Archiv, Berlin.

RANH: Rijksarchief in Noord-Holland, Haarlem.

SUB: Staats- und Universitätsbibliothek, Göttingen.

TKC: Theodore von Kármán Collection, Pasadena.

References

Anderson, John D. 2005. Ludwig Prandtl's boundary layer. *Physics Today*, **58** 42–48.

Batchelor, G.K. 1946. Double velocity correlation function in turbulent motion. *Nature*, **158** 883–884.

Battimelli, Giovanni. 1984. The mathematician and the engineer: statistical theories of turbulence in the 20's. *Rivista di storia della scienza*, **1** 73–94.

Blasius, Heinrich. 1908. Grenzschichten in Flüssigkeiten bei kleiner Reibung. *Zeitschrift für Mathematik und Physik*, **56** 1–37.

Blasius, Heinrich. 1913. Das Ähnlichkeitsgesetz bei Reibungsvorgängen in Flüssigkeiten. *Forschungsarbeiten auf dem Gebiete des Ingenieurwesens*, 131.

Boussinesq, M.J. 1897. *Theorie de l'Ecoulement Tourbillonnant et Tumultueux des Liquides dans les Lits Rectilignes á Grandes Sections*. Gauthiers-Villars et fils, Paris.

Burgers, J.M. 1925. The motion of a fluid in the boundary layer along a plane smooth surface. *Proceedings of the First International Congress for Applied Mechanics, Delft, 1924*, C.B. Biezeno and J.M. Burgers (eds), 113–128.

Collar, A.R. 1978. Arthur Fage. 4 March 1890–7 November 1977. *Biographical Memoirs of Fellows of the Royal Society*, 33–53.

Comte-Bellot, G. and Corrsin, S. 1966. The use of a contraction to improve the isotropy of grid-generated turbulence. *J. Fluid Mech.*, **25** 667–682.

Darrigol, Olivier. 2005. *Worlds of Flow*. Oxford University Press, Oxford.

Dryden, Hugh L. and Kuethe, A.M. 1929. The measurement of fluctuations of air speed by the hot-wire anemometer. *NACA Report*, **320** 357–382.

Dryden, Hugh L., Schubauer, G.B., Mock, W.C. and Skramstadt, H.K. 1937. Measurements of intensity and scale of wind-tunnel turbulence and their relation to the critical Reynolds number of spheres. *NACA Report*, **581** 109–140.

Dryden, Hugh L. 1938. Turbulence investigations at the National Bureau of Standards. *Proceedings of the Fifth International Congress on Applied Mechanics, Cambridge Mass.*, J.P. Den Hartog and H. Peters (eds), Wiley, New York, 362–368.

Dryden, Hugh L. 1955. Fifty years of boundary-layer theory and experiment. *Science*, **121** 375–380.

Eckert, Michael. 2006. *The Dawn of Fluid Dynamics*. Wiley-VCH, Weinheim.

Eckert, Michael. 2008. Theory from wind tunnels: empirical roots of twentieth century fluid dynamics. *Centaurus*, **50** 233–253.

Eckert, Michael. 2010. The troublesome birth of hydrodynamic stability theory: Sommerfeld and the turbulence problem. *European Physical Journal, History*, **35**(1) 29–51.

Eiffel, Gustave. 1912. Sur la résistance des sphères dans l'air en mouvement. *Comptes Rendues*, **155** 1597–1599.

Eisner, F. 1932a. Reibungswiderstand. *Werft, Reederei, Hafen*, **13** 207–209.

Eisner, F. 1932b. Reibungswiderstand. In *Hydromechanische Probleme des Schiffsantriebs. Hamburg*, G. Kempf and E. Foerster (eds), 1–49.

Fage, A. and Townend, H.C.H. 1932. An examination of turbulent flow with an ultra-microscope. *Proc. Roy. Soc. Lond. A*, **135** 656–677.

Flüigge-Lotz, Irmgard and Flügge, Wilhelm. 1973. Ludwig Prandtl in the nineteen-thirties: reminiscences. *Ann. Rev. Fluid Mech.* **5** 1–8.

Föppl, Otto. 1912. Ergebnisse der aerodynamischen Versuchsanstalt von Eiffel, verglichen mit den Göttinger Resultaten. *Zeitschrift für Flugtechnik und Motorluftschiffahrt*, **3** 118–121.

Fritsch, Walter. 1928. Der Einfluss der Wandrauhigkeit auf die turbulente Geschwindigkeitsverteilung in Rinnen. *Abhandlungen aus dem Aerodynamischen Institut der Technischen Hochschule Aachen*, **8**.

Gilbert, Martin. 2006. *Kristallnacht: Prelude to Destruction*. Harper Collins Publishers.

Görtler, H. 1942. Berechnungen von Aufgaben der freien Turbulenz auf Grund eines neuen Näherungsansatzes. *Zeitschrift für Angewandte Mathematik und Mechanik (ZAMM)*, **22** 44–254.

Goldstein, Sydney. 1969. Fluid mechanics in the first half of this century. *Ann. Rev. Fluid Mech.*, **1** 1–28.

Grossmann, Siegfried, Eckhardt, Bruno and Lohse, Detlef. 2004. Hundert Jahre Grenz-schichtphysik. *Physikjournal*, **3** (10), 31–37.

Gruschwitz, E. 1931. Die trubulente Reibungsschicht bei Druckabfall und Druck-anstieg. *Ingenieur-Archiv*, **2** 321–346.

Gruss, Peter and Rürup, Reinhard (eds). 2011. *Denkorte. Max-Planck-Gesellschaft und Kaiser-Wilhelm-Gesellschaft, Brüche und Kontinuitäten 1911–2011.* Sandstein Verlagr., Dresden.

Hager, W.H. 2003. Blasius: A life in research and education. *Experiments in Fluids*, **34** 566–571.

Hahn, Hans, Herglotz, Gustav and Schwarzschild, Karl, 1904. Über das Strömen des Wassers in Röhren und Kanälen. *Zeitschrift für Mathematik und Physik*, **51** 411–426.

Hamel, H. 1943. Streifenmethode und Ähnlichkeitsbetrachtungen zur turbulenten Be-wegung. *Abh. Preuss. Akad. Wiss. Physik.–Math.*, **8**.

Heim, Susanne, Sachse, Carola and Walker, Mark (eds). 2009. *The Kaiser Wilhelm Society under National Socialism.* Cambridge University Press, Cambridge.

Heisenberg, W. 1948. Zur statistischen Theorie der Turbulenz. *Zeitschrift für Physik*, **124** 628–657.

Heisenberg, W. 1958E. On the statistical theory of turbulence. NACA-TM-1431, 1958.

Hensel, Susann. 1989. *Mathematik und Technik im 19. Jahrhundert in Deutschland. Soziale Auseinandersetzungen und philosophische Problematik*, chapter Die Au-seinandersetzungen um die mathematische Ausbildung der Ingenieure an den Technischen Hochschulen in Deutschland Ende des 19. Jahrhunderts, 1–111. Van-denhoeck und Ruprecht, Göttingen.

Hopf, Ludwig. 1910. *Hydrodynamische Untersuchungen: Turbulenz bei einem Flusse. Über Schiffswellen. Inaugural-Dissertation.* Barth, Leipzig.

Jakob, M. and Erk, S. 1924. Der Druckabfall in glatten Rohren und die Durchflusszif-fer von Normaldüsen. *Forschungsarbeiten auf dem Gebiete des Ingenieurwesens*, **267**.

Kármán, von, Theodore. 1921. Über laminare und turbulente Reibung. *ZAMM*, **1** 233–252. CWTK 2, 70–97.

Kármán, von, Theodore. 1924. Über die Oberflächenreibung von Flüssigkeiten. In *Vorträge aus dem Gebiete der Hydro- und Aerodynamik (Innsbruck 1922).* Theodore von Kármán und T. Levi-Civita (eds). Berlin: Springer, 146–167. CWTK 2, 133–152.

Kármán, von, Theodore. 1930a. Mechanische Ähnlichkeit und Turbulenz. *Nachrichten von der Gesellschaft der Wissenschaften zu Göttingen, Mathematisch–Physikalische Klasse*, 58–76. CWTK 2, 322–336.

Kármán, von, Theodore. 1930b. Mechanische Ähnlichkeit und Turbulenz. In *Pro-ceedings of the Third International Congress of Applied Mechanics, Stockholm.* CWTK 337–346.

Kármán, von, Theodore. 1932. Theorie des Reibungswiderstandes. In *Hydro-mechanische Probleme des Schiffsantriebs. Hamburg*, G. Kempf, E. Foerster (eds), 50–73. CWTK 2, 394–414.

Kármán, von, Theodore. 1934. Turbulence and skin friction. *Journal of the Aeronauti-cal Sciences*, **1** 1–20. CWTK 3, 20–48.

Kármán, von, Theodore. 1957. *Aerodynamics: Selected Topics in the Light of their Historical Development*. Cornell University Press, Ithaca, New York.

Kármán, von, Theodore (with Lee Edson). 1967. *The Wind and Beyond*. Little, Brown and Company, Boston.

Kármán, von, Theodore and Leslie Howarth. 1938. On the statistical theory of isotropic turbulence. *Proc. Roy. Soc. Lond. A*, **164** 192–215.

Kempf, Günther. 1929. Neue Ergebnisse der Widerstandsforschung. *Werft, Reederei, Hafen*, **10** 234–239.

Kempf, Günther. 1932. Weitere Reibungsergebnisse an ebenen glatten und rauhen Flächen. *Hydromechanische Probleme des Schiffsantriebs*. Hamburg, G. Kempf, E. Foerster (eds), 74–82.

Klein, Felix. 1910. Über die Bildung von Wirbeln in reibungslosen Flüssigkeiten. *Zeitschrift für Mathematik und Physik*, **59** 259–262.

Kuethe, A.M. 1988. The first turbulence measurements: a tribute to Hugh L. Dryden. *Ann. Rev. Fluid Mech.*, **20** 1–3.

Lorentz, Hendrik Antoon. 1897. Over den weerstand dien een vloeistofstroom in eene cilindrische buis ondervindt. *Versl. K. Akad. Wet. Amsterdam*, **6** 28–49.

Lorentz, Hendrik Antoon. 1907. Über die Entstehung turbulenter Flüssigkeitsbewegungen und über den Einfluss dieser Bewegungen bei der Strömung durch Röhren. *Hendrik Antoon Lorentz: Abhandlungen über theoretische Physik*. Teubner, Leipzig, **1** 43–71.

Lumley, John Leask and Panofsky, Hans A. 1964. *The Structure of Atmospheric Turbulence*. Wiley, New York.

Maier, Helmut (ed.). 2002. *Rüstungsforschung im Nationalsozialismus. Organisation, Mobilisierung und Entgrenzung der Technikwissenschaften*. Wallstein, Göttingen, 2002.

Maier, Helmut. 2007. *Forschung als Waffe. Rüstungsforschung in der Kaiser-Wilhelm-Gesellschaft und das Kaiser-Wilhelm-Institut für Metallforschung 1900–1945/48*. Wallstein, Göttingen, 2007.

Manegold, Karl-Heinz. 1970. *Universität, Technische Hochschule und Industrie: Ein Beitrag zur Emanzipation der Technik im 19. Jahrhundert unter besonderer Berücksichtigung der Bestrebungen Felix Kleins*. Duncker und Humblot, Berlin.

Meier, Gerd E.A. 2006. Prandtl's boundary layer concept and the work in Göttingen. *IUTAM Symposium on One Hundred Years of Boundary Layer Research. Proceedings of the IUTAM Symposium held at DLR-Göttingen, Germany, August 12–14, 2004*. G.E.A. Meier and K.R. Sreenivasan (eds). Springer, Dordrecht, 1–18.

Mises, von, Richard. 1921. Über die Aufgaben und Ziele der angewandten Mathematik. *Zeitschrift für Angewandte Mathematik und Mechanik (ZAMM)*, **1** 1–15.

Motzfeld, H. 1938. Vorträge aus dem Gebiet der Aero- und Hydrodynamik. Frequenzanalyse turbulenter Schwankungen. *Zeitschrift für Angewandte Mathematik und Mechanik (ZAMM)*, **18** 362–365.

Munk, Max. 1917. Bericht über Luftwiderstandsmessungen von Streben. Mitteilung 1 der Göttinger Modell-Versuchsanstalt für Aerodynamik. *Technische Berichte. Herausgegeben von der Flugzeugmeisterei der Inspektion der Fliegertruppen*. Heft Nr. 4 (1. Juni 1917), 85–96, Tafel XXXX–LXIII.

Nagib, Hassan M., Chauhan, Kapil A. and Monkewitz, Peter A. 2007. Approach to an asymptotic state for zero pressure gradient turbulent boundary layers. *Phil. Trans. R. Soc. A*, **365** 755–770.

Nikuradse, Johann. 1926. Untersuchungen über die Geschwindigkeitsverteilung in turbulenten Strömungen. *Forschungsarbeiten auf dem Gebiete des Ingenieurwesens*, **281**.

Nikuradse, Johann. 1930. Über turbulente Wasserströmungen in geraden Rohren bei sehr grossen Reynoldsschen Zahlen. *Vorträge aus dem Gebiete der Aerodynamik und verwandter Gebiete (Aachen 1929)*. A. Gilles, L. Hopf and Th. v. Kármán (eds). Springer, Berlin, 63–69.

Nikuradse, Johann. 1932. Gesetzmässigkeiten der turbulenten Strömung in glatten Rohren. *Forschungsarbeiten auf dem Gebiete des Ingenieurwesens*, 356.

Nikuradse, Johann. 1933. Strömungsgesetze in rauhen Rohren. *Forschungsarbeiten auf dem Gebiete des Ingenieurwesens*, 3611.

Noether, Fritz. Das Turbulenzproblem. *Zeitschrift für Angewandte Mathematik und Mechanik*, **1** 125–138, 218–219.

O'Malley Jr., Robert E. 2010. Singular perturbation theory: a viscous flow out of Göttingen. *Ann. Rev. Fluid Mech.*, **42** 1–17.

Oswatitsch, K. and Wieghardt, K. 1987. Ludwig Prandtl and his Kaiser-Wilhelm-Institut. *Ann. Rev. Fluid Mech.*, **19** 1–26.

Prandtl, Ludwig. 1905. Über Flüssigkeitsbewegung bei sehr kleiner Reibung. In *Verhandlungen des III. Internationalen Mathematiker-Kongresses, Heidelberg*, 484–491. LPGA 2, 575–584.

Prandtl, Ludwig. 1910. Eine Beziehung zwischen Wärmeaustausch und Strömungswiderstand der Flüssigkeiten. *Physikalische Zeitschrift*, **11** 1072–1078.

Prandtl, Ludwig. 1914. Der Luftwiderstand von Kugeln. *Nachrichten der Gesellschaft der Wissenschaften zu Göttingen, Mathematisch–Physikalische Klasse*, 177–190. (LPGA 2, 597–608).

Prandtl, Ludwig. 1921a. Bemerkungen über die Entstehung der Turbulenz. *Zeitschrift für Angewandte Mathematik und Mechanik (ZAMM)*, **1** 431–436.

Prandtl, Ludwig. 1921b. *Ergebnisse der Aerodynamischen Versuchsanstalt zu Göttingen*. Oldenbourg, Berlin, München.

Prandtl, Ludwig. 1922. Bemerkungen über die Entstehung der Turbulenz. *Physikalische Zeitschrift*, **23** 19–25.

Prandtl, Ludwig. 1925. Bericht über Untersuchungen zur ausgebildeten Turbulenz. *ZAMM*, **5** 136–139. LPGA 2, 714–718.

Prandtl, Ludwig. 1925E. Aufgaben der Strömungsforschung: Tasks of air flow research. *Die Naturwissenschaften*, **14** (16) 355–358, 1925; NACA–TN–365, 1926.

Prandtl, Ludwig. 1926a. Bericht über neuere Turbulenzforschung. *Hydraulische Probleme*. VDI-Verlag, Berlin, 1–13. LPGA 2, 719–730.

Prandtl, Ludwig. 1926b. Klein und die Angewandten Wissenschaften. *Sitzungsberichte der Berliner Mathematischen Gesellschaft*, 81–87. LPGA 2, 719–730.

Prandtl, Ludwig. 1927a. Über den Reibungswiderstand strömender Luft. *Ergebnisse der Aerodynamischen Versuchsanstalt zu Göttingen*. Oldenbourg, Berlin, München, **3** 1–5. LPGA 2, 620-626.

Prandtl, Ludwig. 1927b. Über die ausgebildete Turbulenz. *Verhandlungen des II. Internationalen Kongresses für Technische Mechanik*. Füssli, Zürich, 62–75. LPGA 2, 736–751. NACA-TM-435, version in English.

Prandtl, Ludwig. 1930. Vortrag in Tokyo. *Journal of the Aeronautical Research Institute*, Tokyo, Imperial University, **5** (65) 12–24. LPGA 2, 788–797.

Prandtl, Ludwig. 1931. *Abriss der Strömungslehre*. Vieweg, Braunschweig.

Prandtl, Ludwig. 1932. Zur turbulenten Strömung in Rohren und längs Platten. In *Ergebnisse der Aerodynamischen Versuchsanstalt zu Göttingen*, **4** 18–29. LPGA 2, 632–648.

Prandtl, Ludwig. 1933. Neuere Ergebnisse der Turbulenzforschung. *Zeitschrift des Vereines Deutscher Ingenieure* **77** 105–114. LPGA 2, 819–845.

Prandtl, Ludwig. 1933E1. Neuere Ergebnisse der Turbulenzforschung: Recent results of turbulence research. *Zeitschrift des Vereines deutscher Ingenieure*. **7** (5) 105–114. NACA-TM-720. 1933.

Prandtl, Ludwig. 1933E2 Herstellung einwandfreier Luftstrome (Windkanale): Attaining a steady air stream in wind tunnels. *Handbuch der Experimentalphysik*. Vol. 4, (2) 65–106, 1932; NACA-TM-726, 1933.

Prandtl, Ludwig. 1934. Anwendung der turbulenten Reibungsgesetze auf atmosphärische Strömungen. In *Proceedings of the Fourth International Congress of Applied Mechanics, Cambridge*, 238–239. LPGA 3, 1098–1099.

Prandtl, Ludwig. 1938. Beitrag zum Turbulenzsymposium. In *Proceedings of the Fifth International Congress on Applied Mechanics, Cambridge MA*, J.P. Den Hartog and H. Peters (eds), John Wiley, New York, 340–346. LPGA 2, 856–868.

Prandtl, Ludwig. 1942a. *Führer durch die Strömungslehre*. Vieweg, Braunschweig.

Prandtl, Ludwig. 1942b. Zur turbulenten Strömung in Rohren und längs Platten. *Ergebnisse der Aerodynamischen Versuchsanstalt zu Göttingen*, **4** 18–29. LPGA 2, 632–648.

Prandtl, Ludwig. 1942c. Bemerkungen zur Theorie der Freien Turbulenz. *Zeitschrift für Angewandte Mathematik und Mechanik (ZAMM)* **22** 241–243. LPGA 2, 869–873.

Prandtl, Ludwig. 1945. Über die Rolle der Zähigkeit im Mechanismus der ausgebildete Turbulenz: The role of viscosity in the mechanism of developed turbulence. *GOAR* 3712, DLR Archive.

Prandtl, Ludwig. 1948a. Turbulence. *FIAT Review of German Science 1939–1946: Hydro- and Aero-dynamics*, Albert Betz (ed.), Office of Military Government for Germany Field Information Agency Technical, 55–78.

Prandtl, Ludwig. 1948b. Mein Weg zu den Hydrodynamischen Theorien. *Physikalische Blätter*, **3** 89–92. LPGA 3, 1604–1608.

Prandtl, Ludwig. 1948c. *Führer durch die Strömungslehre*. Vieweg, Braunschweig.

Prandtl, Ludwig. 1949E. Bericht über Untersuchungen zur ausgebildeten Turbulenz: Report on investigation of developed turbulence. NACA-TM-1231, 1949.

Prandtl, Ludwig and Eisner, Franz. 1932. Nachtrag zum 'Reibungswiderstand'. In *Hydromechanische Probleme des Schiffsantriebs. Hamburg*, G. Kempf, E. Foerster (eds), 407.

Prandtl, Ludwig et al. 1932. Erörterungsbeiträge. *Hydromechanische Probleme des Schiffsantriebs. Hamburg*, G. Kempf, E. Foerster (eds), 87–98.

Prandtl, Ludwig and Schlichting, Hermann. 1934. Das Widerstandsgesetz rauher Platten. *Werft, Reederei, Hafen*, **21** 1–4. LPGA 2, 649–662.

Prandtl, Ludwig and Reichardt, Hans. 1934. Einfluss von Wärmeschichtung auf Eigenschaften einer turbulenten Strömung. *Deutsche Forschung*, **15** 110–121. LPGA 2, 846–855.

Prandtl, Ludwig and Wieghardt, Karl. 1945. Über ein neues Formelsystem für die ausgebildete Turbulenz. *Nachrichten der Akademie der Wissenschaften zu Göttingen, Mathematisch–Physikalische Klasse*, 6–19.LPGA 2, 874–887.

Rayleigh, Lord. 1887. On the stability or instability of certain fluid motions II. *Proceedings of the London Mathematical Society*, **19** 67–74. Reprinted in *Scientific Papers by John William Strutt, Baron Rayleigh, vol. III (1887–1892)*. Cambridge University Press, Cambridge. 17–23.

Reichardt, H. 1933. Die quadratischen Mittelwerte der Längsschwankungen in der turbulenten Kanalströmung. *Zeitschrift für Angewandte Mathematik und Mechanik (ZAMM)*, **3** 177–180.

Reichardt, H. 1934. Berichte aus den einzelnen Instituten. Physikalisch-Chemisch-Technische Institute. *Naturwissenschaften*, **22** 351.

Reichardt, H. 1935. Die Torsionswaage als Mikromanometer. *Zschr. f. Instrumentenkunde*, **55** 23–33.

Reichardt, H. 1948E. The torsion balance as a micromanometer. NRC-TT-84, 1948-11-13.

Reichardt, H. 1938a. Vorträge aus dem Gebiet der Aero- und Hydrodynamik. Über das Messen turbulenter Längs- und Querschwankungen. *Zeitschrift für Angewandte Mathematik und Mechanik (ZAMM)*, **18** 358–361.

Reichardt, H. 1938b. Messungen turbulenter Spannungen. *Naturwissenschaften*, **26** 404–408.

Reichardt, H. 1941. Über die Theorie der freien Turbulenz. *Zeitschrift für Angewandte Mathematik und Mechanik (ZAMM)*, **21** 257–264.

Reichardt, H. 1942. Gesetzmässigkeiten der freien Turbulenz. *VDI Forschungsheft*, **414**, 22 pages.

Reichardt, H. 1944. Impuls- und Wärmeaustausch in freier Turbulenz. *Zeitschrift für Angewandte Mathematik und Mechanik (ZAMM)*, **24** 268–272.

Reichardt, H. 1951E. On the recording of turbulent longitudinal and transverse fluctuations. NACA-TM-1313, 1951.

Rotta, Julius C. 1990. *Die Aerodynamische Versuchsanstalt in Göttingen, ein Werk Ludwig Prandtls. Ihre Geschichte von den Anfängen bis 1925*. Vandenhoeck und Ruprecht, Göttingen.

Rotta, Julius C. 2000. Ludwig Prandtl und die Turbulenz. In *Ludwig Prandtl, ein Führer in der Strömungslehre: Biographische Artikel zum Werk Ludwig Prandtls*, Gerd E.A. Meier (ed), 53–123.

Sachse, Carola and Walker, Mark (eds). 2005. *Politics and Science in Wartime: Comparative International Perspectives on Kaiser Wilhelm Institutes*. University of Chicago Press.

Saffman, P.G. 1992. *Vortex Dynamics*. Cambridge University Press, Cambridge.

Schiller, Ludwig. 1921. Experimentelle Untersuchungen zum Turbulenzproblem. *Zeitschrift für Angewandte Mathematik und Mechanik (ZAMM)*, **1** 436–444.

Schlichting, Hermann. 1933. Zur Entstehung der Turbulenz bei der Plattenströmung. *Nachrichten der Gesellschaft der Wissenschaften zu Göttingen*, 181–208.

Schlichting, Hermann. 1949. Lecture series "boundary layer theory", part ii: Turbulent flows. *NACA-TM-1218*, 1949. Translation of "Vortragsreihe" W.S. 1941/42, Luftfahrtforschungsanstalt Hermann Göring, Braunschweig.

Schmaltz, Florian. 2005. *Kampfstoff-Forschung im Nationalsozialismus. Zur Kooperation von Kaiser-Wilhelm-Instituten, Militär und Industrie.* Wallstein, Göttingen.

Schuh, H. 1945. Die Messungen sehr kleiner Windschwankungen (Windkanalturbulenz). *Untersuchungen und Mitteilungen der Deutschen Luftfahrtforschung,* **6623**.

Schuh, H. 1946. *Windschwankungsmessungen mit Hitzdrähten.* AVA Monographien, D1, Chapter 4.3.

Schultz-Grunow, Fritz. 1940. Neues Reibungswiderstandsgesetz für glatte Platten. *Luftfahrtforschung,* **17** 239–246.

Schultz-Grunow, Fritz. 1941E. New frictional resistance law for smooth plates. *NACA-TM-986*, 1941.

Simmons, L.F.G., Salter, C. and Taylor, G.I. 1938. An experimental determination of the spectrum of turbulence. *Proc. Roy Soc. Lond. A*, **165** 73–89.

Sommerfeld, Arnold. 1935. Zu L. Prandtls 60. Geburtstag am 4. Februar 1935. *ZAMM*, **15** 1–2.

Spalding, D.B. 1991. Kolmogorov's two-equation model of turbulence. *Proc. Roy. Soc. Math. Phys. Sci.*, **434** 211–216.

Tani, Itiro. 1977. History of boundary-layer theory. *Ann. Rev. Fluid Mech.*, **9** 87–111.

Tietjens, Oskar. 1925. Beiträge zur Entstehung der Turbulenz. *ZAMM*, **5** 200–217.

Taylor, G.I. 1929. Stability of a viscous liquid contained between two rotating cylinders. *Phys. Trans. Roy. Soc.*, **223** 289–343.

Taylor, G.I. 1935a. Turbulence in a contracting stream. *ZAMM*, **15** 91–96.

Taylor, G.I. 1935b. Statistical theory of turbulence. *Proc. Roy. Soc. Lond. A* **151** 421–444.

Taylor, G.I. 1935c. Statistical theory of turbulence II. *Proc. Roy. Soc. Lond. A* **151** 444–454.

Taylor, G.I. 1935d. Statistical theory of turbulence. III. Distribution of dissipation of energy in a pipe over its cross-section. *Proc. Roy. Soc. Lond. A* **151** 455–464.

Taylor, G.I. 1935e. Statistical theory of turbulence. IV. Diffusion in a turbulent air stream. *Proc. Roy. Soc. Lond. A* **151** 465–478.

Taylor, G.I. 1936. Correlation measurements in a turbulent flow through a pipe. *Proc. Roy. Soc. Lond. A* **157** 537–546.

Taylor, G.I. 1938a. The spectrum of turbulence. *Proc. Roy. Soc. Lond. A* **164** 476–490.

Taylor, G.I. 1938b. Some recent developments in the study of turbulence. In *Proceedings of the Fifth International Congress on Applied Mechanics, Cambridge MA,* J.P. Den Hartog and H. Peters (eds), John Wiley, New York, 294–310.

Tollmien, Walter. 1926. Berechnung turbulenter Ausbreitungsvorgänge. *ZAMM*, **6** 468–478.

Tollmien, Walter. 1929. Über die Entstehung der Turbulenz. *Nachrichten der Gesellschaft der Wissenschaften zu Göttingen*, 21–44.

Trischler, Helmuth. 1994. Self-mobilization or resistance? Aeronautical research and National Socialism. In *Science, Technology and National Socialism*, Monika Renneberg and Mark Walter (eds), Cambridge University Press, Cambridge, 72–87.

Vogel-Prandtl, Johanna. 1993. *Ludwig Prandtl. Ein Lebensbild. Erinnerungen. Doku-mente*. Max-Planck-Institut für Strömungsforschung, Göttingen, 1993. (– Mitteilungen aus dem MPI für Strömungsforschung, Nr. 107).

Weizsäcker, von, C.F. 1948. Das Spektrum der Turbulenz bei grossen Reynoldsschen Zahlen. *Zeitschrift für Physik*, **124** 614–627.

Wieghardt, Karl. 1941. Zusammenfassender Bericht über Arbeiten zur statistischen Turbulenztheorie. *Luftfahrt-Forschung*, FB 1563.

Wieghardt, Karl. 1942. Erhöhung des turbulenten Reibungswiderstandes durch Oberflächenstörungen. *ZWB*, FB 1563.

Wieghardt, Karl. 1942E. Correlation of data on the statistical theory of turbulence. NACA-TM-1008, 1942.

Wieghardt, Karl. 1943. Über die Wandschubspannung in turbulenten Reibungsschichten bei veränderlichem Aussendruck. *ZWB*, UM 6603.

Wieghardt, Karl. 1944. Zum Reibungswiderstand rauher Platten. *ZWB*, UM 6612.

Wieghardt, Karl. 1947. *Der Rauhigkeitskanal des Kaiser Wilhelm-Instituts für Strömungsforschung in Göttingen*. AVA-Monographien D1 3.3.

Wieghardt, Karl and Tillmann, W. 1944. Zur turbulenten Reibungsschicht bei Druckanstieg. *ZWB*, UM 6617.

Wieghardt, Karl and Tillmann, W. 1951E. On the turbulent friction layer for rising pressure. NACA-TM-1314, 1951.

Wieselsberger, Carl. 1914. Der Luftwiderstand von Kugeln. *Zeitschrift für Flugtechnik und Motorluftschiffahrt*, **5** 140–145.

3

Theodore von Kármán

A. Leonard and N. Peters

3.1 Introduction

Theodore von Kármán, distinguished scientist and engineer with many interests, was born in Budapest on 11 May 1881. His father, Maurice von Kármán, a prominent educator and philosopher at the University of Budapest, had a significant influence over his early intellectual development. After graduating from the Royal Technical University of Budapest in 1902 with a degree in mechanical engineering, von Kármán published in 1906 the first of a long string of papers concerning solid mechanics problems outside the domain of linear elasticity theory, in this case on the compression and buckling of columns. In that same year, apparently at the urging of his father, von Kármán left Hungary for graduate studies at Göttingen. For his 1908 PhD, supervised by Ludwig Prandtl, he developed the concepts of reduced-modulus theory and their application to column behavior such as buckling. Later, with H.-S. Tsien and others, he developed a nonlinear theory for the buckling of curved sheets. His final work in solid mechanics was on the propagation of waves of plastic deformation published as a classified report in 1942 and in the open literature in 1950. In von Kármán's words:

> It was another version of the problem I had solved for my doctor's thesis, in which I had extended Euler's classical theory of buckling to a situation beyond the elastic limit. (von Kármán and Edson, 1967, p. 248)

Von Kármán stayed at Göttingen for another four years as a Privat-docent, achieving a number of significant advances in diverse fields. He collaborated with Max Born on the vibrations of crystal lattices and the theory of specific heats. His forays into aeronautics and fluid mechanics began during this period also, apparently inspired by witnessing an airplane flight for the first time in Paris in 1908. He contributed to the literature on turbulent skin friction and, of

Figure 3.1 Von Kármán as a lieutenant in the Austro–Hungarian Army. (Courtesy
of the Archives, California Institute of Technology.)

course, made his famous stability analysis of a vortex street. Von Kármán also
computed the drag of a cylinder with a vortex-street wake by use of a control-
volume method, apparently the first appearance of such an approach in the
literature (Vincenti, 1993). Again, and on a number of other occasions during
his young career, the influence of Prandtl greatly benefitted von Kármán.

In 1912 von Kármán accepted an invitation to organize an Aerodynamical
Institute at the Technical University of Aachen as director and Professor of
Aerodynamics and Mechanics. His work there was interrupted during World
War I when he served in the Austro–Hungarian Army (see Figure 3.1). Recognizing his talents as an aeronautical engineer, the military authorities assigned
to him and his collaborators the task of developing the first stable, captive
(tethered) hovering helicopter to be used as a battlefield observation post. The
effort was successful. After the war he continued building a highly respected,
internationally known reputation for himself and the institute (Figure 3.2). In
1922 von Kármán invited leading researchers in fluid mechanics to meet in
Innsbruck to discuss progress in that field, the first of many instances where
he showed his leadership in scientific cooperation. It was decided then to have

Figure 3.2 Von Kármán lecturing at Aachen, *c.* 1922. (Reprinted with permission of the Institute of Aerodynamics at the RWTH Aachen University.)

regular, somewhat extended, meetings on a regular basis, first known known as International Congresses of Applied Mechanics and then, starting in 1948, meetings of the IUTAM, the International Union of Theoretical and Applied Mechanics. Von Kármán was Honorary President of the latter until his death. It was also during von Kármán's Aachen period that he developed his well-known logarithmic law for turbulent boundary layers. Here again interactions with Prandtl play a major role as discussed in the next section.

In 1929 von Kármán decided to accept Robert A. Millikan's offer to become director of the Guggenheim Aeronautical Laboratory at the California Institute of Technology and Professor of Aeronautics. Three years earlier von Kármán had advised Caltech on revisions to their aeronautics curriculum and on the design of the 10 ft wind tunnel. His mother and sister, Josephine (Pipö) de Kármán, moved to Pasadena with him, providing warm hospitality to the many students, faculty, and distinguished visitors that were guests at their home. On one occasion Enrico Fermi wanted to see a Hollywood studio and von Kármán was able to oblige (Figure 3.3) with a lunch with some of the stars at a studio because Pipö knew several Hungarian-born actors such as Bela Lugosi and

Figure 3.3 Edward Teller, Enrico Fermi, and von Kármán in Hollywood, 1937. (Courtesy of the Archives, California Institute of Technology.)

Paul Lukas (von Kármán and Edson, 1967, p. 178). Von Kármán's interest in rocket research began during his first decade at Caltech and led to the formation of the Jet Propulsion Laboratory and later to the establishment of the Aerojet General Corporation. Von Kármán was director of the former from 1938 to 1945. It was also during this period that von Kármán made considerable contributions to the theory of isotropic turbulence with notable interactions with G.I. Taylor as described in Section 3.3.

3.2 The logarithmic law of the wall

Von Kármán's interest in turbulence became manifest in 1921, when he compared friction coefficients in laminar and turbulent boundary layers and pipe flows (von Kármán, 1921). He follows a suggestion by Prandtl to deduce the mean velocity profile close to the wall in a turbulent pipe flow from the friction coefficient that Blasius had determined earlier by careful measurements. He states that "the publication occurs with Prandtl's approval, whereas my derivation is a little different from his". He uses dimensional analysis, where the friction velocity

$$u_\tau = \sqrt{\frac{\tau_w}{\rho}} \qquad (3.1)$$

already appears as a parameter, to derive the 1/7-law for the velocity profile from the $-1/4$ power Reynolds number dependence of the friction coefficient that Blasius had reported. He extends the 1/7-law to the turbulent boundary layer of a flat plate and finds a $-1/5$ power Reynolds number dependence. He remarks that Prandtl had found the same Reynolds number dependence for the friction coefficient earlier. This "cooperative competition", as von Kármán called the interaction between his group in Aachen and Prandtl's group in Göttingen later, stimulated his further ideas in developing a "theory of turbulence" (von Kármán, 1924). In order to estimate the energy balance in a turbulent shear flow he assumes periodic fluctuations of the longitudinal and vertical velocity components, and formulates a maximum principle to calculate the mean wavelength of the fluctuations. As a side condition he uses the energy budget. Using scaling arguments in the spirit of Prandtl (1921) and Tietjens (1925), he then derives the friction coefficient. In his autobiography (von Kármán and Edson, 1967), where he entitles Chapter 17 as 'Turbulence', he qualifies his paper (von Kármán, 1924), presented at the first Congress of Applied Mechanics in Delft, as the one that set "some foundations for a theory". But he "was aware that it would take a really 'happy thought' to regiment turbulence into a workable theory". He mentions the "outstanding contribution" of Prandtl in introducing the mixing length concept in 1926 at the 2nd International Congress of Applied Mechanics in Zürich (Prandtl, 1927), and considers the search for a universal law of turbulence his goal, to be pursued in competition with his former professor. For the next Congress in Stockholm 1930 both players were invited to give papers on turbulence. In order to compete with Prandtl, von Kármán knew that he first had to "find a method of developing a simplified physical concept" to arrive "at the universal formula that would fit the experimental data then available" (von Kármán and Edson, 1967).

On 12 December 1929 he writes a letter to Burgers in Delft where he sketches the logarithmic law for the maximum velocity in turbulent tube and channel flows that will later appear in the Stockholm paper. He remarks:

> This is not a theory but it shows that one can do a theory of turbulence without friction and what remains of the physics when one replaces the distribution by mean values.

There is a fascinating personal anecdote in von Kármán and Edson (1967) about von Kármán's interaction with Frank Wattendorf, an American student from MIT, who helped him in finding that universal formula during an inspired evening at von Kármán's home in Vaals, a small town in the Netherlands across the border from Aachen. Had the story not been confirmed by Wattendorf himself, as stated in the biographical memoir by H.L. Dryden (1965), one might have thought that it was too beautiful to not have been invented by von Kármán himself. The story goes that both passed that evening by plotting on different kinds of logarithmic graph paper the experimental data that Wattendorf had

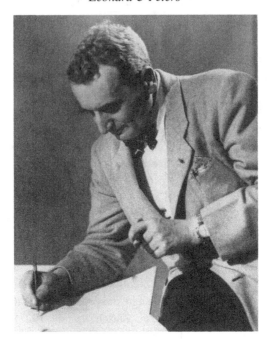

Figure 3.4 Von Kármán studying his lecture notes at Caltech, *c*. 1943. (Courtesy of the Archives, California Institute of Technology.)

brought from Göttingen. When the last streetcar from Vaals was to leave to Aachen at midnight, von Kármán escorted Wattendorf to the station, where he began writing his newly developed theory into the dirt on the side of the waiting streetcar. At first the conductor waited patiently, then looked repeatedly at his watch, but finally became insistent that the car must leave. In order to save the formulas from disappearance, Wattendorf had to jump from the car to the street at each stop, copy a few lines, and jump back on the car as it moved on. That night, back in his own room, he boiled the experimental data points down to one master plot, which he brought to von Kármán the next day.

The outcome of that night was published in two papers (von Kármán, 1930, 1931), the first one presented before the Society of Sciences at Göttingen on Prandtl's invitation and the second one at the Stockholm meeting. The Göttingen paper (von Kármán, 1930) focuses on channel flows and derives a log law for the centerline velocity as a function of the channel width. He first defines a new mixing length, later called the 'von Kármán length', which is proportional to the ratio of the first to the second derivative of the mean axial velocity profile. He then compares it to Prandtl's mixing length and

Figure 3.5 Hugh Dryden, Ludwig Prandtl, von Kármán, and Hsue-Shen Tsien in Göttingen 1945. Dryden, von Kármán, and Tsien were on a US Air Force mission to assess the state of German aeronautics technology. (Courtesy of the Archives, California Institute of Technology.)

argues that his mixing length provides more information because it contains two different derivatives of the axial mean flow. Introducing it into the expression for the turbulent shear stress he obtains a second-order differential equation which he solves to obtain the axial mean velocity profile across the channel. This expression contains the log of a function of the normalized distance from the centreline. He also shows that his mixing length scale is proportional to the distance from the channel wall y for small values of y. He determines the constant of proportionality κ by comparison with the Göttingen experiments by Nikuradse (1926) as 0.36 and mentions that it is "probably" universal.

The Stockholm paper (von Kármán, 1931), having the same title as the Göttingen paper, repeats the previous results but in addition presents an expression for the mean velocity $\bar{u}$ as a function of the distance from y the wall:

$$\bar{u}/u_\tau = \frac{1}{\kappa}\left(\ln\frac{yu_\tau}{\nu} + C\right). \tag{3.2}$$

Here ν is the kinematic viscosity. He discusses and rejects the possibility of a power-law dependence of the mean velocity on y and expresses his conviction that the logarithmic form is the final one. He concedes, however, that at not so large Reynolds numbers there is an influence of viscosity, which he hopes to be able to explain by looking closer at the region in the vicinity of the wall.

In his book of 1931 entitled *Abriß der Strömungslehre* Prandtl (1931) presents von Kármán's derivation of the log law and remarks that the constant should be 0.4 rather than 0.36 "according to recent findings", probably new experiments by Nikuradse done in Prandtl's laboratory in Göttingen. These were published in 1932 (Nikuradse, 1932) and contained a section 'Universal velocity distribution' which starts:

> In a new statement of his ideas, Prandtl … assumes only that the velocity in the vicinity of the wall is dependent only upon the physical quantities that are valid near the wall.

Nikuradse then derives the linear velocity profile in the viscous layer. Then he presents many experimental results and Kármán's theory of the Göttingen paper (von Kármán, 1930) in detail. At the end of the report he enters into the 'Similarity hypothesis according to Prandtl'. With u_τ as the only scaling parameter available he shows that

$$u_\tau = \kappa y \frac{\partial \bar{u}}{\partial y} \qquad (3.3)$$

with κ being a universal constant. After integration he reproduces (3.2) and states that the measurements reproduce $\kappa = 0.4$.

In a review article Prandtl (1933) presents this derivation, but then acknowledges that the "theory of Kármán" leads to the same result under the assumption of constant shear stress. He notes that both his mixing length, and that of von Kármán, lack a convincing foundation. (For more details on the competition with Prandtl see Section 2.6, 'Skin friction and turbulence II. The logarithmic law and beyond' in Chapter 2.)

One may ask why Prandtl published a different derivation of a result that von Kármán had already published. Was it the simplicity of the derivation? Or was it that he had sensed that there was more to it, namely a general scaling principle? We will see below that there is a striking similarity between the procedure that led to the logarithmic law and to the 4/5-law of isotropic turbulence.

The question whether there is a logarithmic or a power law dependence has been raised again by Barenblatt (1993), who argues in favour of a power law

$$\bar{u}/u_\tau = C\left(\frac{yu_\tau}{\nu}\right)^\alpha \qquad (3.4)$$

where both the prefactor C and the power α should depend on the log of the Reynolds number. The logarithmic law would then represent the envelope of a family of power-type curves, each corresponding to a fixed Reynolds number. A different point of view is taken by George et al. (1992); see also Wosnik et al. (2000). They used an asymptotic invariance principle for the zero-pressure-gradient boundary layer to suggest that the profiles in an overlap region between the inner and the outer regions are power laws. In the limit of infinite Reynolds numbers the log law was recovered in the inner region. Finally, Oberlack (2001), using Lie-group analysis, showed that for turbulent shear flows both power laws and logarithmic laws are admissible. From experimental data plotted in a semi-log plot he observes a logarithmic scaling region, the extent of which depends on the Reynolds number.

3.3 Isotropic turbulence

3.3.1 Kármán–Howarth paper

Leslie Howarth Von Kármán made significant contributions to the theory of homogeneous, isotropic turbulence. After the concept of isotropic turbulence was first introduced by G.I. Taylor (1935), von Kármán was quick to see the potential in investigating this case of turbulence for making inroads into the mysteries of turbulence. A significant collaborator of von Kármán's in the early phases of this work was Leslie Howarth. Howarth, a new PhD from the University of Cambridge, came to Caltech in the summer of 1936 for a year as a King's College Research Fellow. The visit was arranged with von Kármán by Sydney Goldstein of Cambridge (CIT von Kármán collection, Box 11, Folder 20), who was his advisor there as well as at Manchester, where Howarth was an undergraduate. Howarth returned to Cambridge in 1937 and became well known for his work in boundary layer theory and computation, among other areas of fluid mechanics. In 1949 he moved to Bristol University as Professor of Applied Mathematics where he built a strong research group.

Correlation tensor The assumptions of rotational and reflexion invariance as well as homogeneity led von Kármán and Howarth (1938) to the following result for the correlation coefficient for fluid velocities separated by a distance r:

$$\frac{\overline{pq}}{\overline{u^2}} = [f(r, t) \cos \alpha \cos \beta + g(r, t) \sin \alpha \sin \beta] \sin \gamma, \qquad (3.5)$$

where, referring to Figure 3.6, p is the velocity component at point P in direction PP', q is the velocity component in direction QQ' at point Q, a distance r from P, and the angles α, β, and γ are also defined in Figure 3.6. The vector QQ'' is the orthogonal projection of QQ' onto the plane formed by PP' and PQ. The vector $Q''Q'$ is therefore perpendicular to this plane.

The function $f(r, t)$ is the correlation coefficient for velocities unidirectional along the line PQ and $g(r, t)$ is the correlation coefficient for velocities unidirectional but perpendicular to the line PQ (see Figure 3.7). Using (3.5), von

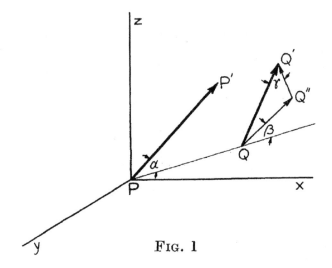

FIG. 1

Figure 3.6 Geometry of two velocity vectors (von Kármán and Howarth, 1938).

Kármán and Howarth construct the correlation tensor, $R_{ij} = \overline{u_i u'_j}/\overline{u^2}$, to be

$$R = \frac{f(r,t) - g(r,t)}{r^2} rr + g(r,t)I, \qquad (3.6)$$

where $r = x' - x = (\xi_1, \xi_2, \xi_3)$ is the spatial vector separating u' at x' and u at x, and I is the unit tensor. Application of the continuity equation then gives the following relation between f and g:

$$g = f + \frac{r}{2}\frac{\partial f}{\partial r}. \qquad (3.7)$$

As demonstrated by von Kármán and Howarth, using (3.6) one can compute the correlation of velocity derivatives such as $\overline{\frac{\partial u_k}{\partial x_i}\frac{\partial u_l}{\partial x_k}}$ or even $\overline{\frac{\partial^2 u_k}{\partial x_i \partial x_j}\frac{\partial^2 u_l}{\partial x_r \partial x_s}}$ with relative ease.

The above results (3.5–3.7) were in the original version of the Kármán–Howarth paper, submitted to G.I. Taylor by von Kármán on 9 January 1937 for publication in the *Proc. Roy. Soc.* (CIT von Kármán collection, Box 29, Folder 34) and essentially those same results appeared in von Kármán (1937a), the February issue, in which von Kármán acknowledged Howarth's "cooperation" in the development of the theory. In a letter dated 24 January 1937 to von Kármán (CIT von Kármán collection, Box 29, Folder 34) Taylor notes that because of (3.6–3.7) above

> I can analyse by your method some sets of observations of R made between pairs
> of points along lines at various angles to the direction of the axis of the tunnel.

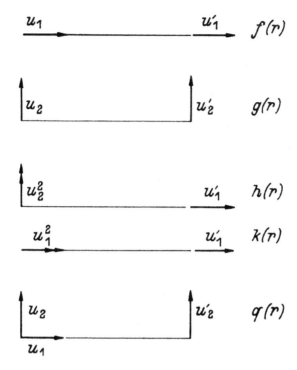

Figure 3.7 Definition of the correlation coefficients f, g, h, k, and q (von Kármán and Howarth, 1938).

Indeed, Taylor (1937) shows that (3.6–3.7) are well satisfied by data taken by L.F.G. Simmons of the National Physical Laboratory (NPL).

Triple correlation function Von Kármán and Howarth next consider the triple correlation function for two velocities at x and one at x', separated by $\xi = x' - x$ as above, and obtain the result

$$\overline{u_i u_j u'_k} = \overline{(u^2)}^{3/2} \left[\xi_i \xi_j \xi_k \frac{k - h - 2q}{r^3} + \delta_{ij} \xi_k \frac{h}{r} + \left(\delta_{ik} \xi_j + \delta_{jk} \xi_i \right) \frac{q}{r} \right], \quad (3.8)$$

where $k(r,t)$, $q(r,t)$, and $h(r,t)$ are basic triple correlations defined in Figure 3.7. Imposing continuity on this result led von Kármán and Howarth to the following:

$$
\begin{aligned}
k &= -2h, \\
q &= -h - \tfrac{r}{2} \tfrac{\partial h}{\partial r}.
\end{aligned}
\quad (3.9)
$$

Thus, as in the case of the double correlations, the triple correlations for isotropic, homogeneous turbulence can be expressed in terms of a single function of the distance between the points in question.

Kármán–Howarth equation and applications After showing by symmetry and continuity that pressure–velocity correlations are zero, von Kármán and Howarth apply their results to the Navier–Stokes equations to obtain the well-known Kármán–Howarth dynamical equation relating f and h,

$$\frac{\partial \overline{f u^2}}{\partial t} + 2\overline{(u^2)}^{3/2}\left(\frac{\partial h}{\partial r} + \frac{4}{r}h\right) = 2\nu \overline{u^2}\left(\frac{\partial^2 f}{\partial r^2} + \frac{4}{r}\frac{\partial f}{\partial r}\right), \tag{3.10}$$

which they call "the fundamental equation for the propagation of the correlation function f".

By examining (3.10) as $r \to 0$ and using $f(0, t) = 1$, von Kármán and Howarth obtain

$$\frac{d\overline{u^2}}{dt} = 10\nu\frac{\partial^2 f}{\partial r^2}(0, t)\overline{u^2} \tag{3.11}$$

which, they show, recovers Taylor's equation (Taylor, 1935) for the decay of energy behind a grid

$$\frac{d\overline{u^2}}{dt} = U\frac{d\overline{u^2}}{dx} = -10\nu\frac{\overline{u^2}}{\lambda^2} \tag{3.12}$$

by using their result (3.7), Taylor's definition of $\lambda(\lambda^{-2} = -\partial^2 g/\partial r^2(0, t)/2)$, and $d/dt = U d/dx$.

Von Kármán and Howarth also consider the case of "small Reynolds number", neglecting the triple correlation function h and assuming that f is a function of $\chi = r/\sqrt{\nu t}$ only. They find that $f(\chi)$ satisfies

$$\frac{d^2 f}{d\chi^2} + \left(\frac{4}{\chi} + \frac{\chi}{4}\right)\frac{df}{d\chi} + 5\alpha f = 0 \tag{3.13}$$

where

$$\alpha = -\left(\frac{d^2 f}{d\chi^2}\right)_{r=0}.$$

As von Kármán and Howarth note, the solution to (3.13) is given in terms of a confluent hypergeometric function as follows:

$$f(\chi) = \exp(-\chi^2/8)M[10\alpha - \frac{5}{2}, \frac{5}{4}, \frac{\chi^2}{8}], \tag{3.14}$$

where $M[a, b, z]$ is the Kummer's hypergeometric function. When $\alpha = 1/4$

they obtain the solution $f(\chi) = \exp(-\chi^2/8)$ which later was found to give a good fit to the experimental data of Batchelor and Townsend (1948).

Of course the Kármán–Howarth equation has been used by many other authors as the basis for new results on isotropic turbulence. We give three examples: two from the period just after the publication of the 1938 paper and one from recent times.

Loitsyanski (1939) multiplied (3.10) by r^4 and integrated over r from 0 to ∞ to find

$$\frac{d}{dt}\left[\overline{u^2}\int_0^\infty r^4 f(r,t)dr\right] = \left[\overline{u^2}^{3/2} r^4 h(r,t)\right]_{r\to\infty} + 2\nu\left[\overline{u^2}r^4\frac{\partial f}{\partial r}(r,t)\right]_{r\to\infty}. \quad (3.15)$$

Assuming that the correlation functions f and h vanish sufficiently rapidly at large separation, as Loitsyanski did, one finds an integral of the motion,

$$I = \overline{u^2}\int_0^\infty r^4 f dr = \text{constant}, \quad (3.16)$$

known as Loitsyanski's invariant. For a number of years this result was generally accepted and incorporated into, for example, the analysis of the low-wavenumber limit of the energy spectrum of isotropic turbulence. See Lin (1949b) and further discussion below. Then, especially after Batchelor and Proudman's (1956) analysis of the effect of pressure fluctuations on velocity correlations and Saffman's (1967) analysis of the dynamics of the large scales of turbulence, it was widely thought that Loitsyanski's integral in (3.16) did not exist. But recent numerical simulations have shown that if the initial condition for homogeneous, isotropic decay is impulse-free, then the integral exists and, after an initial transient, it remains constant. See Davidson (2009) and Pullin and Meiron (2011) for more discussion and analysis.

Kolmogorov apparently used the Kármán–Howarth paper to derive his 4/5 law. In Kolmogorov (1941a) he considers the dynamics of the second-order structure function $B_{\ell\ell}(r,t)$ in terms of the third-order structure function $B_{\ell\ell\ell}(r,t)$ defined respectively by

$$B_{\ell\ell}(r,t) = \overline{(\boldsymbol{\ell}\cdot\boldsymbol{u}(\boldsymbol{x}+r\boldsymbol{\ell},t) - \boldsymbol{\ell}\cdot\boldsymbol{u}(\boldsymbol{x},t))^2},$$

$$B_{\ell\ell\ell}(r,t) = \overline{(\boldsymbol{\ell}\cdot\boldsymbol{u}(\boldsymbol{x}+r\boldsymbol{\ell},t) - \boldsymbol{\ell}\cdot\boldsymbol{u}(\boldsymbol{x},t))^3}, \quad (3.17)$$

where $\boldsymbol{\ell}$ is an arbitrary unit vector. Presumably he introduced the relations

$$B_{\ell\ell}(r,t) = 2\overline{u^2}(1 - f(r,t)),$$

$$B_{\ell\ell\ell}(r,t) = -12\overline{u^2}^{3/2} h(r,t), \quad (3.18)$$

into (3.10) to obtain, with an additional steady-state assumption for $B_{\ell\ell}$, the equation

$$4\epsilon + \left(\frac{dB_{\ell\ell\ell}}{dr} + \frac{4}{r}B_{\ell\ell\ell}\right) = 6\nu\left(\frac{d^2B_{\ell\ell}}{dr^2} + \frac{4}{r}\frac{dB_{\ell\ell}}{dr}\right). \tag{3.19}$$

Here $\epsilon = -(3/2)\,d\overline{u^2}/dt$ is the energy dissipation rate. He integrates the equation twice and shows that there is a viscous layer for very small r where

$$B_{\ell\ell}(r) = \frac{1}{15}\frac{\epsilon r^2}{\nu}. \tag{3.20}$$

For larger values of r he neglects the viscous term and obtains

$$B_{\ell\ell\ell}(r) = -\frac{4}{5}\,\epsilon r. \tag{3.21}$$

In a recent use of the Kármán–Howarth equation, Lundgren (2002) employed the technique of matched asymptotic expansions and Lin's (1949a) modified version of the von Kármán self-preservation hypothesis (see (3.30) below) on (3.18) to determine explicit Reynolds-number corrections to the 4/5 law.

There is a striking analogy between the log law including the viscous sublayer as it was derived by Prandtl and Kolmogorov's derivation of the 4/5 law. Both start from a one-dimensional differential equation, the former in terms of the physical distance from the wall, the latter in terms of the correlation coordinate r. Both contain an external scaling parameter appearing as a constant in the equation, namely the shear stress τ_w leading to the friction velocity u_τ in the log law and the energy dissipation rate in the 4/5 law. Dimensional analysis using the viscosity ν as additional parameter in both cases then leads to the viscous length scale $\ell_\nu = \nu/u_\tau$ for the log law and to the viscous scale $\eta = (\nu^3/\epsilon)^{1/4}$ in Kolmogorov's derivation. At larger values of the respective coordinate, when viscous effects vanish, there remains only a single scaling parameter, u_τ and ϵ, respectively. This is related to symmetry-breaking in the chapter on the Russian school in this book (Falkovich, 2011). There is an important difference, however, between in the derivation of the log law and the 4/5 law when it comes to the 'universal' constant. While the log law assumes a proportionality between the mean velocity gradient and u_τ/y leading to an empirical constant that must be determined experimentally (the von Kármán constant) no such proportionality appears in Kolmogorov's derivation. Therefore its coefficient 4/5 is exact. This is not the case for Kolmogorov's 2/3 law where again an empirical constant, the Kolmogorov constant, appears.

It is interesting that von Kármán, having set the basis for both derivations, did not make the final steps to derive these scaling laws. It looks as if he were more interested in the solution of the entire or nearly entire problem, the mean velocity profile in the case of channel flow and the two-point correlation function for the energy-containing range of scales together with the scales in the inertial subrange for the case of isotropic turbulence.

Kármán–Taylor controversy During the first half of 1937, before the revised version of the Kármán–Howarth paper was submitted, von Kármán and G.I. Taylor had a serious dispute concerning their respective theories of the decay of grid turbulence. In Taylor (1935) it is assumed that the large-eddy lengthscale ℓ in the expression for dissipation $\epsilon = C\overline{u^2}^{3/2}/\ell$ is equal to the meshlength M of the grid. Combining this result with Taylor's other expression for dissipation, $\epsilon = 15\nu\overline{u^2}/\lambda^2$ led Taylor to the relation between λ and $\overline{u^2}$

$$\frac{\lambda}{M} = A \sqrt{\frac{\nu}{M \sqrt{\overline{u^2}}}} \tag{3.22}$$

and, thence, to the decay law

$$\frac{U}{\sqrt{\overline{u^2}}} = \frac{5x}{A^2 M} + \text{const.} \tag{3.23}$$

where A is "an absolute constant for all grids of a definite type" (Taylor, 1935). At the time and for the next few years after that the above expression fit the experimental data very well. On the other hand, von Kármán (1937a, 1937b) derives a one-parameter family of decay laws in which, as von Kármán claims, Taylor's is a special case. In these derivations, however, von Kármán assumes that the vortex stretching term in the enstrophy equation is zero in the mean or, equivalently, that the triple correlation term in the Kármán–Howarth equation is zero. Von Kármán (1937b) states:

> Now it can be shown that because of the isotropic feature of the correlation between the velocity components and their derivatives, [the vortex stretching term] vanishes in mean value. The proof will be given elsewhere. As a matter of fact, any value of the [stretching term] different from zero would mean that the vortex filaments had a permanent tendency stretched or compressed in the direction of the vorticity axis.

This assumption led von Kármán to a closed evolution equation for $f(r, t)$ and, assuming similarity or self-preservation for the correlation function f in terms of the variable $\chi = r/\sqrt{\nu t}$, von Kármán (1937a) derived the following decay law with parameter α:

$$\frac{U}{\sqrt{\overline{u^2}}} = \frac{U}{\sqrt{\overline{u_0^2}}} \left(1 + \frac{x}{U t_0}\right)^{5\alpha}, \tag{3.24}$$

and where $f(\chi)$ satisfies (3.13). Von Kármán notes that his result yields Taylor's result as an exact result if $\alpha = 1/5$ and that Taylor obtained (3.23)

> by assuming rather arbitrarily a relation between the length λ and the linear dimension, e.g. the spacing M at the grid or mesh.

A similar derivation of the decay law with corresponding remarks must have appeared in the original version of the Kármán–Howarth paper. In a letter of 24 January 1937 (CIT von Kármán collection, Box 29, Folder 34) to von Kármán, acknowledging receipt of the original manuscript, Taylor states:

> with regard to my linear law [(3.23) above] I think your statement is a little misleading. My assumption is that the scale of turbulence remains constant but the curvature of R (or f) continually decreases. Thus the R curve is not self-preserving. [A corresponding sketch is provided by Taylor.] The fact that you find a self-preserving law that also gives a linear law is an accident. Thus it is misleading to say that my non-self preserving assumption is a limiting case of a self-preserving theory.
>
> I did once think of trying a self-preserving distribution but came to the conclusion that it would be necessary for Re to be constant. Thus if $r/\sqrt{\nu t}$ is constant then u' must decrease such that ru'/ν is constant therefore $u' \sim 1/\sqrt{t}$ (not $1/t$). ... If you alter very slightly your remarks about the linear law I would be grateful.

Taylor also encloses a plot of decay data taken at NPL that clearly supports his linear law. Von Kármán responds in a letter dated 21 February 1937 (CIT von Kármán collection, Box 29, Folder 34) to Taylor:

> I gladly agree to any changes you suggest concerning the linear law of the decay of turbulence (3.23). However, I am afraid there is a basic difference between your viewpoint and mine, as far as the scale of turbulence is concerned. ... you assume that a certain scale of turbulence remains practically invariable during the process of decay. I could not see the justification for such an assumption because I believe that if the correlation function f is given at $t = 0$ then the process of decay is given for all subsequent time.

Although von Kármán, at this point in time, assumes that the triple correlation term is negligible, he seems open to the possibility that the self-similar solutions to (3.10) with $h = 0$ will not be realized in experiments; for example, in von Kármán (1937b) he states "In general the shape of the correlation curve will vary with time".

But in a telegram dated 15 April 1937 (CIT von Kármán collection, Box 29, Folder 34) to Taylor, Howarth states:

> Hold up publication of joint Royal paper. Essential changes necessary, especially points relating to your theory.

Thus, by early April, von Kármán with possible input from Howarth and/or Clark Millikan must have realized his error in neglecting the vortex stretching term or, equivalently, the triple correlation term. And with Howarth he began development of a correspondingly revised theory. In fact, in a letter dated 30 June 1937 (CIT von Kármán collection, Box 13, Folder 38) to von Kármán written aboard the *Queen Mary*, Howarth, on his way back to England, describes his efforts to determine the various terms in the dynamical equation for the *triple* correlation. An appendix to the letter gives his results obtained so far with the comment "did not complete... too many functions with few relations between them".

It won't be until October that von Kármán submits a new version of the Kármán–Howarth manuscript. Meanwhile, in Taylor (1937), submitted 1 May for the June issue, Taylor seems particularly intent on showing that his theory is *not* a special case of von Kármán's general theory and that von Kármán's theory is, in fact, flawed. After showing that (3.7) is verified experimentally as mentioned above, Taylor recalls von Kármán's (1937a) derivation of the dynamical equation for $\overline{u^2}f$ (i.e. (3.10) without the triple correlation term) and states:

> It is difficult to see any *a priori* reason why these inertia terms [corresponding to the triple correlations]... should be zero.... von Kármán promises to give the proof that the mean values of the inertia terms... vanish.

To further support his view, Taylor recounts his results with Green (Taylor and Green, 1937) on the increase in vorticity with time in three-dimensional flow with Taylor–Green initial conditions demonstrating the importance of the vortex stretching term for that problem. Taylor had, in fact, given von Kármán a preview of those results in a letter dated 12 March 1936 (CIT von Kármán collection, Box 29, Folder 33):

> With a student I have just finished an amusing piece of work. Do you recall that in my statistical theory of turbulence that I get a formula [(3.22)]?

In that letter, Taylor continues by describing their power series solution to t^5, the fact that $\overline{\omega^2}$ increases to a maximum and then decreases, that the maximum $\overline{\omega^2}$ increases with Re, and the connection with Taylor's formula (3.22). He concludes:

> [This] puts my dissipation theory of turbulence on a sounder basis.

Taylor (1937) then proceeds to show that von Kármán's self-similar form of the solution to the dynamical equation for $\overline{u^2}f$, i.e. the solution to (3.13), does not agree with experiment in two ways. First, Taylor argues that because $1/\lambda^2 \sim U$ (see (3.12)) the shape of the curve of f near $r = 0$ will be affected by a change in U but experiment shows little change in the curve away from $r = 0$ with variations in U. A self-similar solution does not have this freedom and, thus, is not in accordance with experimental observation. Next, Taylor solves von Kármán's self-similar form of the evolution equation for f, (3.13), by power series and shows that its solution is in significant disagreement with experiment.

With the publication of the Kármán–Howarth paper, however, the dispute between von Kármán and Taylor is essentially over. In a footnote in that paper von Kármán acknowledges his previous error in assuming that the triple correlation term would vanish in isotropic turbulence and, when von Kármán and Howarth revisit the small Re approximation, they note that $\alpha = 1/5$ (for (3.13) and (3.24)) gives Taylor's result but the coincidence "is rather formal".

Section 11 of the paper, written by von Kármán "especially after reading Taylor's (1937) contribution", explores possible solutions for large Re. Von Kármán finds a one-parameter family of self-similar solutions (see §3.3.2 below) and, *again*, Taylor's result is a special case. "It appears that further experimental results will decide whether Taylor's assumptions are not too narrow and whether [von Kármán's more general solution] correspond more closely to the experimental facts." This time von Kármán is eventually proven to be closer to the truth. For example, Batchelor (1953) presents grid turbulence data that closely follow the decay law,

$$\frac{U}{\sqrt{\overline{u^2}}} = (C + Dx)^\beta \tag{3.25}$$

with $\beta = 1/2$ rather than 1. Batchelor (1953) states:

> The earliest measurements of $\sqrt{\overline{u^2}}$ at different stages of the decay were a little
> confused by inadequate corrections ..., by turbulence, from sources other than
> the grid, and by the inclusion of observations [too close to the grid].

In fact the decay law (3.25) with $\beta = 1/2$ had been proposed by Dryden in a letter dated 16 September 1939 to von Kármán (CIT von Kármán collection, Box 7, Folder 26) and published in Dryden (1943). Recall, even earlier, Taylor had derived and then discarded the same law; see *Kármán–Taylor controversy* above. More recent data taken at higher Re are satisfied by (3.25) with $\beta \approx$ 0.65; see, for example, Pope (2000, p. 160).

In spite of the sometimes testy exchanges in 1937 described above, von Kármán and Taylor, in general, had a very cordial, amiable relationship extending over several decades. There is correspondence between them starting in 1928 and lasting for 30 years and a number of mutual visits were undertaken. For example, in a note of 27 March 1930 (CIT von Kármán collection, Box 29,

Folder 33) to Pipö, von Kármán's sister, Taylor thanks her for her hospitality while he was in Aachen on a visit and continues:

> Please remember me kindly to your mother and tell her that I hope for her sake and for the sake of European Science that your brother will not go to Pasadena permanently.

And in the letter dated 9 January 1937 to Taylor (CIT von Kármán collection, Box 29, Folder 34) cited above, von Kármán states:

> I am looking forward with great expectations to my visit in Cambridge next spring. I hope that Mussolini and Hitler will wait with their big war until after my lecture.

3.3.2 Self-similar solutions and spectral closures

C.C. Lin Following the Kármán–Howarth paper and continuing into the early 1950s, von Kármán's contributions to the understanding of isotropic turbulence concentrated on the development of self-similar solutions to the Kármán–Howarth equation and on spectral closures for the dynamical equation for the energy spectrum function. A frequent collaborator of von Kármán during this period was C.C. Lin.

Chia-Chiao Lin was born in Beijing, China, in 1916. He completed a BSc in physics from Tsinghua University in 1937 and an MA in applied mathematics from the University of Toronto in 1941. It is quite likely that John L. Synge, who was at Toronto at the time, suggested that von Kármán consider Lin as a PhD student. Indeed, Lin received a PhD in aeronautics from Caltech in 1944 under the supervision of von Kármán. He then did postdoctoral work at the Jet Propulsion Lab before joining the faculty at Brown University in 1945. He joined the MIT Applied Mathematics faculty as associate professor in 1947. In 1966, Lin was selected to be an Institute Professor. Lin is also very well known for his work on hydrodynamics stability, an early example being his insightful analysis of the stability of two-dimensional parallel shear flows as a part of his PhD dissertation (see Lin, 1944). Later, he worked on significant problems in the hydrodynamics of superfluid helium and in astrophysics, where he worked on gravitational collapse and star and galaxy formation. Early in his career he frequently looked to von Kármán for advice. At one point he asked for von Kármán's opinion on his job opportunities at Brown, Ohio State, and the Courant Institute (CIT von Kármán collection, Box 18, Folder 23). About this same time, 6 February 1946, Lin asked von Kármán for help because he was in trouble with US immigration and was close to being sent back to China. Von Kármán, Clark Millikan, and, ultimately, Brown University officials were able to solve this problem. About a year earlier, Lin told von Kármán in a letter dated 28 February 1945 (CIT von Kármán collection, Box 18, Folder 22) that he was to report for a pre-induction exam for the military draft. This time von Kármán had a Caltech official write to the local draft board urging reclassification for Lin because Caltech was involved with "high priority weapons research".

Kármán's self-preservation hypothesis In the final section of the Kármán–Howarth paper, von Kármán proposes a solution to (3.10) for large Re. He first argues that f, for example, should be of the form

$$f\left(\frac{r}{\sqrt{\nu t}}, \frac{r}{M}, \frac{Ut}{M}\right)$$

and considers only the the range of r away from viscous effects, i.e. those values of r such that $r/\sqrt{\nu t} \to \infty$ as $Re \to \infty$. Next he assumes that f and h preserve their shape with only a change in scale with time. This assumption yields

$$f = f(\psi), \tag{3.26}$$
$$h = h(\psi),$$

with $\psi = r/L$ and L a function of M and Ut. Substituting these expressions for f and g into the Kármán–Howarth equation and using (3.12), von Kármán obtains an equation relating $f(\psi)$ and $h(\psi)$ with the proviso that

$$\frac{\sqrt{\overline{u^2}}\lambda^2}{L\nu} = A, \tag{3.27}$$

$$\frac{dL}{dt} = B\sqrt{\overline{u^2}}, \tag{3.28}$$

where A and B are constants. The functions $f(\psi)$ and $h(\psi)$ are not determined separately by this similarity hypothesis – the usual closure problem – but using (3.12) with (3.27) and (3.28) above, von Kármán finds the general solution for the decay of turbulent energy (with correction as noted by Kolmogorov, 1941b)

$$\frac{U}{\sqrt{\overline{u^2}}} = \frac{U}{\sqrt{\overline{u_0^2}}}\left(\frac{t}{t_0}\right)^{\frac{5}{5+AB}}. \tag{3.29}$$

As von Kármán notes, for the particular solution $L = M =$ a constant or $B = 0$, (3.27) becomes Taylor's result (3.22) for λ and the corresponding solution for $\sqrt{\overline{u^2}}$ is Taylor's decay law (3.23). Von Kármán remarks that "Taylor found [this solution] with a remarkable vision for the relations between the quantities involved".

Refined similarity hypotheses The similarity hypothesis (3.27) is known as "Kármán's self-preservation hypothesis" and has been revisited and refined by a number of authors since it first appeared in 1938. One notable refinement is due to C.C. Lin (1949a). He assumed self-similar forms for $1 - f$ and h, "suggested by Kolmogoroff's concept of turbulence at high Reynolds numbers", as follows:

$$\overline{u^2}(1 - f) = v^2\beta_2(r/\eta), \tag{3.30}$$
$$(\overline{u^2})^{3/2}h = v^3\beta_3(r/\eta),$$

where v and η are velocity and length scales, respectively. Substituting the above into the Kármán–Howarth equation and requiring that similarity extend to all values of r, including the viscous range, Lin finds the Kolmogorov relations,

$$\eta = (v^3/\epsilon)^{1/4}, \quad v = (v\epsilon)^{1/4}, \tag{3.31}$$

and the decay law

$$\overline{u^2} = \alpha(t - t_0)^{-1} + \gamma. \tag{3.32}$$

According to Lin, the decay law (3.32) is an improvement over (3.25) with $\beta = 1/2$, because of the additional factor γ. Indeed it fits the decay data available at the time very well. In addition, Lin's theory also leads to a formula for λ^2 that is quadratic in $t - t_0$ which also was in good comparison with experimental data of Townsend, supplied to Lin by Batchelor.

However, the decay law (3.32) is not compatible with the existence of Loitsyanski's invariant (3.16). In fact Kolmogorov (1941b) had shown that Kármán's self-preservation hypothesis (3.27) on $1 - f$ rather than f with (3.16) leads to the decay law

$$\overline{u^2} = \alpha(t - t_0)^{-\frac{10}{7}}. \tag{3.33}$$

As a consequence, von Kármán and Lin (1949, 1951), the latter being von Kármán's last published effort on isotropic turbulence, proposed a revised theory in which it is assumed that during the early stage of decay all but the largest scales participate in an equilibrium and Lin's decay law (3.32) will hold. For this early stage they are led to the existence of a medium range of eddy sizes that yield a corresponding portion of the energy spectrum that is 'linear', i.e. $E(\kappa) \sim \kappa$. In the intermediate stage, at high enough Re and before the final period of decay, the 'linear' regime is gone and one obtains an equilibrium between the largest scales and the inertial subrange scales (Kármán similarity) and another equilibrium between inertial subrange scales and the smallest scales (Kolmogorov similarity).

Spectral closures In 1948 von Kármán made brief but notable efforts in the theory of spectral closures for isotropic turbulence (1948a, 1948b, 1948c). Lin had given him a running start by informing him in a letter dated 6 June 1947 (CIT von Kármán collection, Box 18, Folder 23) that he had determined that existence of Loitsyanski's invariant leads to $E(\kappa) \sim C\kappa^4$ for small κ (Lin, 1949b) and, in the same letter, by pointing out Heisenberg's (1948) formula relating 1D spectra to 3D spectra. Lin had also sent to von Kármán a copy of his letter of 26 June 1945 to Lars Onsager (CIT von Kármán collection, Box 18, Folder 22) containing his result relating the spectral transfer term $T(\kappa)$ in the evolution equation for the turbulence energy spectrum to the first and second derivatives of the sine transform of $h(r)$ (Lin, 1949b). As to the latter von Kármán states: "Unfortunately this relation does not help, as far as determination of f and h is concerned" (von Kármán, 1948b).

Von Kármán (1948b) postulates that the transfer function has the following form:

$$T(\kappa) = \int_0^\infty \Theta[E(\kappa), E(\kappa'), \kappa, \kappa'] \, d\kappa' \tag{3.34}$$

with the symmetry constraint

$$\Theta[E(\kappa), E(\kappa'), \kappa, \kappa'] = -\Theta[E(\kappa'), E(\kappa), \kappa', \kappa]. \tag{3.35}$$

He then proposes the specific form for Θ:

$$\Theta = -CE^\alpha(\kappa)E^{3/2-\alpha}(\kappa')\kappa^\beta\kappa'^{1/2-\beta} \quad \kappa > \kappa', \tag{3.36}$$

that satisfies dimensional constraints. As von Kármán shows, (3.36) is consistent with $E(\kappa) \sim \kappa^{-5/3}$ in the inertial subrange for any α and β. Various choices of these parameters correspond to some of the other proposals put forward during this era. For example, with $\alpha = 1/2$ and $\beta = -3/2$ one has the theory proposed by Heisenberg (1948). Subsequently, generalizations of (3.36) have been proposed; for example, Kraichnan and Spiegel (1962) modify (3.36) by multiplying the right-hand side by an arbitrary function $\phi(\kappa/\kappa')$ satisfying $\phi(1) = 1$.

Assuming self-preservation of the energy spectrum during decay and the above for $T(\kappa)$, von Kármán derives a nonlinear integro-differential equation for the spectrum function, which "can be solved numerically" for any definite choice of α and β. Rather, von Kármán proposes "for the time being" the interpolation formula

$$E(\kappa) = \frac{C\overline{u^2}}{\kappa_0} \frac{(\kappa/\kappa_0)^4}{[1 + (\kappa/\kappa_0)^2]^{17/6}}, \tag{3.37}$$

where he assumes that "for the time being, Loitsyanski's result is correct". The proposed spectrum (3.37) is known as the *von Kármán spectrum*.

3.4 Epilogue

After 1951 von Kármán's research interests centered on a new field he termed aerothermochemistry, and on magnetofluidmechanics. All the while his non-research efforts were considerable and included organizing and chairing in 1944 a Scientific Advisory Group for the US Air Force and forming in 1951 AGARD, the Advisory Group for Aeronautical Research and Development Group for NATO, the North Atlantic Treaty Organization, the latter being another demonstration of his leadership in international cooperation. Because of the demands of these and other such responsibilities von Kármán resigned from his directorship at Caltech and became Professor Emeritus in 1949.

However, at least in 1955, von Kármán still had turbulence on his mind and was optimistic about our ability in the future to understand the same. He was asked to gather his thoughts about the future in an article entitled 'The next fifty years' (von Kármán, 1955), or as he puts it, to "scan the coffee grounds in order to find out what will become of aviation and the aeronautical sciences during the next half-century". In the last paragraph he comes to turbulence:

> Finally, for the relief of the aerodynamicist, I wish to mention a problem related to fluid mechanics – incidentally, an essentially 'fluid' problem – which proved to be insoluble by scientific means during the first half-century of modern aero-dynamics: I mean the problem of turbulence, which borders on fluid mechanics as well as statistics. Many years ago, the lamented Arthur Sommerfeld, the prominent German specialist of theoretical physics, told me in confidence that in his lifetime he would like to be able to solve two problems related to theoretical physics: true significance of quantum mechanics and the actual mechanism of turbulence.[1] I did not have the opportunity to ask him whether he felt that the present interpretation of the quantum theory (complementarity instead of causality) was of such a nature that it satisfied the spirit. Without a doubt, turbulence will have to disclose its secret in the next fifty years.

In a 1959 tribute to von Kármán, Hugh Dryden (CIT von Kármán collection, Box 7, Folder 26) said:

> A notable feature of von Kármán's scientific and engineering work is the breadth of field covered. His interests extend from mathematics to rocket propulsion, from fluid mechanics to solid mechanics, from the crystal to the sand dune, from

---

[1] Nearly the same thoughts are attributed to Horace Lamb by Sydney Goldstein (Davidson, 2004, p. 24).

the helicopter to the space vehicle. This breadth of interest is not limited to science and engineering but spills over into politics, music, art, and to every activity of the human spirit.

Theodore von Kármán was awarded the first US Medal of Science by President John F. Kennedy in February 1963. He died on 7 May 1963, four days before his eighty-second birthday.

References

Barenblatt, G.I. 1993. Scaling laws for fully developed turbulent shear flows. *J. Fluid Mech.*, **248**, 513–520.

Batchelor, G.K. 1953. *The Theory of Homogeneous Turbulence*. Cambridge University Press, xi+197 pp.

Batchelor, G.K., and Proudman, I. 1956. The large-scale structure of homogeneous turbulence. *Phil. Trans. Roy. Soc. A*, **248**, 369–405.

Batchelor, G.K., and Townsend, A.A. 1948. Decay of turbulence in the final period. *Proc. Roy. Soc. A*, **194**, 527–543.

CIT von Kármán collection. *Papers of Theodore von Kármán, 1881–1963*. Archives of the California Insitute of Technology.

Davidson, P.A. 2004. *Turbulence*. Oxford University Press, xix+657 pp.

Davidson, P.A. 2009. The role of angular momentum conservation in homogeneous turbulence. *J. Fluid Mech.*, **632**, 329–358.

Dryden, H.L. 1943. A review of the statistical theory of turbulence. *Quart. Appl. Math.*, **1**, 7–42.

Dryden, H.L. 1965. Theodore von Kármán. *National Academy of Sciences*, Biographical Memoirs, **38**, 344–384.

Falkovich, G. 2011. The Russian school. In *A Voyage Through Turbulence*. Cambridge University Press.

George, W.K., Knecht, P., and Castillo, L. 1992. Zero-pressure gradient boundary layer revisited. In X. B. Reed, ed., *13th Symposium on Turbulence*, Rolla, MO.

Heisenberg, W. 1948. Zur statistischen Theorie der Turbulenz. *Z. Physik*, **124**, 628–657.

Kármán, von T. 1921. Über laminare und turbulente Reibung. *ZAMM*, **1**, 233–252.

Kármán, von T. 1924. Über die Stabilität der Laminarströmung und die Theorie der Turbulenz. *Proc. First Internat. Congr. Appl. Mech., Delft.*

Kármán, von T. 1930. Mechanische Ähnlichkeit und Turbulenz. *Gött. Nachr.*, 58–76.

Kármán, von T. 1931. Mechanische Ähnlichkeit und Turbulenz. *Proc. Third Internat. Congr. Appl. Mech. Stockholm*, **1**, 85–93.

Kármán, von T. 1937a. The fundamentals of the statistical theory of turbulence. *J. Aero. Sci.*, **4**, 131–138.

Kármán, von T. 1937b. On the statistical theory of turbulence. *Proc. Nat. Acad. Sci.*, **23**, 98–105.

Kármán, von T. 1948a. Progress in the statistical theory of turbulence. *Proc. Nat. Acad. Sci.*, **34**, 530–539.

Kármán, von T. 1948b. Progress in the statistical theory of turbulence. *J. Marine Res.*, **7**, 252–264.

Kármán, von T. 1948c. Sur la théorie statistique de la turbulence. *C.R. Acad. Sci. Paris*, **226**, 2108–2111.

Kármán, von T. 1955. The next fifty years. *Interavia*, **10**, 20–21.

Kármán, von T., and Edson, L. 1967. *Wind and Beyond*. Little, Brown and Co., 376pp.

Kármán, von T., and Howarth, L. 1938. On the statistical theory of isotropic turbulence. *Proc. Roy. Soc. A*, **164**, 192–215.

Kármán, von T., and Lin, C.C. 1949. On the concept of similarity in the theory of isotropic turbulence. *Rev. Mod. Phys.*, **21**, 516–519.

Kármán, von T., and Lin, C.C. 1951. On the concept of similarity in the theory of isotropic turbulence. Pages 1–19 of *Adv. Appl. Mech.*, vol. 2. Academic Press, New York.

Kolmogorov, A.N. 1941a. Dissipation of energy in the locally isotropic turbulence. *Dokl. Akad. Nauk SSSR*, **32**, 19–21.

Kolmogorov, A.N. 1941b. On decay of isotropic turbulence in an incompressible viscous fluid. *Dokl. Akad. Nauk SSSR*, **31**, 538–540.

Kraichnan, R.H., and Spiegel, E.A. 1962. Model for energy transfer for isotropic turbulence. *Phys. Fluids*, **5**, 583–588.

Lin, C.C. 1944. On the stability of two-dimensional parallel flows. *Proc. Nat. Acad. Sci.*, **30**, 316–324.

Lin, C.C. 1949a. Note on the law of decay of isotropic turbulence. *Proc. Nat. Acad. Sci.*, **34**, 540–543.

Lin, C.C. 1949b. Remarks on the spectrum of turbulence. Pages 81–86 of *Proceedings of Symposia in Applied Mathematics, Vol. 1 (Brown U., 1947)*. Amer. Math. Soc.

Loitsyanski, L.G. 1939. Some basic laws for isotropic turbulent flow. *Trudy Tsentr. Aero.-Giedrodin. Inst.*, **440**, 3–23.

Lundgren, T. S. 2002. Kolomogorov two-thirds law by matched asymptotic expansions. *Phys. Fluids*, **14**, 638–642.

Nikuradse, J. 1926. Untersuchungen über die Geschwindigkeitsverteilung in turbulenten Strömungen, Diss. Göttingen. *VDI-Forschungsheft*, **281**.

Nikuradse, J. 1932. Gesetzmäßigkeit der turbulenten Strömung in glatten Rohren. *Forsch. Arb. Ing.-Wes.*, **Heft 356**.

Oberlack, M. 2001. A unified approach for symmetries in plane parallel turbulent shear flows. *J. Fluid Mech.*, **427**, 299–328.

Pope, S.B. 2000. *Turbulent Flows*. Cambridge University Press, xxxiv+771 pp.

Prandtl, L. 1921. Bemerkungen zur Entstehung der Turbulenz. *ZAMM*, **1**, 431–436.

Prandtl, L. 1927. Über die ausgebildete Turbulenz. *Verh. des 2. Internationalen Kongresses für Technische Mechanik 1926, Zürich*, 62.

Prandtl, L. 1931. *Abriß der Strömungslehre*. Vieweg Verlag Braunschweig, 1. Auflage.

Prandtl, L. 1933. Neuere Ergebnisse der Turbulenzforschung. *VDI*, **77**, 105–114.

Pullin, D. I., and Meiron, D. I. 2011. Philip G. Saffman. In *A Voyage Through Turbulence*. Cambridge University Press.

Saffman, P.G. 1967. The large-scale structure of homogeneous turbulence. *J. Fluid Mech.*, **27**, 581–593.

Taylor, G.I. 1935. Statistical theory of turbulence. *Proc. Roy. Soc. A*, **151**, 421–444.

Taylor, G.I. 1937. The statistical theory of isotropic turbulence. *J. Aero. Sci.*, **4**, 311–315.

Taylor, G.I., and Green, A.E. 1937. Mechanism of the production of small eddies from large ones. *Proc. Roy. Soc. A*, **158**, 499–521.

Tietjens, O. 1925. Articles on the formation of turbulence. *ZAMM*, **5**, 200–217.

Vincenti, W. G. 1990. *What Engineers Know and How They Know It*. The Johns Hopkins University Press. vii + 326 pp..

Wosnik, M., Castillo, L., and George, W.K. 2000. A theory for turbulent pipe and channel flows. *J. Fluid Mech.*, **421**, 115–145.

4

G.I. Taylor: the inspiration behind the Cambridge school

K.R. Sreenivasan

4.1 Opening remarks

4.1.1 Taylor's reputation

The obituary of G.I. Taylor (7 March 1886–27 June 1975), written by Sir Brian Pippard[1] in 1975, begins thus: "Sir Geoffrey Ingram Taylor, who died at the age of 89, was one of the great scientists of our time and perhaps the last representative of that school of thought that includes Kelvin, Maxwell and Rayleigh, who were physicists, applied mathematicians and engineers – the distinction is irrelevant because their skill knew no such boundaries. Between 1909 and 1973 he published voluminously, and in a lifetime devoted to research left his mark on every subject he touched and on every one of his colleagues ... his outgoing manner and complete lack of pomposity conveyed, as no formal exposition could have done, the enthusiasm and intuitive understanding that informed all his work." These words, taken together with Pippard's closing sentence, "To his many friends he was an inspiration, at once a profound thinker and, it seemed, a truly happy man", summarize the essential G.I. Taylor. Goldstein (1969) had this to say: "By the end of the first half-century there was a stronger and more widespread element of physics in thought and research on fluid mechanics than in the first twenty or thirty years, and this is much more so now. Several factors and several research workers contributed to this, but the greatest influence has been the example of G.I. Taylor." Binnie (1978) regarded Taylor's death as marking the end of a golden age, referring to fluid mechanics

---

[1] Pippard was the Cavendish Professor when Taylor died. The other Cavendish professors with whom he overlapped are J.J. Thomson, Lord Rutherford (who arrived at Cambridge slightly after Taylor's own return in 1919), William Lawrence Bragg and Nevill Mott – all of whom were Nobel Laureates.

research, in general – and to that in Cambridge, in particular. It is obvious that Taylor was regarded as very special by those who knew him well.

To fluid dynamicists in the world at large, Taylor undoubtedly remains a giant. The centrality of fluid mechanics in Taylor's research cannot be debated – recall the important associations such as Taylor–Couette flow, Taylor number, Taylor vortex, Taylor column, Taylor–Proudman theorem, Taylor's vorticity transfer theory, Taylor–Green problem, Taylor microscale (and the microscale Reynolds number), Taylor's frozen-flow hypothesis, Rayleigh–Taylor instability, Saffman–Taylor fingering, Taylor cone, etc.[2] – but his contributions to solid mechanics were no less. Bell (1995) summarized this sentiment eloquently as: "[Taylor's] stature as an experimentist in Solid Mechanics will, in time, exceed his stature in all of his other research endeavors". Indeed, for a good part of his scientific life, Taylor seems to have thought about mechanics of fluids and solids with equal flair and ease.

Abstracting the accomplishments of this extraordinary man while not ignoring his peculiarities is fraught with many difficulties – especially because this article is meant to be centered mostly on his turbulence work. But that is what I shall attempt.

4.1.2 The dilemmas of writing about Taylor

Among the older accounts of Taylor's background are Southwell (1956) and Spalding (1962), along with the autobiographical notes (Taylor 1970, 1974). There has been a growing literature on Taylor since his death in 1975. Batchelor (1996) distilled Taylor's life in his definitive biography, *The Life and Legacy of G.I. Taylor* – henceforth abbreviated as *The Legacy*. His earlier accounts (Batchelor 1975, 1986), especially the extensive biographical memoir prepared for the Royal Society of London (Batchelor 1976), are of immense value. Turner's (1997) illuminating article is about the impressions that Taylor made on the Cambridge applied mathematics students of the 1950s; see also Townsend (1990). Even if we ignore a number of other minor accounts, the first dilemma is how to say anything new about Taylor.

The second dilemma relates to the fact that Taylor's scientific career extended over 60 or so productive years. During this time, his methods of research and his approach to his peers changed somewhat. In earlier years (especially those that concern us here the most, see below), Taylor organized an

---

[2] Lighthill (1956) also introduced the names 'Taylor thickness' and 'amended Taylor thickness' in the context of shocks under specific non-equilibrium conditions, but the names do not seem to have caught on. The shock structure in strongly non-equilibrium conditions is quite involved – as discussed, for example, by Zel'dovich & Raizer (1969).

important meteorological expedition, learned to fly and parachute, made theories and inspired others to do experiments to test his theories or used other people's data for the same purpose – and was in constant correspondence with many people, often benefitting from such interactions just as much as others undoubtedly benefitted from him. However, Taylor as a person and researcher is most on record[3] (and best known to some of those now alive) during the period following his official retirement; comments on his "originality, independence, insight, and the ability to recognize significance" (see *The Legacy*), his "lateral modes of thinking" on scientific problems (Turner 1997), and the extraordinary self-contained nature of his life relate to that later period. People who knew him in the last twenty-five years of his life – during which he worked on bench-top experiments almost by himself, except for one technician, Walter Thompson, who seemed to have been able to translate even rough drawings of Taylor's into finely working apparatuses – might find it hard to understand why Taylor is described by Rouse & Ince (1957) as a 'meteorologist', or to appreciate that he was perfectly capable of organizing a complex scientific expedition requiring much planning and coordination of effort. How to perceive the persona of Taylor's earlier years without being swamped by numerous impressions of his later life becomes an issue.

Added to these considerations, Taylor was not demonstrative in print, and the lucidity and simplicity of his prose hid the sense of excitement he seemed to have had at times. Taylor never seemed to have pontificated on broader issues of science and society. Burgers, in a letter to Batchelor, written on 14 June 1977 not too long after Taylor's death, got it right: "I would say that the strong scientific interest suppressed something of humanitarian awareness in Taylor or perhaps more to the point, it did not help him to express concerns about humanitarian aspects." Southwell (1956) had already remarked that Taylor's "inattention may be expected if the conversation veered towards problems of philosophy or politics..." Yet, some of Taylor's actions belie this straitjacket description, however appropriate in a coarser sense.

Altogether, it becomes evident that one has to explore sources beyond the obvious in order to get a fuller picture of Taylor. In this regard, Batchelor's (1960) compilation of Taylor's scientific work in four volumes of *Collected Works* is indispensable, as is his monumental effort in organizing Taylor's correspondence, which has been deposited in fourteen boxes in the Trinity College Library at Cambridge. Much of that material would have remained inaccessible without Batchelor's efforts, for Taylor was not especially organized in filing his scientific and personal papers. Part of the archives consists also of the

---

[3] Those of us who have seen Taylor's beautiful movie on low Reynolds number flows, made in the late 1960s, are grateful that he has been preserved in action for posterity.

Figure 4.1 The young and dashing Taylor at aged 39. From Batchelor (1996).

correspondence between Taylor and Batchelor. Finally, Taylor carried on significant scientific correspondence with a number of well-known fluid dynamicists of his time such as W.M. Orr (1866–1934), V.W. Ekman (1874–1954),
L.V. King (1886–1956), L. Prandtl (1875–1953), Th. von Kármán (1881–1963),
L.F. Richardson (1881–1953), J. Proudman (1888–1975), J.M. Burgers (1895–
1981) and H.L. Dryden (1898–1965). I shall provide some hints on these interactions to illustrate Taylor as a scientist and as a person during the years of
particular interest to us. Even the two obituaries that he authored on Kármán
(Taylor 1973a,b) reveal some of Taylor's charm. I believe that Taylor allowed
himself to express more freely in correspondence with friends and contemporaries than in published work. It is for this reason that citing from his letters
where possible seems useful in interpreting his work and attitude to it. In any
case, an assessment of a scientific life is incomplete if it focuses only on the
long-term scientific flow while ignoring short-range human interactions.

Figure 4.2 Taylor in 'retirement' in 1955 (age 69) in the Cavendish Laboratory with his assistant Walter Thompson. From Batchelor (1996).

4.2 Brief chronological account, focusing mostly on scientific career

Taylor's life history has been recorded extensively in *The Legacy*, and only a cursory summary will be presented here. After graduating from Cambridge with a BA in 1908[4], Taylor spent some four years there doing research, first

---

[4] In Part I of Mathematical Tripos, which Taylor read in 1907, he was the 22nd wrangler. Considering that he was independent-minded and the Tripos examinations usually required a lot of routine practice, the placement is probably not a surprise. As an aside, some senior wranglers (i.e. persons to hold the first place) were George Airy (1823), J.C. Adams (1843) of Neptune fame, G.G. Stokes (1841), P.G. Tait (1852), Lord Rayleigh, then John Strutt (1865), Arthur Eddington (1904), J.E. Littlewood (1905) and G.N. Watson (1907), the date in parenthesis indicating the year they passed the Tripos. Lord Kelvin, then William Thomson (1845), J.C. Maxwell (1854), Horace Lamb (1872), J.J. Thomson (1880) and James Jeans (1898) were among the famous second wranglers. A few others who shone in their careers but did less well in the Tripos were Thomas Robert Malthus (9th, 1788), Osborne Reynolds (7th, 1867), William Henry Bragg (3rd, 1885), Bertrand Russell (7th, 1893), G.H. Hardy (4th, 1898) and John Maynard Keynes (12th, 1905). Cambridge replaced the enormously competitive Wrangler system in 1909 by a softer version.

on showing that Young's slit diffraction experiment produced fringes even by feeble light sources emitting less than one photon at a time, and later on shock structure. According to Batchelor's letter of 4 March 1976 to Burgers, Taylor worked on the quantum problem at the suggestion of J.J. Thomson but soon decided that "he did not feel a call to a career in pure physics". About Taylor's work on the shock structure, Lighthill (1956), in his delightful essay on finite amplitude sound waves, had this to say:

> neither author [Rankine and Hugoniot] can have appreciated clearly how the irre-versible processes are brought into play, since they did not exclude the possibility of discontinuous rarefaction waves. Taylor (1910) first indicated why these could never occur, showing ... that if one formed it would instantly become continuous. Conversely, in a compressive wave, he pointed out that, as the velocity and tem-perature gradients increase, more and more importance must be attached to the penetration, between collisions, of molecules with one mean velocity and tem-perature among those with different mean velocity and temperature – in other words, to the phenomena of viscosity, heat conduction and (as he would have added had he been writing later) relaxation of internal degrees of freedom.

Taylor was barely 24 at the time of this masterly writing.

The result of the work on shock structure won him a Trinity Fellowship[5] to do research in Cambridge, until the 26-year-old Taylor, past experience nil, was appointed to the Schuster Readership in Dynamical Meteorology, which he chose to hold in Cambridge. He used this opportunity to study the transport of heat, humidity and momentum by making measurements using specially de-signed wind vanes, but the Readership was truncated by the opportunity in the early part of 1913 to sail as a resident meteorologist on the ship *Scotia*. This expedition, organized in response to the sinking of the large passenger liner, *Titanic*, in 1912, was meant to study the problems of icebergs on Atlantic ship-ping routes. In the six months spent near the Grand Banks of Newfoundland – a place where cold water adjoins the warmer land and near where the *Titanic* had sunk – Taylor found the opportunity to measure not only vertical distribu-tions in the atmosphere of temperature, humidity and wind, but also to infer the eddy diffusivities of heat and humidity. The work of this expedition was to be a source material for several papers which emerged later.

Soon after, in early 1915, Taylor was drawn into military work on early aeronautics.[6] He joined the Royal Aircraft Factory at Farnborough. There, he

---

[5] Around the same time, L.F. Richardson's application for a Fellowship at King's College, Cambridge, was turned down; see Hunt (1998).

[6] In the Trinity Archives, one can find letters addressed to 'Major Taylor' during this period, in-cluding one from Richardson in 1919. One of Lord Rayleigh's technical papers for the *Advisory Committee for Aeronautics*, number 1296, authored in 1919, is entitled *Remarks on Major G.I. Taylor's papers on the distribution of air pressure*. A letter from W.M. Orr, dated as late as 9 April 1923, also addresses 'Major Taylor'. I may note in passing that Orr's work on stability,

continued to author papers on meteorology-related work – some in response to specific questions apparently raised by his co-workers, others self-motivated. But it was also the time that set the stage for his fruitful career that advanced both fluid and solid mechanics. This phase lasted until his appointment in 1919 to a Fellowship and Lectureship at Trinity. The lecturing career was short-lived. In 1923, Taylor was appointed as the Yarrow Research Professor of the Royal Society with no teaching and administrative responsibilities. War work entrained his attention yet again during most of the 1940s, and he was working, until his retirement in 1951, on problems either directly related to the war or on problems that arose from it. Taylor's later reflection (Batchelor 1975) that "the course of my scientific career has been almost entirely directed by external circumstances" suggests that he became his own master mostly after retirement, which provided him greater freedom to choose and pursue a number of unconventional research problems in fluid mechanics. Taylor was highly productive in this 'retired phase' of some twenty years, writing about as many papers as he did in the previous twenty, including some great ones. It is fair to think that this period transformed him from one of the best scientists to a giant.

Taylor married Stephanie Ravenhill in August 1925 and remained married to her until her death in June 1967. They shared many interests but had no children. By all accounts, Taylor's personal life was on an unperturbed and even keel. In his long life, well lived, forty-two years of them with Stephanie, Taylor stayed active in Cambridge from 1905 until his death in 1975 – excepting roughly between 1913 and 1919, and the visits abroad, including a few weeks at a time to Los Alamos in the atomic bomb era. From April 1972, he was incapacitated by a stroke affecting the left half of his body. He tried heroically to keep up with some correspondence even then (see later for partial descriptions).

4.3 Ideas originated in the period 1915–1921

4.3.1 Historical context and background

At the time Taylor published his first paper on turbulence in 1915, knowledge of the subject was in its infancy, yet a fair bit was known. Boussinesq (1870) had postulated that the momentum transfer caused by turbulent eddies could be modelled by an 'eddy viscosity', and the germs of the idea were already apparent to Saint-Venant (1851). Frisch (1995), who provides excerpts of the

published in 1907 in *Proc. Roy. Irish Acad.*, was ignored in Germany until about 1921 and, somewhat surprisingly, also in England. I could not determine the date of Taylor's military appointment, or find good pictures of him in military uniform.

French originals with English translations, notes that the ideas of Saint-Venant and Boussinesq came from observations of water flow in canals which, though important at the time, fell into oblivion until some of the same ideas were rediscovered in the newly emerging science of aerodynamics; by that time our understanding of momentum transport in kinetic theory had also advanced to a level of maturity and directly spurred people like Prandtl. The French work did not seem to have reached Taylor.

In England, Reynolds (1883) had observed turbulence and the transition between laminar and turbulent states and had set Rayleigh (1892), Orr (1907) and Sommerfeld (1908) on a course for investigating the stability of fluid flows. In 1895, in a paper that was poorly understood then but has turned out to be an inspiration for enormous activity ever since, Reynolds had also formulated his 'Reynolds averaging' and 'Reynolds stresses'. See Chapter 1 by Launder & Jackson for a description of Reynolds' contributions. Turbulence in boundary layers was mentioned by Reynolds (1874), Lanchester (1907) and by Prandtl[7]. Eiffel (1912) had demonstrated that the drag on a sphere could be reduced by introducing turbulence, and Prandtl (1914) had offered the right explanation for it. Blasius (1913) had provided his famous 1/4th power law for forecasting frictional drag. The second edition of Lamb's *Hydrodynamics*, in which he first lamented about turbulence as "the chief outstanding difficulty of our subject", had appeared in 1895; the section on turbulence was less than seven pages long and referred primarily to Reynolds' (1883) work on transition, and Rayleigh's work on inviscid instability (with a few remarks on Lord Kelvin's tentative efforts to include viscosity). The next edition of the book, issued in 1906, added significant material from Reynolds (1895) on his decomposition method[8] and the equation for the energy integral with explicit forms of the dissipation and production integrals. Taylor was exposed to Lamb's *Hydrodynamics* while still at school, in his 'Uncle Walter Scott's library' (see *The Legacy*). The mathematical treatment of eddy motion in the atmosphere was regarded as a problem of great difficulty and, as Taylor (1915) remarked, this was because attention was chiefly directed "to the behaviour of eddies considered as individuals rather than to the average effect of a collection of eddies". This dichotomy of treatment is still quite alive in turbulence research.

By my count, Taylor wrote some thirty papers on topics related to turbulence. They spanned two roughly five-year segments of his life (*c.* 1915–1921

---

[7] Some crucial ideas attributed to Prandtl by both Karman and Nikuradse were published by Prandtl in 1927.

[8] Horace Lamb was, in fact, one of the two referees of the paper; the second was George Gabriel Stokes. Its editor was Lord Rayleigh. See Chapter 1 by Launder & Jackson.

and 1935–1940). The first group of papers was written while he was still a young researcher, while the second was written some dozen years into his Royal Society Research Professorship. Most papers in the earlier period were on specific aspects of meteorology and ocean dynamics, incorporating the *Scotia* expedition, but included the fundamental paper on diffusion by continuous movements, which he wrote after he moved back to Cambridge. The second group is focused on basic aspects of turbulence in which Taylor introduced isotropic turbulence and its methods. In the intervening years, Taylor worked and wrote on a wide range of topics such as soap-film methods in elasticity, physics of the large deformation of solids (almost certainly one of Taylor's greatest achievements), aeronautical problems including compressible and supersonic flows, rotating flows and several other problems of fluid mechanics not directly linked to turbulence, including the path-breaking work on Taylor–Couette flow (another of Taylor's major accomplishments, about which more will be said in Section 4.4.1). The temporal gap of about fifteen years between the two 'turbulence' periods seems to have made no difference to the continuity of his thoughts. In the first of his famous series on the statistical theory of turbulence (Taylor 1935a), for example, Taylor took off exactly from where he had left the subject some fifteen years earlier in his diffusion paper of 1921; one finds this continuity in his mixing length work as well. Taylor (1971) himself stressed this continuity. He chose not to write about turbulence beyond about 1938, the exceptions being Taylor & Batchelor (1949) and Taylor (1954) on the dispersion of matter in turbulent flow through a pipe.

4.3.2 Mixing length

The sponsors of the *Scotia* expedition required Taylor to measure vertical distributions of mean temperature, humidity and wind speed at different locations presumably because of their relevance to the formation of fog and icebergs, and probably expected a report. Taylor wrote up such a report in 1914 and later published a paper on related topics (Taylor 1917a). He pointed out that fog formation over the sea (produced by warm air passing over a cold sea or by cold air blowing over a warmer sea) is simpler to understand than that over the land (the cooling of the ground at night by radiation to the sky). The latter requires not only predicting the amount by which the temperature is likely to fall at night but also assessing if the drop in temperature is large enough to condense the water vapor of the atmosphere. Our knowledge of supersaturation and condensation of water vapor has advanced sufficiently beyond Taylor's considerations of more than ninety years ago, so this topic is better left alone at this point.

The scientific goal of the *Scotia* expedition that Taylor seems to have set for himself was the quantitative description of turbulent processes in the atmosphere. In *Eddy motion in the atmosphere* (Taylor 1915), he used his own kite measurements as well as data interpolated from other ship measurements to infer vertical distributions of humidity and temperature in six instances. He showed by heuristic arguments that the turbulent transport needed to account for the vertical transport of these quantities was of the form $w'L$ where w' is a measure of the fluctuation in the vertical component of the velocity and L is a length connected with turbulence – later called the 'mixing length' – analogous to the mean free path of molecules ("the average height through which an eddy moves from the layer at which it was at the same temperature as its surroundings, to the layer with which it mixes"). This mechanism to mimic molecular motion was regarded as "a purely hypothetical process". Taylor deduced the rough numerical value for the effective thermal diffusivity in the turbulent wind over the sea to be on the order of $3 \times 10^3 \, \text{cm}^2 \, \text{s}^{-1}$, independent of height for a range of wind speeds. Later (Taylor 1917b), using the measurements made earlier at different heights on the Eiffel tower, he estimated the eddy diffusivity to be of the order $10^5 \, \text{cm}^2 \, \text{s}^{-1}$, a much higher value than that over the sea (though it varied with the season and the height). Needless to say, the eddy diffusivity is many orders of magnitude larger than the molecular diffusivity.

Taylor then considered the vertical transport of momentum. He supposed that momentum cannot remain constant over the path of the eddy but postulated that vorticity might. He conjectured that the eddy viscosity is of the same order as the eddy diffusivity, although there were no measurements to back it up. This is in the spirit of the so-called Reynolds analogy (more later).

The vorticity transport theory, according to which the dynamics of turbulent motion can be represented as the diffusion of vorticity rather than as a diffusion of momentum, had its origins in Taylor's Adams Prize essay,[9] and was put forward first explicitly in Taylor (1915). Taylor regarded his 1915 description to have been so brief that it appeared to him "to have escaped notice" (Taylor 1932b). He had become well aware of Prandtl's momentum transport theory soon after 1925 and, in the 1932a paper, made a test of the comparative merits of the two theories. He noted that "if the motion is limited to two dimensions the local differences in pressure do not affect the vorticity of an element, whereas Prandtl has to neglect them or to assume arbitrarily that

---

[9] Among the winners of the Adams Prize, of interest to fluid dynamicists, are James Clerk Maxwell, Joseph Proudman, Harold Jeffreys, Sydney Chapman, Sydney Goldstein, S. Chandrasekhar, George Batchelor, Leslie Howarth, James Oldroyd, Timothy Pedley, Michael McIntyre and, most recently, Jacques Vanneste.

they do not affect the mean transfer of momentum even though they certainly affect the momentum of individual elements of fluid". In the same paper, Taylor showed that the momentum and vorticity transport theories yielded velocity distributions that were different from temperature distributions in the wake of a heated obstacle (as expected from the vorticity transport theory). It was also in this paper that Taylor extended his theory to three-dimensional motion. In the appendix to this same paper, A. Fage and V.M. Falkner reported temperature distributions in the wake of a heated cylinder, and claimed that the results were in better agreement with the vorticity transport theory. In Taylor (1935f), it was concluded that, in the annular gap between rotating cylinders (the inner one heated), Prandtl's momentum transport theory does well to account for the distribution of the velocity near the surface but the vorticity transport theory does better in the bulk.

This comparison was revisited in Taylor (1937a), especially when it appeared that Prandtl (1925), Kármán (1930) and Goldstein (1937) had pressed ahead with Prandtl's theory rather than with that of Taylor, who thought that his theory had something superior to offer. Here, he compared the predictions of a 'modified' vorticity transport theory with Nikuradse's mean velocity distributions in pipe flow and claimed "very good agreement with observation over the whole range except perhaps very close to the wall". (When this conclusion was later questioned by Prandtl in a private letter, Taylor accepted the criticism and sent him the revised figures.)

Despite considerable effort (and comparable anguish expressed over it in the correspondence with Prandtl and Ekman; see later), Taylor did not consider – at least in hindsight – any of the transport theories satisfactory enough. His objections were not about the use of mixing length as a practical device but about the underlying physics. Indeed, in Taylor (1970), he reflected that he

> was never satisfied with the mixture-length theory, because the idea that a fluid mass would go a certain distance unchanged and then deliver up its transferable property, and become identical with the mean condition at that point, is not a realistic picture of a physical process.

4.3.3 Turbulent diffusion

The molecular diffusion of dynamically passive matter into its surroundings obeys the well-known advection–diffusion equation

$$D\theta/Dt = \kappa\nabla^2\theta, \tag{4.1}$$

where κ is the diffusivity for θ, and D/Dt is the material derivative. Throughout much of the nineteenth century, scientists debated the molecular nature of matter but the experimental observations of Brownian motion suggested that molecular agitation was the underlying cause of the diffusion. Einstein had shown that the same diffusion coefficient appears in the stochastic problem at the microscopic level as in the diffusion problem on the macroscopic level. The next qualitative breakthrough in the problem occurred in Taylor (1921) where we find the following sentences as the motivating salvo:

> turbulent motion is capable of diffusing heat and other diffusible properties through the interior of a fluid in much the same way that molecular agitation gives rise to molecular diffusion...On the other hand, nothing appears to be known regarding the relationship between the constants which might be used to determine any particular type of turbulent motion and its 'diffusing power'. The propositions set down in the following pages are the result of efforts to solve this problem.

This is the genesis of turbulent diffusion (Taylor 1921), where, for the first time, an effort was made to link statistically a continuous and stochastic velocity function to the displacement of a fluid element – an effort towards describing the effects of turbulence by means other than invoking the mixing length. The most important result of the paper is the now-famous formula

$$\frac{d\langle X^2\rangle}{dt} = 2\langle u^2\rangle \int_0^t Q(\tau)d\tau, \tag{4.2}$$

where $\langle X^2\rangle$ is the mean-square displacement of the fluid element in one direction and Q is the Lagrangian correlation function of the velocity component u in the same direction at two different times separated by the time interval $\tau : Q(\tau) = \langle u(t)u(t+\tau)\rangle/\langle u^2\rangle$. The velocity u is assumed to be statistically stationary. The integral from 0 to ∞ of $Q(\tau)$ gives the Lagrangian integral time scale, T_L.

For small times $t \ll T_L$, we have $Q \approx 1$, so the result is that

$$\langle X^2\rangle = \langle u^2\rangle t^2, \tag{4.3}$$

reflecting the fact that a fluid element simply moves, for some small time, with its initial velocity. For large time, on the other hand, the integral in (4.2) is simply T_L, so we have the result

$$\langle X^2\rangle = 2\langle u^2\rangle T_L t, \tag{4.4}$$

akin to Brownian motion. The combination $2\langle u^2\rangle T_L$ is the eddy diffusivity.

In thinking about possible applications, Taylor regarded his theory to be capable of predicting the mean shape for a smoke plume.[10] He thus sought clues in the form of smoke trails leaving a chimney. In a paper inspired by Taylor (1915), Richardson (1920) had published several photographs resulting from long-term exposure of the smoke trail coming off a mast in a field. In at least one of them (his Figure 2), it appeared that the initial spread was conical (consistent with (4.3)) whereas the spread was paraboloidal beyond a certain distance (consistent with (4.4)). While "opinions may differ" according to Richardson (1920), Taylor declared that both these observations were "in good agreement" with his equations. The evidence is somewhat flimsy and it is not trivial to relate the spread of a contaminant with the spread of a smoke from a chimney, but there is no doubt that Taylor felt that the theory[11] accomplished a lot.

Taylor knew that the role of viscosity in diffusion would be interesting to understand but did not dwell on it by noting that "the 'diffusing power' of any of turbulence appears to depend ... little on the molecular conductivity and viscosity of the fluid". Later, Townsend (1954) and Saffman (1960) attempted to clarify the role of molecular diffusion. Their first-order corrections add identical amounts to the total dispersion thickness at time t by the factor $2\kappa t$, but the sign and magnitude of the next order corrections are different; indeed, this issue remains to be settled even after so many years. A further remarkable result of Taylor (1954) is that, in capillary flows involving the interaction of convection and molecular diffusion, the effective axial diffusion is inversely proportional to molecular diffusivity! This result is the basis, for instance, of modern liquid chromatography.

In a related development, Richardson (1926) showed that the spread of a cloud of particles (as opposed to how the center of gravity of the cloud drifts, which was Taylor's concern) proceeds according to the third power of time (though Richardson's empirical basis for this claim was less than stellar). See Section 4.6.5 for a brief account of the interaction between Taylor and Richardson; and Ashford (1985), Hunt (1998) and Chapter 5 by Benzi for more details of Richardson's life and work. It is accurate to say that Richardson's 1926 paper on diffusion is the starting point for our understanding of scale-dependent

---

[10] In later recollections, Taylor (1970), we find the following statement: "While thinking of these things [making mixing length concepts more definitive], I became interested in the form which a smoke trail takes after leaving a chimney ... This led me to think of other ways than mixture-length theory to describe turbulent diffusion. The result was my paper, *Diffusion by continuous movements* ..."

[11] Kampé de Fériet (1939), who put Taylor's theory on a more formal mathematical footing, also reported measurements, by his associates, of the diffusion of tiny soap bubbles in a turbulent air stream and claimed the results to be "in complete accordance" with the theory.

eddy diffusivity. Vast avenues of work were thus opened up by these two remarkable people who laid the foundations for the statistical theory of turbulent
diffusion.

4.3.4 Other work from that period

The papers written up until Taylor's appointment in October 1919 to a Fellowship and Lectureship in mathematics at Trinity were quite specific, many
related to the work of his colleagues in England, mostly on meteorology and
physical oceanography. For instance, he was able (Taylor 1916) to obtain reasonable estimates for the skin friction coefficient in atmospheric boundary layers. A more serious paper on the subject of friction on solid surfaces is Taylor
(1918) in which Pitot tube measurements on a long board mounted in a wind
tunnel were reported. This report is interesting because it sheds some light
on the confusion that prevailed, even as late as 1916, on the proper boundary
conditions to be used on a solid surface. Taylor also initiated a closer study
of the so-called Reynolds analogy which Reynolds (1874) had proposed some
nine years before publishing in 1883 his famous paper on 'direct and sinuous
motion'. In this far-sighted but short paper, Reynolds had noted that heat is removed from a solid surface by the eddies in the flow, which mixed hotter fluid
with cooler fluid. The Reynolds analogy was taken up again in Taylor (1930) –
essentially commenting on the previous work by Eagle & Ferguson (1930) –
wherein he came out in favor of the analogy.

  Several fundamental ideas expressed in qualitative terms in these papers
lay dormant for some time. Of greater interest to this readership is Taylor
(1917c), *Observations and speculations on the nature of turbulent motion*,
which summarized the experiments carried out in 1913 before *Scotia*. Three
very important ideas can be found in this paper. First, Taylor showed by direct observation that the velocity fluctuations in the atmosphere are approximately isotropic about 3 m above the ground ("there is approximately an equipartition of eddy energy among the various directions"). Second, he argued
towards the end of the paper that only this approximate isotropy enables finite energy dissipation to take place in a fluid of infinitesimal viscosity. This
is what is now called the 'zeroth' law of turbulence. Third, in the very next
paragraph, Taylor stated the principal conjecture of what is now known as
the small-scale intermittency: "the eddy motion must tend to produce small
whirls or discontinuities where a very large vorticity occurs in a very small
volume". We will have more to say about this aspect later. He was interested in
the root-mean-square pressure fluctuation even then, though it was only later
(Taylor 1936c) that he estimated it for a special velocity field to be $\frac{1}{2}K\rho\overline{u^2}$

where u is the eddy velocity and K lies between 1 and $\sqrt{2}$. These are remarkable insights for someone at the beginning of his research career in a difficult field!

4.4 The intervening period

I must now briefly call attention to a few aspects of Taylor's work in the period leading up to the next 'turbulence' episode of his scientific life. I shall say nothing about his work on the micromechanics of plastic flow of single crystals and polycrystalline materials or on his triumphant theory of dislocations, except to recall Bell's remark on its enormous value in mechanics and material science, but will focus on three problems of fluid mechanics that bear on turbulence.

4.4.1 Taylor–Couette flow

Taylor's investigation of the stability of the Taylor–Couette flow – as it has come to be known – has not only been an important cornerstone in the development of hydrodynamic stability theory, but a paradigm for flows in which successive increases in spatio-temporal complexity eventually lead to turbulence through several, non-unique routes. To quote from *The Legacy*:

> The notion of stability of fluid flow systems at that time was imprecise, and, although the experiments of Reynolds and others had shown that under certain conditions some theoretical steady flow systems could not be realized and were replaced in practice by a permanently fluctuating turbulent flow, there was no case of steady flow for which a critical condition for stability had been both calculated and established from observation.... [Taylor] perceived intuitively that steady flow between concentric cylinders, for which an algebraic criterion for stability to inviscid small disturbances had recently been given by Rayleigh, would allow a normal-mode analysis of the behaviour of small disturbances, and that the interesting critical disturbance that neither grows nor decays would have the simple property of not propagating in the direction of the cylinder axis. He also anticipated correctly that the role of viscosity would not be singular in any way.

In the event, Taylor (1923) found remarkable agreement with his stability theory and observations. Indeed, I know some fluid dynamicists who regard this excellent agreement as the first foolproof confirmation of the correctness of the no-slip boundary condition in viscous flows.

As is now well known, the Taylor–Couette flow is steady and purely azimuthal for low angular velocities. Taylor showed that when the angular velocity of the inner cylinder is increased above a certain threshold, the basic flow becomes unstable and a secondary steady state, characterized by axisymmetric toroidal vortices, known as Taylor vortices, emerges. Further increases in the angular speed result in a progression of instabilities and lead to greater complexity. Taylor's work has subsequently paved the way for a rich study of phenomena associated with states of varying degrees of complexity (see, for example, Andereck et al. 1986), and I shall cite only two major milestones. One of them is the experimental observation that multiple flow regimes are possible for the same external conditions depending on the path followed in the parameter space (Coles 1965). Coles also discovered a second kind of transition when the outer cylinder has a larger angular velocity than the inner one. In this case, the fluid is divided into distinct regions of laminar and turbulent flow spiralling without any change in shape except an occasional shift from a right-hand to a left-hand pattern. Gollub & Swinney (1975) showed that, as the rotation rate of the inner cylinder increases, three distinct transitions appear, each of which adds a new discrete frequency to the velocity spectrum, followed by a sharply defined Reynolds number at which discrete spectral peaks suddenly disappear. They speculated that this observation signalled the Ruelle–Takens scenario for the onset of turbulence, according to which turbulence is possible only after a small number of bifurcations.

4.4.2 Flow in a curved pipe and relaminarization

White (1929) had observed that flow through curved pipes can be maintained laminar for substantially higher Reynolds numbers than is possible in straight pipes (excluding the case of unusually smooth inlet conditions). Taylor (1929) was intrigued by this observation and studied the problem of transition to turbulence in curved pipes. He visualized the nature of secondary flow by careful dye injection studies and determined the conditions at which turbulence appears. A more detailed study has been presented by Sreenivasan & Strykowki (1983). Taylor reasoned that turbulence, which may have already appeared in a straight pipe owing to the fact that the Reynolds number exceeded the Reynolds criterion of about 2300, might revert to a laminar state if the pipe was subsequently curved. This process of relaminarization of an initially turbulent flow has been studied now in a large number of contexts, as described by Narasimha & Sreenivasan (1979).

4.4.3 Stratified flows

One thread of work, begun again in the Adams Essay of 1914 but continued through the 1930s, deserves special mention. This has to do with the effects of stratification in the atmosphere. Taylor knew well that the fluctuations in the wind increased or decreased depending on the difference between the speeds at two heights and on the difference between temperatures at these two heights. Encouraged by S. Goldstein, who was beginning to work on a similar topic, he published these results without any experimental support – contrary to his habit. Prandtl and Richardson were also working on this same topic. Taylor (1931a) assumed an exponential drop in density and a linear velocity distribution and showed that a steady stream of uniformly sheared flow in stratified environment is unstable to small disturbances if

$$\frac{g\Delta\rho}{\rho H (dU/dz)^2} \tag{4.5}$$

exceeds a critical value of order unity. Here $\Delta\rho$ is the decrease in density over a layer with a height H in the vertical direction, and dU/dz is the mean velocity gradient above the ground. The numerical value of the constant itself depends on the assumptions made. This non-dimensional number is now known as the (gradient) Richardson number. The next paper (Taylor 1931b) considered a related problem of determining the critical value of the density gradient above which the mean flow cannot persist in a uniformly sheared state. Taylor was searching particularly for internal wave structure here. In this paper, Taylor did make comparisons with experiments. Details of Taylor's stability calculations have been superseded, but there is no doubt they gave substantial impetus for the researchers who followed.

4.5 Ideas explored in the period 1935–1940

We have already seen that Taylor (1921) introduced, evidently for the first time, the idea of correlation in Lagrangian terms. Taylor (1970) remarked that:

> even then [i.e. in 1921] I realized that Eulerian correlations might furnish a useful means for describing a turbulent field. I did not, however, see any way in which it could be used to connect such a field with measurable mean properties, so I did not see any reason for publishing the idea. Some time, about 1924, I believe that the idea was published in Russia[12] but without predictions which could be tested experimentally. It was not until after 14 years, when improvements in hot-wire techniques enabled Dryden at the National Bureau of Standards and Simmons

[12] The reference is to the work of Keller & Friedman (1924) of which Falkovich in Chapter 6 gives a brief account (see also Monin & Yaglom 1971). Keller & Friedman introduced correlation functions and a particular *ad hoc* method of closure. Their paper was presented at the First International Congress of Applied Mechanics in Delft, which Taylor attended. He thus no doubt knew the existence of this paper, presented in German, but seems to have regarded the idea to have occurred to him earlier; the introduction of isotropy was entirely Taylor's.

at the National Physical Laboratory to measure the decay of turbulence behind regularly spaced grids, that I realized that energy dissipation could be related to the Eulerian definition of correlation. I could only get a simple result, verifiable experimentally, if I assumed that turbulence was isotropic.

This was the genesis of isotropic turbulence – a subject that has consumed the attention of a number of turbulence researchers, occasionally inviting some skepticism.[13] Taylor (1939) paid greater attention to hot-wire developments and mentioned the work of King (1914), van der Hegge Zijnen (1924), the NACA work (Dryden & Kuethe 1929, Mock & Dryden 1932 and Mock 1937), Ziegler (1934), Simmons & Salter (1934), Townend (1934) as well as Schubauer (1935). He wrote to Dryden a longish typewritten letter (rare for Taylor, since handwritten notes were the norm) describing the NPL hot-wire work (without, however, mentioning Simmons and Salter by name).

4.5.1 Basic ideas of isotropic turbulence

Taylor wrote five papers on the *statistical theory of turbulence* (Taylor 1935a–d, 1936a) in quick sequence. To this series must be added Taylor (1936b, 1937b, 1938a,b), as well as the reviews (1938c and 1939). These papers contain his principal results on isotropic turbulence, and will be summarized below. Batchelor remarked in *The Legacy* that "present-day students seldom have need to go back to [Taylor's] papers" because their results have been "exclusively adopted, and greatly extended", but it was Batchelor himself who did much of this adoption and extension in his masterly monograph of 1953, *The Theory of Homogeneous Turbulence*; see Chapter 8 by Moffatt.

In 1935a, Taylor began with a review of his 1921 paper on Lagrangian diffusion and Lagrangian integral scale, introduced for the first time the transverse Eulerian length scale L as the integral of the equal-time transverse correlation function, interpreted it as the "average size of the eddies", discussed how to measure the integral scale, wrote down the general expression for energy dissipation and reduced it to the isotropic form, connected it to the definition of his microscale[14] λ through the curvature at the origin of the transverse correlation function, noted the inviscid scaling for the dissipation rate per unit mass, $\epsilon = O(u'^3/L)$, found the relation between λ and L, and, finally, discussed the

---

[13] This skepticism has been voiced, for example, by Saffman (1968); see Chapter 12 by Pullin & Meiron for a summary of Saffman's major contributions. Lighthill, in an unpublished and elegant paper, entitled *Vorticity balance in isotropic turbulence. I*, probably dating around 1955, also expressed this same view, but more subtly. The chief point of criticism is that the spectral theory of isotropic turbulence does not play explicit attention to vortex dynamics, although it clearly includes it. There is no contradiction of principle but only a divergence of emphasis and the lore that goes with it.

[14] Taylor mistakenly interpreted the microscale to be the length that determines the scale of the eddies responsible for the dissipation of energy; see, especially, Taylor (1935d).

application of this theory to turbulence behind "regular grids or honeycomb". He noted that "if a regular grid or honeycomb is constructed the scale of the turbulent motion produced by it at any distance down-stream beyond the point where the 'wind-shadow' has disappeared will depend only on the form and mesh size of the grid". He used this expectation to deduce the evolution of the microscale, λ, as

$$\frac{\lambda}{M} = A\sqrt{\frac{\nu}{Mu'}}, \qquad (4.6)$$

where the constant A was "found experimentally to be about 2.0", and the decay of ϵ with downstream distance from the grid as

$$\frac{U}{u'} = \frac{5x}{A^2 M} + \text{constant}. \qquad (4.7)$$

Here, x is the distance behind a grid of mesh size M, located in a fluid stream of velocity U, producing a root-mean-square velocity fluctuation u' at x; and ν is the kinematic viscosity of the fluid. Taylor further noted that the

> linear law of increase of U/u' should... apply to all wind-tunnels where the scale of turbulence is controlled by a honeycomb or grid, and the value of the constant A determined experimentally, using (59) [the present (4.7)] should be universal[15] for all square grids. Thus, the turbulence behind a square-section honeycomb with long cells should obey the same law of decay as that produced by a square-mesh grid of flat slats or a square-mesh grid of round bars, and the values of A should be identical in all these cases.

I have chosen to mention Taylor's principal results specifically, albeit briefly, because he put great store in them and repeated them in private correspondence with Prandtl, Kármán, Burgers and Dryden. In fact, in a letter to Kármán, dated 24 January 1937, Taylor referred to an argument he had earlier constructed, but rejected in view of the data, in favor of a half-power dependence on x (this being closer to the truth – as we know it today). Now we know that Taylor's expectation of universality turned out to be optimistic because the conditions at the grid do indeed make a difference (Stewart & Townsend 1951; Barenblatt 1979, 1996; Sreenivasan 1984). In the letter to Kármán just cited, Taylor included a number of plots purportedly verifying the linear decay, but in the words of Batchelor (1953):

> The earliest measurements of u [the root-mean-square velocity] at different stages of the decay were a little confused by inadequate corrections for the errors introduced by the finite length of the recording hot wire, by the existence in the wind tunnel stream of turbulence from sources other than the grid, and by the inclusion of observations at positions in the range $0 < x < 10\,M$.

[15] The word 'universal' means something different today from Taylor's usage. I am unable to determine if he was the first to use the word in this broad context.

Taylor was quite aware of these limitations: he conveyed them in the letter to Kármán: "The results are not bad but the trouble at the NPL is that they have initial turbulence of large scale before getting to the grid." However, the agreement of his theory with measurements, whatever their limitations, seemed to have made a powerful impression on Taylor's thinking.

Whatever reservations can be expressed about these specific results with the hindsight of many years, there can be no question that the paper has served as the source of founding ideas for isotropic turbulence, and, more broadly, for the entire statistical theory of turbulence.

Taylor (1935b,c) makes experimental contact with the ideas expounded in Taylor (1935a):

> At that time measurements of the rate of decay of the energy of turbulence had recently been made by Hugh Dryden at the Bureau of Standards. I therefore asked the National Physical Laboratory to measure Eulerian correlations in turbulence produced by obstacles in a wind stream which had the same geometry as those used by Dryden when he measured the rate of decay of turbulent energy.

Together, these measurements produced results which had direct relevance to Taylor's (1935a) ideas. The measurements were about the correlation functions and the decay formula (4.7) behind grids and honeycombs, at both NACA and NPL. Assuming isotropy of the dissipative scales of motion (except very close to the wall, where Taylor well understood that isotropy would not hold) and using the correlation measurements from the channel data of Reichardt (sent by Prandtl), as well as those of Wattendorf and Kuethe, Taylor (1935c) estimated the dissipation in the outer part of turbulent channel flows. He also estimated the production of turbulence by fitting the velocity distribution to the Prandtl–Kármán log-law and reached the conclusion that the rate of dissipation was greater than the production rate above a certain height from the wall, while the opposite was true at lower heights. These results have been elaborated upon several times by Laufer (1954), Townsend (1956) and others.

Taylor (1935d) is an application of Taylor (1935a) to the diffusion of heat from a localized line source in a wind tunnel. The line source is one of the simplest ways of injecting a passive scalar into a turbulent field. Given the nature of the data available at the time (the measurements of Schubauer (1935) at NACA and of Simmons & Salter at NPL), Taylor deduced a surprising number of characteristics of the spreading of the passive contaminant. Taylor's analysis was somewhat preliminary also because of his assumptions on the constancy of the integral length scale with downstream distance and the self-preservation of the Lagrangian correlation function during the decay of turbulence. Measurements and analyses following Taylor's lead have been done by others (e.g. Uberoi & Corrsin 1953, Townsend 1954, and many others). The spreading of the heated wake has been repeated since then in a number of flows other than grid turbulence, and simulations have also been performed. There is no doubt that all the later developments, which are too extensive to cite here, owe much to Taylor's path-breaking work.

Perhaps the less known part of Taylor (1936c) is the discussion of the pressure microscale λ_p defined through

$$\overline{(\partial p/\partial y)^2} = 2\rho^2(\overline{v^2})^2/\lambda_p^2.$$

This microscale plays a role in the estimates of the pressure–velocity terms in the equation for turbulent kinetic energy. The main result was that the ratio $\lambda_p/\lambda = O(1)$. Batchelor (1951) showed that Taylor's result is essentially correct for low Reynolds number but the ratio increases as $Re_\lambda^{1/2}$ for very high $Re_\lambda \equiv u'\lambda/\nu$. Recent work suggests that Batchelor's result itself may need further modifications.

4.5.2 The Taylor–Green problem

Taylor had concluded that his expressions for the downstream development of his microscale and the linear decay law were correct, but that they were coarse manifestations of

> the fundamental process in turbulent flow, namely the grinding down of eddies produced...into smaller and smaller eddies until these eddies are of so small a scale that they die away owing to viscosity more rapidly than they can be produced by the grinding down process.

In modern parlance, he was trying to understand the energy cascade, an idea that was already apparent in the 1914 Adams Prize essay. Taylor & Green (1937) argued elegantly that:

> we may expect first an increase in $\overline{\omega^2}$ [mean-square-vorticity]...When $\overline{\omega^2}$ has increased to some value which depends on the viscosity, it is no longer possible to neglect the effect of viscosity.

Later work (see, for example, Eyink et al. 2008) has revealed that this behavior has to do with the possible occurrence of inviscid singular behavior. Instead of addressing the problem abstractly, Taylor and Green noted that "It is difficult to express these ideas in a mathematical form without assuming some definite form for the disturbance..." and chose a special form of the initial motion represented by

$$\left. \begin{aligned} u &= A\cos ax \sin by \sin cz \\ v &= B\sin ax \cos by \sin cz \\ w &= C\sin ax \sin by \cos cz \end{aligned} \right\} \tag{4.8}$$

with

$$Aa + Bb + Cc = 0 \qquad\qquad (4.9)$$

to respect continuity. The full three-dimensional equations of motion were solved with these initial conditions, and the solution was written in the form of a triple Fourier series in which the coefficients were powers of time, t. For special subsets of coefficients, the detailed analysis was taken up to t^5 and t^6. The results that Taylor and Green found can be stated qualitatively as follows: even though the mean-square energy decreased, mean-square vorticity increased up to a maximum and then decreased. The ratio of the maximum vorticity to the initial vorticity increased as a function of the Reynolds number (expressible in terms of the coefficients of (4.8)). A comparison with experiments necessitated the identification of the scale of turbulence in the model with the actual one. By taking that lengthscale in measurements to be a constant of the order of half the mesh size, Taylor and Green were able to observe satisfactory correspondence.

There is no doubt that Taylor was excited about these calculations. He shared them with Burgers on 17 November 1935 describing his progress in calculating "terms containing t^2 and some terms in t^3", and with Kármán on 12 March 1936 saying that the "labour is awful but we have managed to carry it to terms t^5". To Kármán, Taylor essentially summarized the main results of the work and drew the inference that they supported his theory. We shall discuss this aspect more in Section 4.6.3.

The Taylor–Green paper has a modern feel to it, especially in its connection with finite time singularities of incompressible Navier–Stokes or Euler equations that may develop from smooth initial conditions. This is one of the Millennium Prize problems of the Clay Mathematics Institute, presumably because its mathematical solution will likely involve inventing completely new tools in the theory of partial differential equations. Singularity formation in the Euler equation is of great relevance in understanding small-scale intermittency, since the central issue there is the self-amplification process and the resulting (near) singular structures. When Taylor was asked at the 1968 Stanford meeting on the computation of turbulent flows as to which problem he would like to see solved on the increasingly powerful computers, he cited the Taylor–Green problem as an example (private correspondence with the late M. van Dyke). His wish has indeed come true, and the initial effort has been superseded by huge calculations by various groups (see, for example, Brachet et al. 1983 and Cichowlas & Brachet 2005). There is no doubt that even the latest calculations will soon be superseded. For example, Cichowlas & Brachet argued that grid resolutions in the range 163843–327683 are needed to really probe such singularities. This open invitation in an age of rapidly improving computing power will not be ignored for long.

I should note the importance of external stimulus Taylor received to work on this aspect, though the broad idea was clear to him since 1915. By 1937, other people were becoming interested in Taylor's theory, and the shield provided by the originality of his research thus far had begun to soften. I have already referred to the NACA measurements which were eloquently summarized by Dryden (1943). An early remark in that paper is this:

> A statistical theory of turbulence...was inaugurated by Taylor and further developed by himself and by von Kármán. It is the object of this paper to give a connected account of the present state of this particular statistical theory of turbulence.

This remark shows that Kármán (with L. Howarth and C.C. Lin, both working with Kármán) had begun to take a major interest in the theory. Indeed, the particular research on vortex stretching was a direct consequence of Taylor's interaction with Kármán. This issue has been discussed also by Leonard & Peters in Chapter 3 and we shall say more about it in Section 4.6.3.

4.5.3 Rapid distortion theory

Another result of Taylor, the scope of which has expanded beyond his specific result, may be mentioned now. For the 60th birthday Festschrift for Prandtl, Taylor (1935e) wrote a paper on the effect of contraction on turbulence, taking cues from earlier arguments of Prandtl (1930) himself. Prandtl's arguments were based on the conservation of circulation and did not contain much appreciation for the circumstances under which they would hold. Taylor recognized that nonlinear effects can be ignored if the contraction is rapid enough, and proceeded to calculate the effects of contraction quantitatively. The premise of the theory, as he explained to Prandtl on 2 March 1935 in response to a question from the latter, is that

> the contraction is so rapid that the relative motion of two neighbouring [points] due to the contraction is large compared to that due to turbulence.[16]

---

[16] This letter is important because it is here that Taylor introduced his theory of isotropic turbulence to Prandtl, described the grid and his linear decay law, and asked: "Have you made any measurements in Göttingen which bear on this point?" He expressed the view that the Reynolds number ought to be high enough, "$u'L/\nu > 100$", for his theory to hold. This is also the place that Taylor discussed his critical Reynolds number theory for the sphere (see Section 4.5.5). Prandtl did not comment on the grid problem but was not particularly impressed by the critical Reynolds number work. This should not surprise us because an immense amount of first-rate work on boundary layer stability had been going on in Prandtl's own institute; see Section 4.5.5.

The velocity field was represented by the sinusoidally varying model of the Taylor–Green problem. Taylor computed the vorticity whose evolution through the contraction he evaluated, after which he inverted the relation to get the velocity field to deduce the effect of contraction. He also presented a comparison with the measurements made by Simmons, Townend and Fage (using different techniques). According to Batchelor (see *The Legacy*, p. 168), Taylor "was held up by not knowing how to 'integrate' the results over all Fourier components of a turbulent, and hence, random fluid velocity";[17] the task was later undertaken successfully by Batchelor & Proudman (1954), whose work brought the subject to the level of having its own acronym RDT, attracting the attention of later researchers towards qualitative evaluation of the effects of a 'mean strain' that is relatively large.

4.5.4 Energy spectrum and the Wiener–Khintchine theorem

Taylor had realized, probably around 1917,

> that if a statistically steady state could be established, with large eddies being supplied at a given rate and their energy finally disappearing owing to viscosity, a definite spectrum would result ...

He cited Lord Rayleigh's work as the source for the harmonic representation of an arbitrary signal and noted that Dryden et al. (1937) at NACA and Simmons & Salter (1938) at NPL had measured the spectrum of turbulence by using band-pass filters. As we already know, Taylor had made Eulerian correlations the basis of his theory of isotropic turbulence. In Taylor (1938b), it was made clear that the spectrum and the correlation functions are Fourier transforms of each other. This is the Wiener–Khintchine theorem. Taylor himself attributed the general mathematical form of the theorem to Norbert Wiener's 1933 book but we know of no concrete examples pre-dating Taylor which showed that the measurements follow the Fourier transform relation.[18] Taylor was to show more. He used the correlation functions of the band-pass velocity and demonstrated that the modified spectrum agreed well with the Fourier transform result. Research of band-passed velocity signals was taken to the next level of

---

[17] Taylor amusingly related (Batchelor 1975) that he was once described as an 'x, y, z' man rather than a modern 'i, j, k' person. Townsend (1990) remarked that, in his lectures to a few graduate students, Taylor derived "in long hand the mean-value ratios of velocity gradients, using the methods of his 1935 paper. Obviously he saw no point in deriving them in another way to get the same result."

[18] Wiener was well aware of Taylor's work and attributed, on more than one occasion (see, for example, Wiener 1939), much of the advances in the statistical theory of turbulence to Taylor. There is, however, no record of any correspondence between them in the Taylor Archives in Trinity, or, according to my brief investigation, in the Wiener Archives at MIT.

sophistication many years later by Comte-Bellot & Corrsin (1971). This paper of Taylor's has had an enormous practical impact on later turbulence research because the two-point velocity correlation could be obtained from the more convenient spectral analysis. This was also the paper in which Taylor argued that the velocity fluctuation changes relatively slowly as it is carried past a point of measurement where the spectral density is measured. This so-called Taylor's frozen-flow hypothesis has been a working tool for hundreds of turbulence researchers (sometimes indiscriminately without realizing its limitations). One can see the rough outlines of the hypothesis in Taylor's earlier papers as well.

4.5.5 Critical Reynolds number of transition

We now turn briefly to Taylor (1936a), which is devoted to the relationship between the level of free turbulence in a wind tunnel stream and the drag on spheres immersed in it. Recall that W. Tollmien and H. Schlichting at Prandtl's institute in Göttingen had already calculated the conditions of linear instability in boundary layers but no one had been able to observe the so-called Tollmien–Schlichting (T–S) waves. Taylor (1923) had by then calculated the instability of the flow between two concentric cylinders and successfully observed them to be in good agreement with the theory (see Section 4.4.1). The enormous success of the stability theory in Taylor–Couette flow and its utter failure, until then, in the boundary layer, and the subtlety of the role assigned to viscosity in the latter, must have led Taylor to disbelieve the very existence of T–S waves in the transition on flat plates (see also Taylor 1932a)[19] and to think that the transition to turbulence in the boundary layer can perhaps be better linked to local separation under the action of fluctuating pressure gradients. (An interesting historical article worth reading is Eckert 2010.) Taylor formulated a parameter $\Lambda = -\frac{\delta^2}{\nu U \rho} \Delta p'$, with dimensions of the square of a length, where $\Delta p'$ is a typical value of the fluctuating pressure gradient and δ is the boundary layer thickness. He was, however, unable to make contact with experiments on flat plates (presumably because data were lacking); instead, he turned attention to the critical Reynolds number experienced by spheres in streams of varying turbulence levels. According to Batchelor (*The Legacy*, p. 166), Taylor "took a

---

[19] The key to observing the T–S waves was to reduce the background noise to low enough levels. They were finally observed, after many years of patient work on controlling background noise, in the momentous measurements at NBS (Schubauer & Skramstadt 1947). The authors described the palpable excitement of the moment: "as the pickup was moved downstream, an almost pure sinusoidal oscillation appeared, weak at first, but with increasing amplitude . . . downstream". The flat plate used for those measurements is now exhibited in the laboratory of W.S. Saric at the Texas A&M University. It is not known if Taylor later reflected on his neglect of the T–S waves.

close interest" in the work of B.M. Jones related to this problem. After some further manipulations using the concepts already summarized, he expressed the critical Reynolds number Re_{cr} as

$$Re_{cr} = f[(u'/U)(D/M)^{1/2}]$$

where D is the sphere diameter, U is the free-stream velocity with the root-mean-square fluctuation of u' and M is the characteristic size of the device that is generating the turbulence in the free stream; f is an unknown function of its argument. He used the data available from Dryden to verify the theory in its broad terms. A better assessment of the theory was made by Dryden et al. (1937) who correlated the critical Reynolds number data very well with the expectations of the theory.

4.5.6 A bridging remark

In the period spanning roughly 1915–1921, Taylor's results on meteorological and oceanographic measurements, those on flat-plate resistance, the analogy between heat and momentum transport, and so forth, were built on previous knowledge or produced in direct reaction to points raised by the research work of colleagues and coworkers. Well into the 1930s, he continued expanding on these themes, as and when external circumstances (such as correspondence, establishing priority) demanded his response. The exception to this statement is his work on diffusion by continuous movements and that on isotropic turbulence; they are both pristine in originality with no precedent. In Batchelor's words, Taylor "took the investigation of turbulence down to a deeper level and revealed the language and the concepts that future research must use". He kept in contact with the measurements made at NACA and at NPL, and brought the theory and experiments together in exemplary ways. The next major development in isotropic turbulence took place in the paper by Kármán & Howarth (1938), in whose publication Taylor played an important role. The interaction with Prandtl was scientifically more substantial. Unfortunately, their relationship soured during the war years. Taylor had declared Prandtl as "our chief" in 1935 but considered him, in 1973, "the poor man ... under pressure from [the Nazi] propaganda machine" who wished to generate a sympathetic image of Hitler. The companion Chapter 2 by Bodenschatz & Eckert sheds some light on the relationship between Taylor and Prandtl, but we shall consider it briefly as well. For a short account of Dryden's important work on turbulence, see Eyink & Sreenivasan (2006). Regretfully, various limitations have prevented a discussion here of Dryden's correspondence with Taylor.

4.6 A window into Taylor's personality through his correspondence

4.6.1 Taylor and Burgers

Since this collection of essays on the history of turbulence does not include a self-contained article on Burgers, my brief summary here exceeds the connection with Taylor (resumed in Section 4.6.2). Burgers' essay (1975) makes delightful reading in this context; also of interest are Burgers' *Selected Works* compiled by Nieuwstadt & Steketee (1995) and his autobiographical notes edited by Sengers & Ooms (2007).

Brief life history of Burgers Johannes Burgers was born in Arnhem, the Netherlands. His father, a post-office clerk, was a self-educated amateur scientist. In 1914, Burgers entered the University of Leiden, where he came to know Hendrik Lorentz, Kamerlingh Onnes, Albert Einstein and Niels Bohr, and was part of a group of students of Ehrenfest. Burgers was, in fact, the first of his students in Leiden to complete a PhD thesis in 1918. His dissertation was on the Rutherford–Bohr model of the atom.

At the age of 23, before receiving his PhD, Burgers was appointed as Professor in the Department of Mechanical Engineering, Shipbuilding and Electrical Engineering at the Technical University in Delft, where he founded the Laboratory of Aero- and Hydro-dynamics. Somewhat reminiscent of the appointments of the young and inexperienced Taylor as Reader of Meteorology and of Reynolds as Professor at Manchester at the age of 26, Burgers was appointed a professor in a field of study in which he had no experience. Burgers (1975) noted that the selection committee "considered it desirable to look for a person with sufficient background in mathematics and its applications who would be prepared to build up the subject from its fundamentals". In his characteristically modest account of his early years in Delft, Burgers wrote that one reason why he accepted the position was his concern of "having insufficient fantasy for making fruitful advances in Bohr's theory".

Sometime after the Second World War, Burgers played an important role in establishing the International Union of Theoretical and Applied Mechanics, which was admitted to the International Council of Scientific Unions in 1947. He served as general secretary of the Union from 1946 to 1952. This was another context through which Burgers and Taylor interacted scientifically.

Towards the end of 1955, at age 60, Burgers left Delft to join the faculty of the University of Maryland. There he developed an interest in the relation of the Boltzmann equation to the equations of fluid dynamics. His book *Flow Equations for Composite Gases*, published in 1969, represents some work of that period. He also published, in 1974, the book *The Nonlinear Diffusion Equation*; he was then 79 years of age.

Burgers' scientific work Once he arrived at Delft, Burgers quickly learned about the work of Eiffel in Paris, of Prandtl and his school in Göttingen, of Reynolds and Taylor in England, and of other leading figures in Europe, and was to become a recognized authority on fluid dynamics.

Burgers' first work was on Oseen's theory of flow at low Reynolds numbers and its connection with Prandtl's work on airfoils. In 1921 he met Kármán, with whom he had a long and close association – both personal and professional. In about 1920, he became aware of L.V. King's work on the hot-wire measurements of the convection of heat from small cylinders in a stream of fluid, and, with the help of his laboratory assistant van der Hegge Zijnen, became a pioneer in using the hot-wire anemometer to probe velocity fluctuations in turbulent flows. These measurements were made in conjunction with vane anemometers and Pitot tubes. Though the measurements were rough, they were adequate to delineate the different regions of the boundary layers as well as the suppression of turbulence in stratified flows.

Obviously because of his background in statistical mechanics, Burgers' work on the theory of turbulence was devoted to developing a statistical theory of turbulence and to treating theoretical models. His best-known work concerns the equation now known after him, which is a one-dimensional version of the Navier–Stokes equations without pressure (Burgers 1939, a more accessible reference being Burgers 1948), and is written as

$$\frac{\partial u(x,t)}{\partial t} + u\frac{\partial u}{\partial x} = \nu\frac{\partial^2 u}{\partial x^2}. \tag{4.10}$$

Burgers himself regarded that this one-dimensional, nonlinear partial differential equation would yield some insights into the fundamental processes of turbulence.[20] The multi-dimensional version of the equation has also been used extensively.

The hopes that this equation would shed any light on turbulence dwindled after it was recognized that the so-called Hopf–Cole transformation (Hopf 1950, Cole 1951) rendered the equation linear. In recent years, beginning in the mid-1980s, the equation has been used extensively in condensed matter physics (for example, in modeling surface growth, where it is known as the so-called KPZ equation; Kardar et al. 1986) and in cosmology (where the three-dimensional versions of the inviscid and viscous Burgers equation – called the Zel'dovich

---

[20] In fact, Bateman (1915) had suggested the same equation for describing the discontinuous motion of a fluid whose viscosity tends to zero.

approximation and the adhesion model – are used to represent the formation of large-scale structures of the Universe; see, for example, the review by Shandarin 1989).

The one-dimensional and multi-dimensional Burgers equation with stochastic initial conditions (e.g. She et al. 1992, Sinai 1992) and stochastic forcing (e.g. Chekhlov & Yakhot 1995, Polyakov 1995) has served in recent years as a testing ground for field-theoretic and advanced probabilistic methods. The latter has been used as a counterpart of turbulence models where the steady state is maintained by external forcing, especially for understanding how an intermediate region, where the system exhibits anomalous scaling (not predictable by dimensional analysis), develops. This anomaly for velocity structure functions can effectively be calculated in the case of the stochastic Burgers equation. Furthermore, the probability density function of the velocity gradients can be calculated essentially exactly. These aspects often provide the point of departure in understanding the structure of the Navier–Stokes turbulence.

In addition to his work on turbulence, Burgers became interested in solid mechanics in his capacity as Secretary of the Joint Committee on Viscosity and Plasticity. Taylor had already made pioneering contributions to dislocation theory in plasticity, and so was a natural correspondent on this topic as well. Burgers collaborated with his brother, Professor W.G. Burgers, in the work on dislocations in crystal lattices and introduced, in 1939, the so-called Burgers vector, which is a measure of the strength of a dislocation in a lattice. He also studied the fluid dynamics of dilute polymer solutions and wrote some of the fundamental papers on the intrinsic viscosity of suspensions.

Burgers carried on elaborate correspondence with many leading scientists of his day (apparently with ease in English, German and Italian as well as his native Dutch). Much of it is preserved in the Burgers Archives placed in the Libraries of the University of Maryland.

4.6.2 Correspondence between Taylor and Burgers

Three principal scientific topics of exchange are relevant to us: hot-wire anemometry, statistical theories of turbulence and the First International Congress of Applied Mechanics held in Delft in 1924. Burgers was some three years younger than Taylor and the correspondence reflects the self-assuredness of Taylor and the slight tentativeness of Burgers. They met for the first time at the 1924 Congress. Burgers (1975) remembered that Taylor "came with a large collection of instruments for demonstrating motions in rotating fluids...". Many letters were exchanged between 1924 and towards the end of 1947, but only a few will be discussed here.

The first letter found in the archives appears to be a response from Burgers to Taylor and is dated 24 January 1924. One may infer from Burgers' letter that Taylor might have inquired about velocity distributions in turbulent flows. Burgers responded with his summary of the contributions of Kármán, Prandtl, Blasius and Taylor himself, mentioned his hot-wire work with van der Hegge Zijnen and added some figures. The letter is addressed formally as "Dear Sir" (which turns into "Dear Taylor" by 2 April 1927 – if not sooner). On 3 May 1935, Taylor began his letter to Burgers, apparently without any prompting from the latter, as follows:

> I have lately been working out a theory of the dissipation of energy in turbulent motion and have got results which agree extremely well with observations. In the course of the work I have come across some general relationships which I have not seen before but which it is very possible that some one has noticed. I am therefore writing to ask you whether you have come across them. They are concerned with the idea of isotropic turbulence, i.e., turbulence which is isotropic statistically so that all statistical averages are independent of the direction of the axis of reference.

He then listed most of the principal results of Taylor (1935a). The measurements with which Taylor claimed excellent agreement were "Dryden's observations of decay of turbulence behind honeycombs, NPL observations of the same type, and recent unpublished observations of Dryden's on decay behind similar grids of varying scales", all of which, he claimed, obeyed the linear law (4.7), with the value of the constant A "between 2.8 and 3.1 whatever the nature of the turbulence producing mechanism i.e., square rods, honeycombs, grids of flat slats arranged in square and grids of round bars arranged in squares".

In the response of 23 May 1935, Burgers enclosed a 27-page typewritten work of his own, entitled *On the mechanism which determines the intensity of turbulence*, mostly related to shear flows. Burgers made the following three points: (a) turbulence generated in one place in the flow could be dissipated elsewhere (he was obviously thinking about inhomogeneous flows such as in pipes, as one can easily see from his notes just mentioned); (b) that "certain phase relationships must be present . . . by which the phases of the various terms are coupled"; and (c) according to his model, the transverse velocity fluctuations would be the largest near the wall – a result from his model that he did not understand. Shear flows were not Taylor's concern of the moment, but he tried to make contact with Burgers' points of view. In his response of 28 May, Taylor agreed with (a) and (b) but pointed out that:

> If turbulent energy is produced asymmetrically by the mean velocity acting on the Reynolds stress near the walls the strong tendency of turbulence to become

isotropic seems to be effective during the transfer of the energy from the place of origin near the walls to the middle part of the pipe.

He also wrote, perhaps somewhat discouragingly: "I would very much like to see a theory of the general type of yours, but I cannot at present see how to accept it in its present form". He promised to send all his five papers to Burgers as soon as he inserted a correction in the first part of the series. Though he noted that the "secretarial dept of the NPL is very slow", he was prompt to follow up only two days later with preprints of his papers.

Burgers began his letter of 8 June 1935 with "Now in your letter you have made a serious criticism against the attempt I made ... May I try to defend myself against the criticisms?" and gave broad statistical mechanical arguments in support. There is no record in Burgers Archives that Taylor responded immediately, but he assured Burgers on 17 November 1935 that he had "been thinking about the relationship that must exist between the problem you discussed ... and the problem I discussed in my statistical theory of turbulence". Taylor again drew attention to his view that, once established by whatever process, the system of eddies will degrade and the question of this degradation is "well worth mathematical study". In his view, the essence of three-dimensional turbulence was the property of vorticity of increasing continually until the resulting small scales die away more quickly than can be produced by the larger ones. He cited his system of three sinusoidally varying velocity components and noted that "so far I have only been able to proceed as far as the terms containing t^2 and some terms in t^3 but there is no doubt about what is happening". He went on to suggest that increasing number of Fourier modes come into play with increasing time, and suggested that if one likened the action of the initial motion of the three Fourier modes to the effect of the mean motion, the problem considered by him should be quite similar to that occupying Burgers' attention. This letter is a preview of the Taylor–Green paper discussed above. It is not clear how influential was Burgers' insistence on the mean flow effects in fixing Taylor's attention on the Taylor–Green problem, but Taylor did try to make a connection with it in his correspondence with Burgers.

On 11 December 1935, Burgers noted that the result of Taylor's previous letter "comes not unexpected to me", but cautioned that the "difficulty of the problem however lies in the circumstance that in reality one does not start with one single term in u, v, or w, but works with a whole 'spectrum' of terms". He again returned to his theme of the "phase relation between various terms, or groups of terms, which are critical to the debate".

My overall sense is that Taylor and Burgers were not on the same wavelength (despite the mutual fondness), which may be one reason why the correspondence went silent even as Taylor was in the middle of his work on isotropic turbulence and was carrying on vigorous correspondence with Prandtl, Kármán and, though less frequently, with Dryden.

Two further letters were sent from Burgers to Taylor immediately towards the end of the war and were promptly reciprocated. One of them has only minor relevance to turbulence but are summarized here because of their human interest. On 16 July 1945 Burgers wrote: "We are anxious to know whether friends and acquaintances in England are still living . . . My family has not suffered much", and inquired after some common friends: he also wondered about the possibility of a visit to England to pursue aeronautical issues, to which Taylor responded on 8 August 1945: "Very glad to know that you have survived the war without too much hardship. With regard to the proposed visit, would be delighted but the security mentality is still with us though it is disappearing in some departments . . . ". Burgers followed up by sending his papers and wrote: "The similarity considerations, which seem to have been used by Kolmogoroff, Onsager and von Weizsäcker, can be applied also to the model [what we now know as the Burgers equation], so that a similar formula for the correlation function (depending on $r^{2/3}$) can be deduced". Taylor knew about Kolmogorov's work already through Batchelor (see Section 4.6.6) and was not particularly impressed by Burgers' work. In a postscript to his response of 1 March 1947, he wrote: "I have now read your paper on the model of turbulence and it certainly seems to give some interesting results. The main objection to it from my point of view seems to be that until many of the problems which are solved in analogue in your paper had been solved with great labour using the actual equations of a fluid I could not see the usefulness of the analogue. One sees the usefulness of the analogue only when it is no longer so necessary." He could very well not have anticipated the resurgence of interest in Burgers' equation. These were the last scientific exchanges between the two men.

In Burgers' words from a letter to Batchelor on 7 October 1975, "I have had relatively little contact with Taylor; most of it was in the early period, of 1924 to about 1934. There has been almost nothing since I moved to the US at the end of 1955."[21] A few brief exchanges did take place sometime after April 1973: Taylor sent a reprint of his SIAM obituary notice of Kármán to Burgers with the handwritten notation, "I hope you do not mind my taking your name in vain!", to which Burgers responded with great fondness, recalling Taylor's stay with Kármán in Burgers' place, and so forth. Taylor wrote back

---

[21] In fact, he came to know of Taylor's death through the obituary by Pippard (1975) but later wrote in Dutch, in response to a request from the Royal Netherlands Academy of Sciences, a six-page obituary of Taylor.

with reminiscences of his own, mentioning Stephanie's death in 1967 and his stroke in April 1972.[22]

4.6.3 Taylor and Kármán

In *The Legacy*, Batchelor recorded that no correspondence between Taylor and Kármán was found among Taylor's papers; indeed, the Taylor Archives in Trinity contain no related documents. It would appear that the papers were misplaced *in toto* because the two of them corresponded from the late 1920s to almost the end of Kármán's life. Fortunately, there is some record of this extensive correspondence in the Kármán archives at Caltech. Among those preserved, Taylor's notes are the more substantial and more numerous. The correspondence was exceptionally cordial through the years, but ceased to be scientifically driven sometime after the publication of Kármán & Howarth (1938) – the only paper in which Kármán signed off as "Theodore de Kármán".

The relevant part of the correspondence occurred during 1935–38. Although they have been cited in Chapter 3 by Leonard & Peters (2011), a partial recapitulation here might not be out of turn.

On 9 January 1937, Kármán sent the first version of the Kármán–Howarth paper to Taylor, with a request to "accept [it] for publication in the *Proceedings of the Royal Society*, and to expedite its publication. He called attention to his earlier paper (Kármán 1937a) analyzing the "simplest case" of "one-dimensional turbulence" using similar symmetry principles. The paper by Kármán and Howarth defined the general isotropic forms of two-point correlation functions of second and third order, and derived the now well-known Kármán–Howarth equation which has played a key role in the theory of turbulence on more than one occasion (e.g. in the derivation of Kolmogorov's famous 4/5th law; Kolmogorov 1941b).

Kármán & Howarth obtained a family of self-preserving solutions, a special case of which is the well-known 5/2-power law for the final period of decay (see Batchelor 1953). As a further special case, Kármán (1937a) obtained the linear power-law (4.7) that Taylor had obtained through other means. Taylor noted in his response on 24 January 1937 that he was "tremendously interested in the paper and [looked] forward to verifying the work in detail when it comes out". This is precisely what he did in his paper published in the *J. Aero*.

---

[22] Taylor also wrote that Enrico Volterra had asked him for a contribution celebrating Tullio Levi-Civita's centennial. In his response, Burgers mistook Enrico Volterra for Enrico Fermi, and Taylor corrected it on August 1973 in his exceedingly scrawly hand. The last letter from Burgers was on 3 October 1973, with apologies for the mix-up and assurance that it didn't go any further.

Soc. (Taylor 1937b).[23] However, he had a "small point" to bring up; this small point is what took up most of the letter. It was that Kármán's attribution that Taylor's linear decay was a special case of the self-preserving solutions could not be correct because self-preservation on all scales was inconsistent with facts as Taylor knew them.[24] He noted that "if you would alter very slightly your remarks about the linear law for $1/u'$ I would be grateful because I don't want to seem, by communicating the paper, to agree with your remarks on all particulars". Taylor pointed out that "if I had tried to apply my ideas to the self-preserving distributions, I should have come to $1/u' \propto \sqrt{t}$ – not to t".

Kármán responded on 21 February 1937:

> I gladly agree to any changes you suggest concerning the linear law of the decay ... in order to accelerate the publication of the paper, please make the corrections to avoid appearance that you agree with all that is said on the matter.

He was clearly anxious to see his paper in print. The suggested change seems to have been implemented; in its printed version we find: "Hence the coincidence between these [self-preserving] equations and Taylor's results is rather formal." In the letter, Kármán went on to point out rather firmly that Taylor's assumption of the constant length scale cannot be correct. In hindsight, we now know that Kármán was right on this point.

There indeed was another important aspect on which Taylor was right, and Kármán not. As far as I can tell, this aspect was not commented on by Taylor in his letter to Kármán but was to occupy a part of one paper (Taylor 1937b) and all of another (Taylor 1938a), for both of which the starting point was a commentary on Kármán's (1937b) paper. Taylor (1937b) first verified the isotropic relation between the longitudinal and transverse correlation functions of isotropic turbulence, earlier discovered by Kármán ("a more general relationship" than his limiting case, according to Taylor 1973a). He then questioned Kármán's assumption of self-similarity of the correlation function and showed why it is inconsistent with observations. Taylor had recognized very early that the small scales and the large scales need to follow two different versions of similarity, and that all the scales cannot be simultaneously self-preserving. As his final point, Taylor (1937b) dealt with the role of the vortex stretching terms in the evolution equation for mean-square vorticity. Kármán (1937b) had assumed that the vortex-stretching term $\overline{\omega_i \omega_k (\partial u_i / \partial x_k)}$ would be

---

[23] In this paper, Taylor cited only the published work of Kármán (1937a,b), and essentially pointed out their errors, as discussed later.

[24] I cannot help admiring Taylor's conscientiousness in writing to Kármán en route to Switzerland, one day after he received the paper; and he had recruited his wife to send the paper off by registered post to the Royal Society the following day.

zero in the equation for the rate of change of mean-square vorticity. His reasoning was that the vortex filaments would otherwise have a permanent tendency to be stretched or compressed along the axis of vorticity. Taylor, who had thought about this matter since his Adams Essay of 1914, went further now and showed that such a tendency does indeed exist on the average. He briefly introduced the Taylor–Green problem, dealt in greater detail in Taylor & Green (1937), and noted that the vortex-stretching terms are large. He clearly understood that the places where vortex filaments are being stretched are also places where high vorticity may be present. Using estimates from available measurements, Taylor & Green (1937) deduced that the term neglected by Kármán was three times as large as the one retained. They also estimated that the vortex-stretching term increases relative to other terms retained by Kármán as a linear power of the large-scale Reynolds number.

Taylor, writing to Kármán on 12 March 1936, had already taken the occasion to mention the Taylor–Green work (still unpublished at that point; see Figure 4.3, which is the reproduction of a page of Taylor's letter), and concluded that "this puts the whole idea of my dissipation theory of turbulence on a sounder basis".

The error in the Kármán–Howarth draft was eventually set right, as described in greater detail in Chapter 3. In the printed version of the paper, a new section was added under the responsibility of the "senior author", acknowledging the previous error as well as the possibility of small and large scales possessing two separate laws of self-preservation, and conforming to Taylor's earlier request for the slight modification in the interpretation of his linear law, with Kármán's addition mentioned earlier: "Hence the coincidence between these equations and Taylor's results is rather formal." All told, Taylor got his way.

I now wish to add a few remarks on the human element of the exchange of letters between Taylor and Kármán. It covered a wide range of topics, often about mutual visits, families and acquaintances, and the topics ranged from pleasantries to serious scientific exchanges. To get a sense of these friendly exchanges, we note that the available correspondence begins with a graceful note from Taylor accepting the award of the honorary degree from the University of Aachen, most likely enabled by Kármán's intervention, an inquiry from Kármán asking for a copy of Heisenberg's 1948 paper on isotropic turbulence, a note from Taylor informally asking for Kármán's thinking on "radio waves coming from beyond our solar system", Kármán's cable to Taylor on the occasion of the latter's 70th birthday: "Dear G.I. First time in life I am ahead of you. Shall be almost 75 as you reach 70. Cordial congratulations", and so forth.

A good sense of this cordiality can be found in Taylor's biographical notes of Kármán (Taylor 1973a,b).[25]

4.6.4 Taylor and Prandtl

The correspondence was almost certainly initiated by Taylor who sent his published papers, most likely in 1923, to Prandtl: we may never know which papers exactly. Prandtl's command of English was minimal at that time, and his awareness of Taylor minimal as well: for instance, he inferred that Taylor was from Cambridge only from the postal seal and even had Taylor's middle initial wrong in his first letter. It is not clear if he read Taylor's papers carefully. However, by the time the two of them met in Cambridge in 1927, on the occasion of the 15th Wilbur Wright Memorial Lecture that Prandtl delivered in London, they were on very good terms and their relationship then warmed up even more; Prandtl's stay in Cambridge was among the "best memories of [his] journey in England". The conversation between them was almost certainly facilitated by Stephanie Taylor since her husband's German was not particularly good.

As is well known, Taylor put forward his vorticity transport theory in 1915 while Prandtl put forward his momentum transport theory in 1925 – independently, as Taylor (and Prandtl) would state (and all the correspondence would seem to suggest). Eddy viscosity or diffusivity does not impress today's turbulence researchers but it was an exciting idea when it emerged as successful in calculating mean distributions of velocity and temperature, in reasonable accord with measurements in several standard flows.

It was thus natural that a serious scientific exchange between them occurred about turbulent transport. Taylor's theory was proposed to hold when the

---

[25] In the *JFM* version (Taylor 1973a), he used the name 'Kármán' consistently – and this was how he addressed all the private letters – but switched to 'Theodore' in the SIAM article (Taylor 1973b), presumably because the talk on which it is based was presented at the meeting by Goldstein in Taylor's absence due to health problems. Considering the enormous regard in which Kármán seemed to have held Taylor through the years, it may sound incongruous that he mentioned Taylor not at all in the *Turbulence* chapter of his autobiography (Kármán 1967). The only place where Taylor's name appears is in the footnote on p. 248. Referring to his work on plasticity, Kármán remarked: "It is interesting to note that at the end of the war I learned that my good friend from Cambridge, Sir G.I. Taylor, working on another problem, had come to the same theory as I had approximately at the same time." The chapter on turbulence is mostly Kármán's rendering of the competition with Prandtl and does not mention other salient aspects such as the Kármán–Howarth equation. Prandtl's daughter, Johanna Vogel-Prandtl, who had read Kármán's autobiography before writing, in 1993, an account of her father's life, had this to say: "Lee Edson published the book on his own only after Kármán's death. In this, certainly, not sufficient care was taken in differentiating what was really said, and what Kármán interpreted with hindsight. In professional circles that have come to know of Kármán's book...the view is held that this...should be read critically....Professor Blenk expressed the view as follows: 'The autobiography...cannot make any claims of historical correctness'..."

motion was predominantly two-dimensional. Prandtl noted on 4 June 1927 that:

> In general, I believe that my previous theory can be supported at least as a good approximation. I have based my consideration on a somewhat different mechanism that is essentially three-dimensional ... I have been recently thinking that perhaps both processes, that studied by you and that studied by me, can occur simultaneously and that one can write for the force per unit volume
>
> $$\frac{\partial \tau}{\partial y} = \alpha \frac{d}{dy}\left(\varepsilon \frac{dU}{dy}\right) + \beta \varepsilon \frac{d^2 U}{dy^2}$$
>
> if one uses the formula $\varepsilon = \rho \ell^2 \frac{du}{dy}$ and assumes ℓ to be proportional to the distance from the wall.

Here, τ is the turbulent stress, which is related to the mean velocity gradient dU/dy through the eddy diffusivity ε; and ℓ is the mixing length. Prandtl noted that the 1/7th power for the velocity distribution would obtain if $\beta = \alpha/3$. Twenty days later, he was already disenchanted with this proposal because:

> the apparent fraction in the inner part of the flow depends only on processes in the laminar layer of the wall and that the contribution of the first term is compensated to a large extent at large Reynolds numbers by the second term, which is negative. This is for me the strangest thing of all and I believe that the formula has no physical justification.

Taylor took nearly three years to answer Prandtl's request for his opinion, and was properly apologetic: He had "thought about the matter but could not come to any definite conclusions". Around the same time, Taylor had been carrying on a correspondence with V.W. Ekman (see later) about the differences between the transport of heat and momentum. It is easy to conjecture that this stimulus must have prompted Taylor to re-examine his transport theory vis-á-vis Prandtl's theory. The result is Taylor (1932a,b, 1935f, 1937a), as already discussed.

Prandtl had great respect for Taylor's work ("if I had known these papers [Taylor 1915 and 1921], I would have found my way to turbulence earlier" – letter of 25 July 1932), but was not hesitant to point out that "The result of your theory [Taylor 1936a] of critical Reynolds number of a ball behind a turbulence grid is very strange" (letter of 12 March 1935). Taylor followed up on 14 April 1935 with a summary of his 1935 papers, to which Prandtl responded with Reichardt's data in the channel. He was enamored by Taylor's work but not entirely sold on isotropic turbulence. On 11 May 1938 (effectively the last serious scientific letter exchanged between them), Prandtl wrote:

> in your new work [Taylor 1938] we were extremely interested in the extremely cunning trick by which you can obtain the correlation curve from the spectrum

and conversely ... Would it be possible to proceed so far for non-isotropic turbulence which is in fact far more important?

In total, Taylor and Prandtl exchanged something like 25 letters between 1923 and 1938, and two more from Prandtl after the war. They had become quite close. The correspondence had started off formally with "Dear Dr. Prandtl" and "Sehr Verehrter Herr Professor Taylor". It evolved through various stages and remained for several years at the level of "Dear Prandtl" and "Dear Taylor". The war intervened in their friendship, after which Taylor wrote two letters without any form of address, and Prandtl reverted to formalities. I will discuss these letters at some length, because they shed important light on Taylor and also the impact that a calamity such as a war can have on personal relationships. This episode needs a more detailed and careful discussion at some point in the future.

While attending the 5th International Congress for Applied Mechanics in Cambridge, MA, Taylor and Prandtl met on 7 September 1938, in J.C. Hunsaker's house, with Stephanie Taylor in attendance. Hunsaker and Taylor remained friends at all times. There is no record of the conversation at the meeting but Vogel-Prandtl (1993) noted that "some personal thoughts were exchanged". One can infer the nature of the conversation by studying the letters subsequently exchanged between the Taylors and Prandtl. After the meeting, Taylor remained in the USA for a little longer. One suspects that the exchanges at the meeting must have festered in his mind, which prompted him to write a testy note to Prandtl from his hotel in Washington, DC. Prandtl's knowledge of English was good enough to read Taylor's letters directly[26] but he had this particular letter translated into German. In this letter, dated 27 September 1938, which was begun without any form of greetings, Taylor said: "Now I must ask you to believe that whatever happens between our countries the friendship and admiration which I, in common with aerodynamical people in all other countries, feel for you will be unchanged. I realised that you know nothing of what the criminal lunatic who rules your country has been doing so you will not be able to understand the hatred of Germany which has been growing for some years in every nation which has a free press ... I hope that by the time I see you again Germany will have developed some method of dealing with her neighbours that will make it possible for the present universal hatred of her rulers to die down."[27]

---

[26] Prandtl was to write on 5 June 1934, while contemplating his visit to Cambridge to receive the honorary degree, that the "scientific discussion should go better than in 1927 because since then I ... understand English rather better".

[27] Prandtl had complained about Taylor's handwriting on 25 January 1933: "Would it be possible for your letters which at the moment are extremely hard to read and cost me a large amount

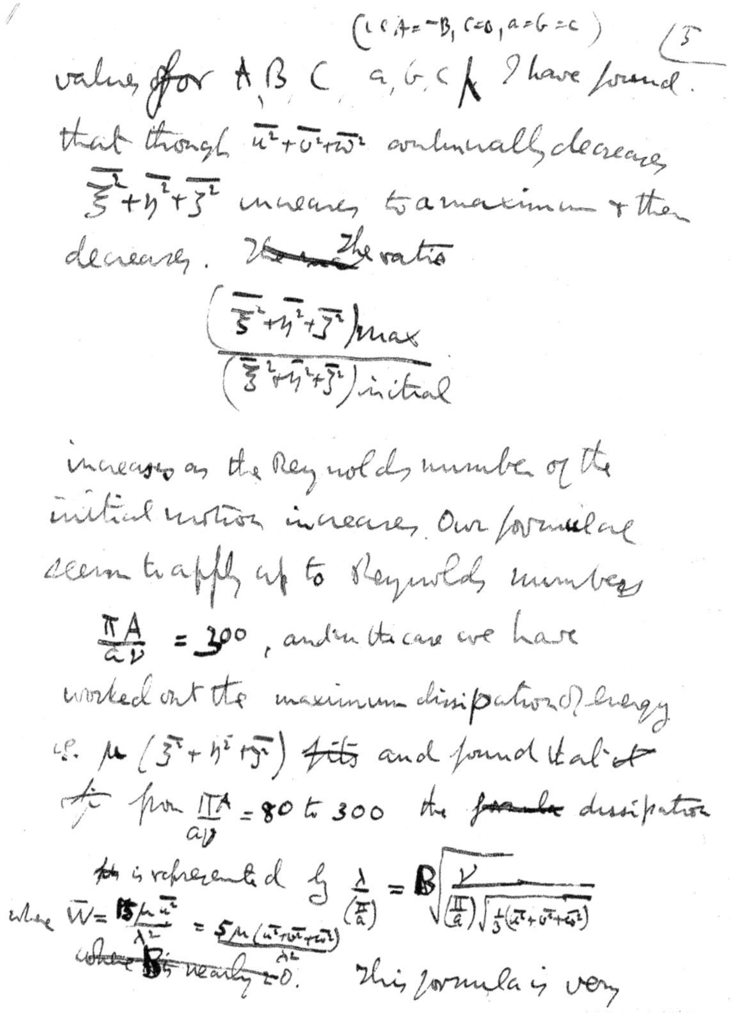

Figure 4.3 Taylor's quite messy handwriting.

Prandtl's response was quite defensive. It is not the appropriate place to evaluate it in detail but the substance is essentially that the so-called free press

of time to be written by somebody else who writes more clearly? I hope that you will not take umbrage against this remark …" Taylor continued to write in his own hand but made a decided effort to be more neat while writing to Prandtl. However, in this and the subsequent letter, he made no such effort and his writing returned to being quite messy.

was giving an incorrect picture of the situation in Germany. Prandtl called attention to the agreement between Hitler and Chamberlain,[28] was thankful that good sense had prevailed on both sides, and added: "You yourself must visit Germany once to convince yourself that we are still governed here very well." Taylor's response was tough. Condemning the atrocities of the Nazis by citing specific examples, he went on to say:

> It seems to me that the only reason for this officially inspired tirade is that the Nazis want war but Chamberlain's visit for the first time penetrated the wall of their censorship and showed them that the people of Germany do not want war. They are therefore doing their best to incite the populace against us in order that the desire for war may spread in Germany. You will see that we are not likely to agree on political matters so it would be best to say no more about them.

He ended the letter and signed it; changing his mind, he crossed it out and continued on the next page: "Fortunately there is no reason why people who do not agree politically should not be the best of friends." This was followed by a question of clarification on a mundane scientific issue.

It appears that this was the last letter from Taylor to Prandtl. But Prandtl, who apparently wished to win Taylor over to his point of view, wrote to Stephanie Taylor a year later (5 August 1939, just before the Second World War officially broke out), essentially seeking an ally in her. He lay the blame for possible war on England, and discussed the political pacts that Germany had closed for peace. If Prandtl had hoped to convince the Taylors about the situation in Germany, he must have failed utterly. There is no record of either of the Taylors having responded to it. There is also no record, either, of any response to two of Prandtl's post-war letters using intermediate carriers. The contents of the letters, as translated by Johanna Vogel-Prandtl, are partially included in the *The Legacy*, p. 187, but a few additional comments may be made.

The letter of 28 July 1945 was basically a post-war status report on Prandtl's Institute, which was then occupied by Allied Officers. The employees of the Institute were allowed only to make some repairs to the building and write up reports to the Allied Committee on the unfinished work done during the war. No new work was allowed. Prandtl was hopeful that things would change for the better but felt that he could not probably last much longer since he was already 70 years old (though in good health). He mentioned the BBC programs in German – some of which he thought were useful, especially those relating to the superiority of the constitutional system of governance – but was less sanguine about other aspects.

---

[28] This agreement occurred on 29 September 1938, a month before Prandtl's letter was written, and is universally regarded as a failed pact.

In the letter of 11 October 1945, the last one in existence, Prandtl reiterated that his Institute was prohibited to do any research, listed areas that "await solution", and said that he had written to the Royal Society, on the advice of the British Resident Scientific Officer, to allow further research to be continued. He attached a copy of that letter, and expressed gratitude if the Royal Society would allow the request. He was clearly seeking Taylor's intervention. It is not known if Taylor did anything. Prandtl also enclosed an improved method for the calculation of mean profiles in turbulent flows (apparently using his new closure method – I have not seen that note). Taylor most likely did not respond to these calculations; his research interests had simply moved on.

4.6.5 Taylor and Richardson

Chapter 5 in this volume, by Benzi, is devoted to Richardson's contributions to turbulence, and my interest here is limited to making a few remarks on his interaction with Taylor. A useful article in this direction is Hunt (1998).

Taylor and Richardson were contrasting figures in some ways. Taylor's career was straightforward and Richardson's convoluted; Taylor's writings were direct and simple, Richardson's pregnant with emergent meaning. Taylor was concrete and Richardson abstract. They had common friends (see Hunt 1998) but seemed not to be close. There are only two letters from Richardson to Taylor in the Trinity Archives, only one substantial. In that letter, dated 17 March 1919, Richardson was responding to Taylor's invitation to join in a meteorological expedition for the summer. This would be about the time when Richardson had returned to the Meteorological Office to work on numerical weather forecasting, just after his period away in France as an ambulance driver during the First World War. Richardson was not sure that he could participate. His question towards the end of the letter, presumably in response to Taylor's narrative of his 1917 work on turbulence near the ground, was this (after one replaces the mathematical symbols by words): "Is there an unshakable definition of the gradient of the wind near the surface?" It reveals the way in which Richardson was thinking of velocity gradients in turbulent flows (and of fractal-like objects in general).

Taylor and Richardson intersected on problems of meteorology, especially on turbulent diffusion and stratification. It was clear that Richardson (1920), which provided useful data on the increase of turbulent diffusivity with height above the ground and the wind speed, was influenced by Taylor (1915). Likewise, Taylor (1921) made a reference to Richardson's work on the diffusion of smoke from a point source a few meters above the ground, cited his photographs to support his predictions on the average growth of the smoke plume, and reproduced some of Richardson's results supplied privately. It is thus clear that each was well aware of the other's work but it is unclear if Taylor was influenced by the famous 'four-thirds' law of Richardson (1926), and its implication that turbulence contains eddies of many length scales, à la his famous rhyme (Richardson 1922) depicting energy cascade – which we may quote here for completeness:

> Big whorls have little whorls,
> Which feed on their velocity;
> And little whorls have lesser whorls,
> And so on to viscosity
> (in the molecular sense).

With respect to the effects of stratification, Taylor had derived for a special case the conditions of stability in stratified flows and was duly appreciative of some specific aspects of Richardson's (1920) contributions.

4.6.6 Taylor and Kolmogorov

The companion Chapter 6 by Falkovich addresses Kolmogorov's far-reaching contributions to turbulence (see also, among others, Yaglom 1994 and Barenblatt 2006). I limit myself to the interaction between Taylor and Kolmogorov. They met at the well-known 1961 meeting at Marseilles but it is almost certain that they did not correspond before or after. In Kolmogorov (1941a), there is considerable reference to Taylor's isotropic turbulence as a point of departure for the locally isotropic turbulence which Kolmogorov invented in that extraordinary paper. As a curious aside, I may note that Kolmogorov (1941a) made a plausible argument for the formation of increasingly smaller eddies, à la Richardson's famous rhyme of 1922 (see above), but was not aware of the latter's work. He invoked Prandtl's mixing length as the first step in a sequential generation of eddies (but referenced Richardson in Kolmogorov 1962).

A piece of lore in Cambridge, which I have verified with two senior people, is that when Batchelor excitedly brought Kolmogorov's 1941 papers to Taylor, the latter remarked that the results were obvious. Later, Batchelor (1975) was to invite Taylor to comment on "some ideas which were expressed in concrete terms [in Taylor's papers] and which have later been reformulated in more abstract and general form and seen to be very important". He particularly had in mind Kolmogorov's work. Taylor's response was that he

> did have ideas about the eddy cascade in 1917 but didn't see how to express them in mathematical form... certainly did not get on to the kind of statistical similarity argument which Kolmogoroff, Heisenberg and Onsager developed independently of one another a few years later.

Taylor had a good idea of the cascade processes even in his Adams essay of 1914, and had more to say about it in his 1937 paper with A.E. Green, but the general picture that Kolmogorov (and later Onsager, Heisenberg and von Weizscäker) brought to bear on the problem had eluded him. In the same way, one can see the stirrings of small-scale intermittency in Taylor (1917c), but its statement as a general principle was either not Taylor's forte – or he was simply loathe to extending himself beyond the concrete. If he momentarily lapsed into thinking that Kolmogorov's results were obvious, we can forgive him for not seeing their full sweep.

4.7 Some reflections

4.7.1 Taylor and the missed Nobel Prize

Taylor was nominated for the Nobel Prize at least once, in 1937. The nominator was Harold Jeffreys.[29] Jeffreys presumed:

> that on this occasion other subjects than atomic physics are to be considered, and in other branches of physics ... Taylor is the strongest candidate. Professors Prandtl, Karman [sic], and Bridgman would be worth much consideration; but Taylor's work covers more ground than any of them and seems to me more satisfactory than that of Karman [sic] and Prandtl.

He presented the highlights of Taylor's work on turbulence, compressible aerodynamics and plasticity, extolling Taylor's expertise in both theory and experiment.

In the meantime, Prandtl was nominated for the Nobel Prize by Sir Lawrence Bragg from Cambridge who, it may be recalled, was the youngest recipient of the Nobel Prize (and his record has not been broken yet) for his joint work with his father, William Bragg, on crystallography. Bragg was not an expert on fluid mechanics and almost certainly motivated by Taylor's initiative in preparing the nomination letter. One can infer it to be so from the similarity of language that Bragg used in his nomination letter to that used by Taylor in a confidential note to Prandtl on 15 November 1935.

To this letter, Prandtl responded revealingly:

> It is extremely friendly of you to be of the opinion that I am a suitable candidate for Nobel Prize. However, I have no hopes in this direction because I cannot consider the work really part of physics. I have also regretted that there is no similar prize foundations for mathematics and for engineering similar to the Nobel

---

[29] On 23 February 1944, Nevil Mott wrote to the Nobel Committee regretting that its letter seeking nominations for the year was received by him too late, but would have liked to have nominated Taylor had there been any time. Whether he did anything in subsequent years is unclear.

foundation. . . . If the Swedes divide science in the same as we do here then I will come into as little consideration as mathematicians and could therefore console myself in the same way mathematicians do.

And the Swedes thought somewhat the same way, as will be seen presently.

The summary is that Taylor and Prandtl were considered together for the Prize by the Nobel Committee for Physics, which concluded thus:

L. Prandtl and G.I. Taylor have both been active in applied mechanics and occupy leading positions in this field. The particular conditions prevailing there are due to the fact that the laws governing the observed phenomena are so complicated that it is impossible to derive the phenomena from them. For those who cannot or will not wait for the mathematical methods to be perfected enough to make today's impossibilities possible, the task must then be to develop approximate methods of calculation, which at least apply in special cases. Typical examples of this kind of approximate methods of calculation are Prandtl's theories for boundary layers (1905) and airplanes (1919). Obviously, activities of this kind cannot in general lead to any 'discovery or invention in the area of physics'. No such discovery or invention has been advanced by the two propositioners who this year recommended Prandtl and Taylor. Therefore, the members of the Committee cannot recommend that either of them be rewarded.

I should note that Prandtl had been nominated for the 1928 Prize as well (by Professors Húckel of Zürich, Schiller of Leipzig and Pöschl of Prague). The conclusion of the Nobel Committee for Physics on that occasion was that the boundary layer theory "provides no true solution to the hydrodynamical problem, but Prandtl's analysis of what actually happens should nevertheless be considered a valuable advance". The key person on both occasions was the committee member C.W. Oseen (1879–1944), who wrote a special opinion in 1928. He stated, in part, that:

The propositioners apparently assume that in order to receive a physics Nobel Prize, a long-standing, dynamic and successful activity within physics, including mechanics, suffices. If this assumption were in agreement with the Charter of the Nobel Foundation, I would have recommended that Professor Prandtl be awarded a physics Nobel Prize. However, since a prerequisite for receiving such a prize, according to the Charter, is a discovery or invention in the field of physics, and since no discovery or invention of the magnitude that would justify a Nobel Prize has been put forward by the propositioners, I conclude that I cannot second the nomination.

As an aside, other nominees who contributed to fluid dynamics were: W. Thomson (later Lord Kelvin, in the years '01, '02, '03. '04, '05, '06, '07), J.W. Strutt (later Lord Rayleigh, nominated in '02, '03, '04, '05, successful in 1904 for his discovery of Argon), Vilhem Bjerknes ('21, '23', '24, '26, '29, '36, '37, ???), Jacob Bjerknes ('28, '36, '37, ???), Martin Knudsen ('14, '16, '17,

'18, '19, '20) and Gustaf de Laval ('08, also for Chemistry Prize in '08). The question marks indicate that there may have been further nominations.

4.7.2 Taylor's school and method of doing science

During his career, Taylor did not make conscious efforts to build a school around his own interests. His sense of 'opportunism' was probably not conducive to such tasks; the war efforts undoubtedly took away too much of his time and attention. In any case, it was Batchelor who built a fluid mechanics school in Cambridge and fashioned it after Taylor's style of doing science, with Taylor a seemingly inspirational presence (Turner 1997, Moffatt 2011). Taylor's memory resides in Cambridge in part due to Batchelor's efforts, among which is the establishment of the G.I. Taylor Professorship. The first person to hold the Chair was G.I. Barenblatt – whose breadth and depth in vistas of mechanics are similar to Taylor's own.

What exactly was Taylor's style of doing science? In his words, for which we have Batchelor (1975) to thank:

> My own method of answering questions of this kind [whether concepts can be given a rational description] is to think of experiments which depend in the simplest possible way on the concept I have in mind, and then set up apparatus to see whether my preconceived idea is likely to be right.

A criticism sometimes leveled against Taylor is that he inelegantly broke up mechanics into small problems without strategic thinking. This criticism has been voiced most strongly by Clifford Truesdell. Again in Taylor's own words: "I do not see how one can plan a 'strategy of research in fluid mechanics' otherwise than by thinking of particular problems." This characterization is consistent with the fact that Taylor's Nobel nomination described a mix of several special problems, instead of a major discovery. There are, of course, different valid ways of doing science, including the formulation of problems in their greatest generality but, in nonlinear science broadly, one builds one's intuition by solving several particular problems – thus eventually strengthening one's understanding that is so important for practical design or fundamental theory.

I should, however, stress that Taylor did not think in incremental terms as a rule. It is clear from the sum-total of the correspondence that Taylor maintained during 1935–39 that he regarded his work on isotropic turbulence as having a holistic coherence to it. Stuart (1986) recalls, in his essay on Keith Stewartson, that "[Stewartson's] philosophy, in a nutshell, was 'to chip away until something gives', but on problems of mathematical interest; if one was lucky, he felt, the contribution might be useful in a physical view also. I recall ... an earlier conversation between Stewartson and Taylor, Stewartson had explained to Taylor his personal philosophy for applied mathematics, as described above. To Keith's chagrin, Taylor's devastating reply had been 'that one did not make progress in science in that way'." However, there is little doubt that Taylor's focus was usually a specific physical question rather than an elegant formulation.

Taylor seems to have placed great value in seeing his theory verified by experiment. Nearly all his papers end with some comparison between his theory and experiment (or suggestions for the same), even if such comparisons were sometimes based on scanty evidence. He didn't seem to have much use for the paradigm of Karl Popper who argued that a scientific idea cannot be proven true by observation. It may be wrong even if a large number of observations agree with it, because there might yet be another that might disprove it: a single contrary experiment can prove a theory incontrovertibly false. Again, Taylor in his own words:

> I have tried to indicate the kind of interplay which I have used between mathematics and experiment. The late Professor G.H. Hardy regarded all applied mathematics as a dull activity, a sort of glorified plumbing, which could not give the kind of satisfaction he found in pure mathematics. My feeling is that I derive a rather similar kind of satisfaction from the interplay between applied mathematics and experiment. It is quite a different kind from the satisfaction one gets in doing something useful, though one derives an added pleasure when anything one does turns out accidentally to be of use in engineering.

In *The Legacy* Batchelor has described Taylor in various places as uncomplicated, modest, lovable, and gentle and whose "simplicity of character and outlook was a source of great scientific strength". There is no reason to think otherwise. But this image of Taylor comes more from his scientific life in the 'retirement phase', during which Taylor worked on problems that did not need large collaborations. Add to this Turner's (1997) description of Taylor as "shy with students, as if he did not know how to talk with them, either professionally or socially (except when he was showing them an experiment he was currently excited about)", and other descriptions that he was not a great conversationalist, never the one to take interest in introspection ("inattention may be expected if the conversation veered towards problems of philosophy or politics" – Southwell 1956) – and we get the impression of a somewhat detached man who lived to satisfy his call, not particularly interested in the world and people around him. There is also the image of Taylor as the scientist's scientist, one who dreamed up his own problems and pursued them without paying much attention to the trends of the day. In commenting on Taylor's tribute to Kármán,

Corrsin (1968) seems to have thought that the following remark of Taylor reflects a characteristic naivete on Taylor's part:

> In California I think less of his [Kármán's] time must have been absorbed in administrative duties than it was in Aachen, for he obviously read very widely and the reviews which he published of the then existing state of knowledge were particularly valuable.

Taylor, Corrsin thought, must have forgotten about the cadre of intelligent and eager assistants with whom Kármán had surrounded himself. One can perhaps speculate that Taylor's uncomplicated nature is consistent with the fact that he was not a particularly abstract thinker.

4.7.3 Taylor's style in the turbulence era

While the ability to discern the core of the problem, the directness of approach and naturalness of the flow of ideas were characteristic of Taylor's research at all points of his life, one may infer that his style of research in the period of interest to us here was perhaps slightly different from that in vintage years. During the latter phase, he was universally respected as an elder scientific statesman, had received many honors including the Order of Merit in 1969[30], and most of his scientific correspondents were by then younger people who adored and admired him. He was no longer working for the government but frequently consulted for the industry. And at no time during this period did Taylor have to fight for his place. That was not necessarily the case during his turbulence phase on which we are focused in this article.

During that period, Taylor took the cause of establishing his scientific reputation quite seriously and was, in fact, combative on occasion. For instance, V.W. Ekman[31] wrote a note to Taylor, on 17 May 1927, saying, "I have criticized your opinion on the relation between eddy conductivity and eddy viscosity, and I hope you will agree with me on this point." Ekman's criticism concerned two aspects. First, in Taylor (1915), of which Ekman had become aware only recently by virtue of a meeting between them in Zürich, Ekman

---

[30] The Order of Merit is perhaps the highest honor bestowed in the United Kingdom (or Commonwealth territories), and is limited to 24 living recipients and a small number of honorary members. This honor no doubt recognized Taylor's many contributions during the two war periods, among which was his exceptional work (Taylor 1950a,b) on self-similar blast waves and uncanny estimates of the tonnage of the 1945 nuclear explosion at Los Alamos, simply by studying published photographs.

[31] Ekman – whom Taylor consistently addressed as Eckman – was a Swedish physical oceanographer, whose work in 1905 had partly anticipated Taylor's results of 1915 on the spiral distribution of wind velocity in the friction layer of the atmosphere due to the effect of Coriolis forces. He was later a Professor of Mechanics in Lund.

inferred that Taylor had regarded the transport mechanisms of heat and momentum transport to be the same – a point of view with which he did not agree; second, Taylor's neglect of viscosity did not sit well with Ekman. Taylor was quick with a reply. While acknowledging that "perhaps I was not very clear in my exposition in 1915"[32], he was adamant that:

> I do not agree with your criticism of my deduction of the connection between heat and momentum transport ... I think that your statement about my method ... is incorrect. I hope that I have made it sufficiently clear that I do not and did not in 1915 state that there is an exact equivalence between the transfer of momentum and that of heat.

Ekman wrote back, in an exquisitely handwritten note of several pages, apologizing for misinterpreting Taylor, but the latter was still not satisfied: "I do not understand your section 6. I still see no reason to believe that your assumption ... is true." The last letter on record is from Ekman: "If you have now weighed the arguments put forth in my last letter of 18 July 1927, ... it might perhaps be better to discuss the matters in a periodical paper. In that case I think the public discussion ought to be opened by you ... "

I have already mentioned that Taylor returned to an examination of the eddy transport theories in the 1930s. It seems to me that this correspondence as well as that with Prandtl must have nudged him in that direction, but one does not get that sense of this external stimulus from reading Taylor's papers.

Also noted earlier is how Taylor did not yield floor to Kármán in his correspondence on self-preservation. Personal respect for Kármán notwithstanding, science was science and he would not allow anyone to misinterpret him. One can cite several other instances from Taylor's correspondence where he firmly stuck to his guns and cared a lot for his reputation. He would sometimes criticize others whom he regarded as having misinterpreted his theories. Taylor was thus an assertive scientist – certainly while he was working on turbulence – corresponding extensively with all the serious people of his time, advancing his ideas firmly, correcting wrong impressions of his work, giving his opinions freely, and making sure that the credit for his work was duly given. He kept up correspondence with other prominent people because it gave him scientific vitality. He also knew well when to give up a field and move on, rather than stay on beyond the period of freshness.

---

[32] It was indeed not clear. For instance, Richardson (1920) inferred that Taylor's suggestion was that they were equal. A study of the the literature, including Taylor's review (1938c), suggests that Taylor viewed somewhat inaccurately the Göttingen school's views on the similarities and differences concerning the momentum and scalar transport mechanisms. This topic, worth a further historical study, requires careful reading of the German originals by Prandtl, Tollmien, Reichardt, Nikuradse, Kármán and others.

It is not entirely accurate to say that Taylor's interests were confined to his science. Recall that he learned both to fly and parachute, no doubt in part because it was easier to make measurements as he flew (rather than have someone else do his bidding), but also because he loved the open air. His love for sailing is indeed legendary (his next love having been, perhaps, playing golf with Rutherford, F.W. Aston and R.W. Fowler – mostly "to hear Rutherford talk", according to *The Legacy*). Indeed, one of his letters to his future wife contains perspicuous comments on the metaphysical poet John Donne, and gives the impression of having derived much pleasure from reading Shakespeare, Milton and Shelley. It would seem that he was reading poetry for pleasure, at least at some point in his life.

It is also not the case that Taylor was indifferent to what others might think of him even by implication. From *The Legacy*, p. 211, we learn how incensed he was when he became aware of a remote resemblance to himself of a character in the fictionalization of the personalities of some scientists in the atom bomb era at Los Alamos: he wrote a strongly worded letter to the publishers asking for an apology. The publisher did tender one.

One further remark: one could not have expected letters of the sort Taylor wrote Prandtl in 1938 unless he was deeply concerned about politics of his day and their implications. It is hard to reconcile the heartfelt agony of those letters with Batchelor's assessment (*The Legacy*, p. 110) that Taylor "was not reflective, and moral or philosophical issues did not often engage his mind". These two letters, from which I have already quoted, show that Taylor was deeply touched by the horrors preceding the Second World War, and that he could express his views strongly even to the person he once held in the highest regard. No one alive then could have escaped the imprint of Second World War, and one may put the fervor of Taylor's remarks to that aberration. But the sheer power of those letters suggests that such thoughts could not have been exceptional to Taylor's mind.

Two other instances are probably worth quoting. Even later in life (12 December 1950) when the Royal Society was considering moving to a new site, he wrote a strong note with: "I most sincerely hope that the matter will be referred to the Fellows before any irrevocable step is taken". When a petition was being prepared in support of Benjamin Levich, who was under persecution by the Soviet authorities of the time, he wrote a letter thus:

> I thought I should say how highly scientists whose work lies in related fields value his contributions to electrohydrodynamics. Many of us regard him as the foremost scientist in the field and would feel that a serious blow has been delivered to science.

Note that this was after Taylor had suffered his stroke, and that he had never even met Levich!

4.7.4 Playful adventurousness as a key to creativity

It would be interesting to understand what in Taylor's youth could have foretold his looming greatness. Both Southwell (1956) and Batchelor (1996) have noted the genetic advantage that he inherited on the side of his mother (who was the second daughter of George Boole, the originator of the backbone of modern digital computers, the so-called Boolean logic). Already in grammar school, Taylor's "ability to work a subject out by himself, in practical work as well as in bookwork" was obvious to the headmaster (despite his poor performance in languages).

A key to his creativity seems to have been his playful adventurousness, in which he did not seem to take himself too seriously. Recall from *The Legacy* how at the start of his research career Taylor adjusted conditions of his first experiment on quantum interference to accommodate his sailing schedule. To quote from Southwell:

> in testing some home-made explosive he and a friend [who were both young lads then] blew a big hole in the garden door, and the wheel-barrow used as a container became a total loss. But this experience, too, seems to have been fruitful of ideas; for I have heard him [i.e. Taylor] recall an adventure at the Round Pond in Kensington Park Gardens, when a fine-ship constructed by the brothers Taylor blew up (by design) amidst a bevy of costly model yachts . . . , and the owners were led away sobbing by indignant nursemaids, gleefully watched by the two small buccaneers from an adjacent clump of shrubs.

He was not above cheerfully admitting to practicing "mild deception" (Taylor 1970) on the captain of the experimental ship *Scotia*, who might otherwise not have allowed meteorological kites to be flown ("well meaning, but oh so stupid!" – see *The Legacy*, p. 55). We have this from Southwell again: "Few men known to me have taken more or better holidays. None known to me has had a bigger output of creative work."

A few words should be said about Taylor's sense of humor, especially because his seriousness as a scientist has been inferred occasionally to preclude it. His correspondence in both directions is occasionally quite humorous. In a letter dated 12 November 1935, Prandtl complained about a paper on explosions, authored by three Englishman at Imperial College and asked Taylor to intervene and correct a mistake: "It would be less unacceptable for the gentlemen concerned, I think, to be corrected by an Englishman rather than by a foreigner". In response, Taylor responded:

I do not not feel competent to open a discussion with such an expert on explosions...feeling that perhaps Cambridge is hardly out of range of the force of such explosions [at Imperial]. I feel however that any detonation arising in the Imperial College, London, would have diminished in intensity by the time it reached Göttingen. I think that a note from you in the subject would be welcomed if you feel disposed to write one.

If one understands that Prandtl had not convinced Taylor that his assessment of the work in question was right, one can infer that the latter was being quite diplomatic here. Recall Taylor's innocent delight, reported in *The Legacy*, in humorously describing how Rutherford could speak no foreign language:

I only remember hearing him speak one sentence in a foreign language. He was showing his apparatus to a distinguished French mathematician who could speak no English. I saw him [Rutherford] point to a certain spot and say "ici les α-particles".

In one of his final letters to Burgers, he mischievously mentioned that his "memory of Tullio Levi-Civita is that he used to come to the congress meetings with a competent wife who was many sizes bigger than Tullio". We may well add the following remark of Binnie (1978):

One of the entertaining stories about him have come down to us by oral tradition. The scene was a London bus, the time March 1918 when the state of the war was desperate. Taylor dressed up as a Major and accompanied by Mrs. (later Lady) Southwell, boarded a bus, but they were unable to obtain adjacent seats, so their conversation had to be conducted in somewhat raised tones. In the course of it, Taylor remarked that the great thing about having two jobs is that, if you are not at one, people think that you are at the other.

This jocularity didn't end particularly well; to continue with Binnie:

At this, the other passengers became much incensed, an old lady struck him with an umbrella, and the conductor rang his bell to stop and said, "I think, Sir, you had better get off".

On 27 March 1949, Taylor humorously wrote to Batchelor, after meeting his parents in Australia, that the father related to the Taylors about "what the Botanists call the 'juvenile form' of the son". It does not seem to have amused the latter, as one can gather by the notation made upon receiving the letter!

4.7.5 A personal remark and closure

My sense is that Taylor, between the ages of thirty and fifty, say, was neither remote from major concerns of his day nor without human passions. He was most of all a serious scientist keen on maintaining his reputation – and this seems to have dwarfed his desire to express other passions. He organized scientific research, often exchanged ideas and happily used others' data to support his theories. He was also not above thinking, occasionally unjustifiably, that some fruitful new ideas were extensions of his own; and that, if an idea had occurred to him earlier in time, he was rightfully its owner, even if others had published it before he did.

These observations do not diminish Taylor's status as a giant. Many enduring scientists accumulate some clay on their feet during a lifetime of discharging their share of responsibilities: this debasing occurs as they wield their academic power (however petit!) and acquire fame through the trappings of committee work, academies and accolades collected through carefully mentored students and post-docs. It is also common for entrenched scientists who do not grow intellectually with time to hang on to their methods and ideas, and become a hindrance to younger scientists who do not toe the line. There are others who, despite living to a ripe old age, are not trapped by these trifles. Such scientists are the salt of the earth, on whose work the dignity of their subjects often rests. Dirac and Feynman come to mind readily, and Taylor clearly belonged to this category: and he escaped many of the trappings by holding, for most of his career, research positions without administrative bureaucracy. This warm feeling does not diminish even with a better understanding of Taylor's foibles.[33] I echo Batchelor (1996) in *The Legacy*: Taylor was indeed

> a happy man who spent a long life doing what he wanted most to do and doing it supremely well. He was a natural scientist whose character and activities were perfectly matched, and that allowed the fullest use of his creative talents.

Acknowledgments Writing this article has been a labor of love but has not been easy. I am grateful to my co-editors for assigning this task to me. The material from which I have cited in the article came from archives at the Libraries of the University of Maryland (Burgers); Caltech (Kármán); Trinity College, Cambridge (Taylor); DLR, Göttingen (Prandtl), and the Nobel Foundation. Professors J.M. Wallace, A. Leonard, H.K. Moffatt, E. Bodenschatz, and the officials of the Swedish Academy of Sciences (Nobel Archives)

---

[33] There are some scientists, especially pure mathematicians and theoretical physicists, who do great work but have short lives. They remain unsullied in our memories because, by the time fate snuffs out their lives prematurely, they have not had the time to fall behind in their science and become obsolete, or to accumulate enough academic follies to be a bane to succeeding generations. No armies of graduate students swarm their labs and no post-docs of theirs are launched into well-placed positions. They are famous simply because their work is glorious. We may cite Evariste Galois (who died at 20), Niels Henrik Abel (who died at 27), Srinivasa Ramanujan (who died at 32), Ettore Majorana (who disappeared and was presumed dead at 32), S. Pancharatnam (who died at 35), etc. They are pure gold and through them shines the untarnished image of science.

have helped me in procuring the relevant documents. I am thankful to them all. The essay was kindly read by Snezhana Abarzhi, Andy Acrivos, Guenter Ahlers, Nadine Aubry, Grisha Barenblatt, Ankit Bhagatwala, Eberhard Bodenschatz, Howard Brenner, Michael Brenner, Peter Davidson, Steve Davis, Russ Donnelly, Diego Donzis, Michael Eckert, Grisha Falkovich, Uriel Frisch, Carl Gibson, Shravan Hansoge, John Hinch, Julian Hunt, Yukio Kaneda, Brian Launder, Charles Meneveau, Keith Moffatt, Roddam Narasimha, Ronnie Probstein, Andrea Prosperetti, Ramakrishna Ramaswamy, Bill Saric, Jan Sengers, Ramamurti Shankar, Bhimsen Shivamoggi, Ladislav Skrbek, Lex Smits, Howard Stone, Harry Swinney, V. Vasanta Ram, James Wallace, John Wettlaufer, Victor Yakhot and P.K. Yeung. Some of them were kind to provide extensive commentaries on the draft, which have been quite helpful in preparing the revised version. Needless to say, the responsibility for all errors of omission and commission rest with me.

References

Andereck, C.D., Liu, S.S. & Swinney, H.L. 1986. Flow regimes in a circular Couette system with independently rotating cylinders. *J. Fluid Mech.* **164**, 155–183.

Ashford O. 1985. *Prophet or Professor: The Life and Work of L.F. Richardson.* Bristol: Hilger.

Barenblatt, G.I. 1979. *Similarity, Self-similarity, and Intermediate Asymptotics.* New York and London: Consultants Bureau, Plenum Press.

Barenblatt, G.I. 1996. *Scaling, Self-similarity, and Intermediate Asymptotics.* Cambridge University Press.

Barenblatt, G.I. 2006. What I remember and will remember forever. In *Kolmogorov in Memories of Disciples*, A.N. Shiryaev (ed.), Moscow: Publishing House MTsNMO, pp. 54–98 (in Russian).

Batchelor, G.K. 1951. Pressure fluctuations in isotropic turbulence. *Camb. Phil. Soc.* **47**, 359–374.

Batchelor, G.K. 1953. *The Theory of Homogeneous Turbulence.* Cambridge University Press.

Batchelor G.K. (ed.) 1960. *The Scientific Papers of Sir Geoffrey Ingram Taylor*, vols. I–IV. Cambridge University Press.

Batchelor, G.K. 1975. An unfinished dialogue with G.I. Taylor. *J. Fluid Mech.* **70**, 625–638.

Batchelor, G.K. 1976. Geoffrey Ingram Taylor. 7 March 1886–27 June 1975. *Biographical Memoirs of Fellows of the Roy. Soc. Lond.* **22**, 565–633.

Batchelor, G.K. 1986. Geoffrey Ingram Taylor. 7 March 1886–27 June 1975. *J. Fluid Mech.* **173**, 1–14.

Batchelor, G.K. 1996. *The Life and Legacy of G.I. Taylor.* Cambridge University Press.

Batchelor, G.K. & Proudman, I. 1954. The effect of rapid distortion on a fluid in turbulent motion. *Quart. J. Mech. Appl. Math.* **7**, 83–103.

Bateman, H. 1915. Some recent researches on the motion of fluids. *Monthly Weather Rev.* **43**, 163–170.

Bell, J.F. 1995. A retrospect on the contributions of G.I. Taylor to the continuum physics of solids. *Exper. Mech* **35**, 1–10. (The paper was presented by Bell as the 15th Sir Geoffrey Taylor Memorial Lecture at the University of Florida on 9 March 1980).

Benzi, R. 2011. Lewis Fry Richardson. Chapter 5 of this volume.

Binnie, A.M. 1978. Some notes on the study of fluid mechanics in Cambridge, England. *Ann. Rev. Fluid Mech.* **10**, 1–11.

Blasius, H. 1913. *Das Ähnlichkeitsgesetz bei Reibungsvorgängen in Flüssigkeiten.* Forsschungsarbeiten des Ver. Deutsch. Ing. No. 131, Berlin.

Bodenschatz, E. & Eckert, M. 2011. Prandtl and the Göttingen school. Chapter 2 of this volume.

Boussinesq, J. 1870. Essai théorique sur les lois trouvées expérimentalement par M. Bazin pour l'écoulment unifrome de l'eau dans les canaux découverts. *C. R. Acad. Sci. Paris* **71**, 389–393.

Brachet, M.E., Meiron, D.I., Orszag, S.A., Nickel, B.G., Morf, R.H. & Frisch, U. 1983. Small-scale structure of the Taylor–Green vortex. *J. Fluid Mech.* **130**, 411–452.

Burgers, J.M. 1939. Mathematical examples illustrating relations occurring in the theory of turbulent fluid motion. *Verhand. Kon. Neder. Akad. Wetenschappen, Afd. Natuurkunde, Eerste Sectie* **17**, 1–53.

Burgers, J.M. 1948. A mathematical model illustrating the theory of turbulence. *Adv. Appl. Mech.* **1**, 171–199.

Burgers, J.M. 1974. *The Nonlinear Diffusion Equation.* Dordrecht: D. Reidel.

Burgers, J.M. 1975. Some memories of early work in fluid mechanics at the Technical University of Delft. *Ann. Rev. Fluid Mech.* **7**, 1–12.

Chekhlov, A. & Yakhot, V. 1995. Kolmogorov turbulence in a random-force-driven Burgers equation. *Phys. Rev. E*, **51**, R2739–R2742.

Cichowlas, C. & Brachet, M.E. 2006. Evolution of complex singularities in Kida–Pelz and Taylor–Green inviscid flows. *Fluid Dyn. Res.* **36**, 239–248.

Cole, J.D. 1951. On a quasi-linear parabolic equation occurring in aerodynamics. *Quart. Appl. Math.* **9**, 225–236.

Coles, D. 1965. Transition in circular Couette flow. *J. Fluid Mech.* **21**, 385–425.

Comte-Bellot, G. & Corrsin, S. 1971. Simple Eulerian time correlation of full and narrow-band velocity signals in grid-generated, 'isotropic' turbulence. *J. Fluid Mech.* **48**, 273–337.

Corrsin, S. 1968. Review of: Theodore von Kármán with Lee Edson: *The Wind and Beyond. ISIS*, **59**, 240–242.

Dryden, H.L., 1943. A review of the statistical theory of turbulence. *Quart. J. Appl. Math.* **1**, 7–42.

Dryden, H.L. & Kuethe, A.M. 1929. *The measurement of fluctuation of airspeed by the hot-wire anemometer.* NACA Tech. Rep. 342.

Dryden, H.L., Schubauer, G.B., Mock, W.C. & Skramstad, H.K. 1937. *Measurements of intensity and scale of wind-tunnel turbulence and their relation to the critical Reynolds number of spheres.* NACA Tech. Rep. 581.

Eagle, A. & Ferguson, R.M. 1930. On the coefficient of heat transfer from the internal surface of tube walls. *Proc. Roy. Soc. Lond.* **127**, 540–556.

Eckert, M. 2010. The troublesome birth of hydrodynamic stability theory: Sommerfeld and the turbulence problem. *Eur. Phys. J. H* **35**, 29–51.

Eiffel, G. 1912. Sur la résistance des sphéres dans l'air en mouvement. *Compt. Rend.* **155**, 1587–1599.

Eyink, G.L. & Sreenivasan, K.R. 2006. Onsager and the theory of hydrodynamic turbulence. *Rev. Mod. Phy.* **78**, 87–135.

Eyink, G.L., Frisch, U., Moreau, R. & Sobolevskii, A. 2008. Euler Equations: 250 Years On. *Physica D* **237**, 1825–2246.

Falkovich, G. 2011. The Russian school. Chapter 6 of this volume.

Frisch, U. 1995. *Turbulence: The Legacy of A.N. Kolmogrov*. Cambridge University Press.

Goldstein, S. 1937. The similarity theory of turbulence, and flow between parallel planes and through pipes. *Proc. Roy. Soc. Lond.* **A159**, 473–496.

Goldstein, S. 1969. Fluid mechanics in the first half of this century. *Ann. Rev. Fluid Mech.* **1**, 1–29.

Gollub, J.P. & Swinney, H.L. 1975. Onset of turbulence in a rotating fluid. *Phys. Rev. Lett.* **35**, 927–930.

Hopf, E. 1950. The partial differential equation $u_t + uu_x = \mu u_{xx}$. *Commun. Pure Appl. Math.* **3**, 201–230.

Hunt, J.C.R. 1998. Lewis Fry Richardson and his contributions to mathematics, meterology, and models of conflict. *Ann. Rev. Fluid Mech.* **30**, xiii–xxxvi.

Kampé de Fériet 1939. Some recent researches on turbulence. In *Proc. 5th Int. Cong. Appl. Mech.* Cambridge, MA, pp. 352–355 .

Kardar, M., Parisi, G. & Zhang, Y.Z. 1986. Dynamical scaling of growing interfaces. *Phys. Rev. Lett.* **56**, 889–893.

Kármán, Th. v. 1930. Mechanische Ahnlichkeit und Turbulenz. *Nach. Ges. Wiss. Gottingen, Math.-Phys.* **Kl**, 58–76.

Kármán, Th. v. 1937a. On the statistical theory of turbulence. *Proc. Nat. Acad. Sci.* **23**, 98–107.

Kármán, Th. v. 1937b. The fundamentals of the statistical theory of turbulence. *J. Aero. Sci.* **4**, 131–138.

Kármán, Th. v. (with Lee Edson) 1967. *The Wind and Beyond*. Boston: Little, Brown and Co.

Kármán, Th. v. & Howarth, L. 1938. On the statistical theory of isotropic turbulence. *Proc. Roy. Soc. Lond.* **A164**, 192–215.

Keller, L.V. & Friedman, A.A. 1924. Differentialgleichung für die turbulente Bewegung einer kompressiblen Flüssigkeit. In *Proc. 1st Intern. Cong. Appl. Mech.* Delft, pp. 395–405.

King, L.V. 1914. On the convection of heat from small cylinders in a stream of fluid: determination of the convection constants of small platinum wires, with applications to hot-wire anemometry. *Proc. Roy. Soc.* **90**, 563–570.

Kolmogorov, A.N. 1941a. The local structure of turbulence in incompressible viscous fluid for very large Reynolds numbers. *Dokl. Akad. Nauk SSSR* **30**, 9–13. Reprinted in *Proc. Roy. Soc. Lond.* **A434**, 9–13.

Kolmogorov, A.N. 1941b. Dissipation of energy in locally isotropic turbulence. *Dokl. Akad. Nauk SSSR* **32**, 16–18. Reprinted in *Proc. Roy. Soc. Lond.* **A434**, 15–17.

Kolmogorov, A.N. 1962. A refinement of previous hypotheses concerning the local structure of turbulence in a viscous incompressible fluid at high Reynolds number. *J. Fluid Mech.* **13**, 82–85.

Lanchester, F.W. 1907. *Aerodynamics*. London: Constable & Co. Ltd.

Laufer, J. 1954. *The structure of turbulence in fully developed pipe flow*. NACA Tech. Rep. 1174.

Launder, B.E. & Jackson, D. 2011. Osborne Reynolds: a turbulent life. Chapter 1 of this volume.

Leonard, A. & Peters, N. 2011. Theodore von Kármán. Chapter 3 of this volume.

Lighthill, M.J. 1956. Viscosity effects in sound waves of finite amplitude. In *Surveys in Mechanics*, G.K. Batchelor & R.M. Davies (eds.), pp. 250–351.

Mock Jr., W.C. 1937. *Alternating current equipment for the measurement of fluctuation of air speed in a turbulent flow*. NACA Tech. Rep. 598.

Mock Jr., W.C. & Dryden, H.L. 1932. *Improved apparatus for the measurement of fluctuation of airspeed in turbulent flow*. NACA Tech. Rep. 448.

Moffatt, H.K. 2011. George Batchelor: the post-war renaissance of research in turbulence. Chapter 8 of this volume.

Narasimha, R. & Sreenivasan, K.R. 1979. Relaminarization of fluid flows. *Adv. Appl. Mech.* **19**, 221–301.

Nieuwstadt, F.T.M. & Steketee, J.A. 1995. *Selected Works of J.M. Burgers*. Dordrecht: Kluwer.

Orr, W.M. 1907. The stability or instability of the steady motions of a perfect liquid and of a viscous liquid. *Proc. Roy. Irish Acad. A* **27**, 9–68; 69–138.

Pippard, B.A. 1975. Obituaries: Sir Geoffrey Taylor. *Phys. Today* **28**, 67.

Polyakov, A.M. 1995. Turbulence without pressure. *Phys. Rev. E* **52**, 6183–6188.

Prandtl, L. 1905. Über Flüssigkeitsbewegung bei sehr kleiner Reibung. In *Verhandlungen des dritten Internationalen Mathematiker-Kongresses in Heidelberg 1904*, A. Krazer (ed.), Leipzig: Teubner, Leipzig, pp. 574–584. English translation in *Early Developments of Modern Aerodynamics*, J.A.K. Ackroyd, B.P. Axcell & A.I. Ruban (eds.), Oxford: Butterworth–Heinemann (2001), pp. 77–87.

Prandtl, L. 1914. Der Luftwiderstand von Kugelin. *Nachrichten der Gesselschaft der Wissenschaften zu Göttingen, Math.-Phys. Klasse*, 177–190.

Prandtl, L. 1925. Bericht über Untersuchungen zur ausgebildeten Turbulenz. *Z. Angew. Math. Mech.* **5**, 136–139.

Prandtl, L. 1927. Über den Reibungswiderstand strömer Luft. In *Ergebnisse der Aerodynamischen Versuchsanstalt zu Göttingen*. Munich, Berlin: Oldenbourg **3**, 1–5.

Prandtl, L. 1930. Attaining a steady air stream in wind tunnels. NACA Technical Memorandum No. 726, 1933. Originally published in *The Physics of Solids and Fluids*, P.P. Ewald, Th. Pöschl, L. Prandtl (eds.). English translation by J. Dougall and W.M. Deans. Glasgow: Blackie and Son, Ltd. (1930), p. 358.

Pullin, D.I. & Meiron, D.I. 2011. Philip G. Saffman. Chapter 12 of this volume.

Rayleigh, Lord 1892. On the question of stability of the flow of fluids. *Phil. Mag.* **34**, 59–70.

Reynolds, O. 1874. On the extent and action of the heating surface for steam boilers. *Proc. Manchester Lit. Phil. Soc.* **14**, 7–12.

Reynolds, O. 1883. An experimental investigation of the circumstances which determine whether the motion of water shall be direct or sinuous, and of the law of resistance in parallel channels. *Phil. Trans. Roy. Soc. Lond.* **174**, 935–982.

Reynolds, O. 1895. On the dynamical theory of incompressible viscous fluids and the determination of the criterion. *Phil. Tran. Roy. Soc. Lond.* **86**, 123–164.

Richardson, L.F. 1920. Some measurements of atmospheric turbulence. *Phil. Trans. Roy. Soc.* **221**, 1–29.

Richardson, L.F. 1922. *Weather Prediction by Numerical Methods*. Cambridge University Press.

Richardson, L.F. 1926. Atmospheric diffusion shown on a distance-neighbour graph. *Proc. Roy. Soc. Lond.* **A110**, 709–737.

Rouse, H. & Ince, S. 1957. *History of Hydraulics*. Dover Publications, Inc.

Saffman, P.G. 1960. On the effect of the molecular diffusivity in turbulent diffusion. *J. Fluid Mech.* **8**, 273–283.

Saffman, P.G. 1968. Lectures on homogeneous turbulence. In *Topics in Nonlinear Physics*, N.J. Zabusky (ed.), Springer, pp. 485–614

Saint-Venant, A.J.C. 1851. Formules et tables nouvelles pour les eaux courantes. *Ann. Mines* **20**, 49.

Schubauer, G.B. 1935. Air flow in a separating laminar boundary layer. NACA Tech. Rep. 527.

Schubauer, G.B. & Skramstad, H.K. 1947. Laminar boundary-layer oscillations and stability of laminar flow. *J. Aero. Sci.* **14**, 69–76.

Sengers, J.V. & Ooms, G. (eds) 2007. Autobiographical notes of Burgers. In *J.M. Burgers Centre at 15 years*, J.M. Burgers Centre, The Netherlands, pp. 20–59; also available at http://www.ipst.umd.edu/aboutus/history.php.

Shandarin, S.F. 1989. The large-scale structure of the universe: turbulence, intermittency, structures in a self-gravitating medium. *Rev. Mod. Phys.* **61**, 185–220.

She, Z.S., Aurell, E. & Frisch, U. 1992. The inviscid Burgers equation with initial data of Brownian type. *Commun. Math. Phys.* **148**, 623–641.

Simmons, L.F.G. & Salter, C. 1934. Experimental investigation and analysis of the velocity variations in turbulent flow. *Proc. Roy. Soc. Lond.* **A145**, 212–234.

Simmons, L.F.G. & Salter, C. 1938. An experimental determination of the spectrum of turbulence. *Proc. Roy. Soc. Lond.* **A165** 73–89.

Sinai, Ya. 1992. Statistics of shocks in solutions of inviscid Burgers equation. *Commun. Math. Phys.* **148**, 601–622.

Sommerfeld, A. 1908. Ein Beitrag zur hydrodynamischen Erklärung der turbulenten Flüssigkeitsbewegungen. *Proc. 4th Internat. Cong. Math. Rome* **3**, 116–124.

Southwell, R.V. 1956. G.I. Taylor: a biographical note. In *Surveys in Mechanics*. G.K. Batchelor & R.M. Davies (eds.), Cambridge University Press, pp. 1–6.

Spalding, D.B. 1962. An interview with Sir Geoffrey Taylor. *The Chartered Mech. Engineer* **9**, 186–191.

Sreenivasan, K.R. 1984. On the scaling of the turbulence energy dissipation rate. *Phys. Fluids* **27**, 1048–1050.

Sreenivasan, K.R. & Strykowski, P.J. 1983. Stabilization effects in flow through helically coiled pipes. *Exp. in Fluids* **1**, 31–36.

Stewart, R.W. & Townsend, A.A. 1951. Similarity and self-preservation in isotropic turbulence. *Phil. Trans. Roy. Soc. Lond.* **A243**, 359–386.

Stuart, J.T. 1986. Keith Stewartson: his life and work. *Ann. Rev. Fluid Mech.* **18**, 1–14.

Taylor, G.I. 1910. The conditions necessary for discontinuous motion in gases. *Proc. Roy. Soc. Lond.* **A84**, 371–377.

Taylor, G.I. 1915. Eddy motion in the atmosphere. *Phil. Trans. Roy. Soc. Lond.* **A215**, 1–26.

Taylor, G.I. 1916. Skin friction of the wind on the Earth's surface. *Proc. Roy. Soc. Lond.* **112**, 196–199.

Taylor, G.I. 1917a. The formation of fog and mist. *Quart. J. Roy. Met. Soc.* **43**, 241–268.

Taylor, G.I. 1917b. Phenomena connected with turbulence in the lower atmosphere. *Proc. Roy. Soc. Lond.* **94**, 137–155.

Taylor, G.I. 1917c. Observations and speculations on the nature of turbulent motion. *Reports and Memoranda of the Advisory Committee for Aeronautics*, **345**.

Taylor, G.I. 1918. Skin friction on a flat surface. *Reports and Memoranda of the Advisory Committee for Aeronautics*, **604**.

Taylor, G.I. 1921. Diffusion by continuous movements. *Proc. Lond. Math. Soc.* **20**, 196–212.

Taylor, G.I. 1923. Stability of a viscous liquid contained between two rotating cylinders. *Phil. Trans. Roy. Soc. Lond.* **A223**, 289–343.

Taylor, G.I. 1929. The criterion for turbulence in curved pipes. *Proc. Roy. Soc. Lond.* **A124**, 243–249.

Taylor, G.I. 1930. The application of Osborne Reynolds's theory of heat transfer to flow through a pipe. *Proc. Roy. Soc. Lond.* **A129**, 25–30.

Taylor, G.I. 1931a. Effect of variation in density on the stability of superposed streams of fluid. *Proc. Roy. Soc. Lond.* **132**, 499–523.

Taylor, G.I. 1931a. Internal waves and turbulence in a fluid of variable density. *Rapports et Procés-Verbaux des Réunions du Council Permanent International pour d'Exploration de la Mer* **76**, 35–42.

Taylor, G.I. 1932a. Note on the distribution of turbulent velocities in a fluid near a solid wall. *Proc. Roy. Soc. Lond.* **A135**, 678–684.

Taylor, G.I. 1932b. The transport of vorticity and heat through fluids in turbulent motion. *Proc. Roy. Soc. Lond.* **A135**, 685–705.

Taylor, G.I. 1935a. Statistical theory of turbulence. I. *Proc. Roy. Soc. Lond.* **A151**, 421–444.

Taylor, G.I. 1935b. Statistical theory of turbulence. II. *Proc. Roy. Soc. Lond.* **A151**, 444–454.

Taylor, G.I. 1935c. Statistical theory of turbulence. III. Distribution of dissipation of energy in a pipe over its cross-section. *Proc. Roy. Soc. Lond.* **A151**, 455–464.

Taylor, G.I. 1935d. Statistical theory of turbulence. IV. Diffusion in a turbulent air stream. *Proc. Roy. Soc. Lond.* **A151**, 465–478.

Taylor, G.I. 1935e. Turbulence in a contracting stream. *Z. Angew. Math. Mech.* **15**, 91–96.

Taylor, G.I. 1935f. Distribution of velocity and temperature between concentric rotating cylinders. *Proc. Roy. Soc. Lond.* **151**, 494–512.

Taylor, G.I. 1936a. Statistical theory of turbulence. V. Effect of turbulence on boundary layer. Theoretical discussion of relationship between scale of turbulence and critical resistance of spheres. *Proc. Roy. Soc. Lond.* **A156**, 307–317.

Taylor, G.I. 1936b. Correlation measurements in a turbulent flow through a pipe. *Proc. Roy. Soc. Lond.* **A157**, 537–546.

Taylor, G.I. 1936c. The mean value of the fluctuations in pressure and pressure gradient in a turbulent stream. *Proc. Camb. Phil. Soc.* **32**, 580–584.

Taylor, G.I. 1937a. Flow in pipes and between parallel planes. *Proc. Roy. Soc. Lond.* **A159**, 496–506.

Taylor, G.I. 1937b. The statistical theory of isotropic turbulence. *J. Aero. Sci.* **4**, 311–315.

Taylor, G.I. 1938a. Production and dissipation of vorticity in a turbulent fluid. *Proc. Roy. Soc. Lond.* **A164**, 15–25.

Taylor, G.I. 1938b. The spectrum of turbulence. *Proc. Roy. Soc. Lond.* **A164**, 476–481.

Taylor, G.I. 1938c. Turbulence. Chapter 5 of *Modern Developments in Fluid Dynamics*, vol. 1, S. Goldstein (ed.), pp. 191–233. Oxford University Press.

Taylor, G.I. 1939. Some recent developments in the study of turbulence. *Proc. 5th Int. Congr. Appl. Mech.*, Cambridge, MA. pp. 294–310, Wiley.

Taylor, G.I. 1950a. The formation of a blast wave by a very intense explosion. I. Theoretical discussion. *Proc. Roy. Soc. Lond.* **A201**, 159–174.

Taylor, G.I. 1950b. The formation of a blast wave by a very intense explosion. II. The atomic explosion of 1945. *Proc. Roy. Soc. Lond.* **A201**, 171–186.

Taylor, G.I. 1954. The dispersion of matter in turbulent flow through a pipe. *Proc. Roy. Soc. Lond.* **A223**, 446–468.

Taylor, G.I. 1970. Some early ideas about turbulence. *J. Fluid Mech.* **41**, 5–11.

Taylor, G.I. 1973a. Memories of Kármán. *J. Fluid Mech.* **16**, 478–480.

Taylor, G.I. 1973b. Memories of von Kármán. *SIAM Rev.* **15**, 447–452.

Taylor, G.I. 1974. The interaction between experiment and theory in fluid mechanics. *Ann. Rev. Fluid Mech.* **6**, 1–16.

Taylor, G.I. & Batchelor, G.K. 1949. The effect of wire gauze on small disturbances in a uniform stream. *Quart. J. Mech. Appl. Math.*, **2**, 1–26.

Taylor, G.I. & Cave, C.J.P. 1916. Variation of wind velocity close to the ground. *Reports and Memoranda of the Advisory Committee for Aeronautics*, **296**, part 1.

Taylor, G.I. & Green, A.E. 1937. Mechanism of the production of small eddies from large eddies. *Proc. Roy. Soc. Lond.* **A158**, 499–521.

Townend, H.C.H. 1934. Statistical measurements of turbulence in the flow through a pipe. *Proc. Roy. Soc. Lond.* **A145**, 180–211.

Townsend, A.A. 1954. The diffusion behind a line source in homogeneous turbulence. *Proc. Roy. Soc. Lond.* **224**, 487–512.

Townsend, A.A. 1956. *The Structure of Turbulent Shear Flow*. Cambridge University Press.

Townsend, A.A. 1990. Early days of turbulence research in Cambridge. *J. Fluid Mech.* **212**, 1–5.

Turner, J.S. 1997. G.I. Taylor in his later years. *Ann. Rev. Fluid Mech.* **29**, 1–25.

Uberoi, M.S. & Corrsin, S. 1953. Diffusion of heat from a line source in isotropic turbulence. NACA Tech. Rep. 1142.

White, C.M. 1929. Streamline flow through curved pipes. *Proc. Roy. Soc. Lond.* **A123**, 645–663.

Wiener, N. 1939. The use of statistical theory in the study of turbulence. *Proc. 5th Int. Congr. Appl. Mech.*, Cambridge, MA. pp. 356–358, Wiley.

van der Hegge Zijnen, B.G. 1924. *Measurements of the velocity distribution in the boundary layer along a plane surface.* Doctoral Dissertation, University of Delft.

Vogel-Prandtl, J. 1993. *Ludwig Prandtl: Ein Lebensbild, Erinnerungen, Dokumente,* Max-Planck Institute Report 107, Göttingen.

Wiener, N. 1933. *The Fourier Integral and Certain of Its Applications,* Cambridge University Press, (see p. 70, Theorem 9.09).

Yaglom, A.M. 1994. A.N. Kolmogorov as a fluid mechanician and founder of a school in turbulence research. *Ann. Rev. Fluid Mech.* **26**, 1–23.

Zel'dovich, Ya.B. & Raizer, Yu.P. 1969. Shock waves and radiation. *Ann. Rev. Fluid Mech.* **1**, 385–412.

Ziegler, M. 1930. The application of hot wire anemometer for the investigation of the turbulence of an airstream. *Versl. d. Kon. Akad. v. Wetensch Amsterdam* **33**, 723-736.

5

Lewis Fry Richardson

Roberto Benzi

5.1 Introduction

The nature of turbulent flow has presented a challenge to scientists over many decades. Although the fundamental equations describing turbulent flows (the Navier–Stokes equations) are well established, it is fair to say that we do not yet have a comprehensive theory of turbulence. The difficulties are associated with the strong nonlinearity of these equations and the non-equilibrium properties characterizing the statistical behaviour of turbulent flow. Recently, as predicted by von Neumann 60 years ago, computer simulations of turbulent flows with high accuracy have become possible, leading to a new kind of experimentation that significantly increases our understanding of the problem. The largest numerical simulations nowadays use a discretized version of the Navier–Stokes equations with several billion variables producing many terabytes of information that may be analyzed by sophisticated statistical tools and computer visualization. None of these tools were available in the 1920s when some of the most fundamental concepts in turbulence theory were introduced through the work of Lewis Fry Richardson (1881–1953). Although his name is not as well-known as other contemporary eminent scientists (e.g. Einstein, Bohr, Fermi) and although his life was spent outside the mainstream of academia, his discoveries (e.g. the concept of fractal dimension) are now universally known and essential in understanding the physics of complex systems.

Detailed biographies of Richardson (Ashford (1985); Hunt (1987)) are available to the interested reader. Our aim here is simply to review his most important scientific ideas and to understand how the concepts that he introduced transformed our view of turbulent flow and, more generally, of complex phenomena. The significance of Richardson's life has been well summarized by Hunt (1987) in the following terms:

Figure 5.1 L.F. Richardson.

His name can be added to the list of inventors of computational mathematics (with von Neumann, Courant, and Turing), of modern meteorology and fluid mechanics (with Bjerknes, Taylor, and Prandtl), of quantitative techniques in psychology and social sciences (with James), and of analysis and modelling of complex systems (with Norbert Wiener).

Richardson was born on 11 October 1881, of Quaker parents, in Newcastle-upon-Tyne. After two years education at Newcastle University, he proceeded in 1900 to King's College, Cambridge, and graduated in 1903 with a first-class degree. From 1903 to 1913, Richardson held a series of short research posts. During this period, in order to study percolation of water through peat, he developed a new method for solving Laplace's equation in complex geometry, based on a suitably discretized form of the continuous equation. Richardson submitted this work as a dissertation for the Fellowship competition at King's College; his ideas were however simply too new for the scientific community; he did not get a Fellowship and never subsequently worked in any of the main centres of academic research.

In 1913 Richardson was appointed Superintendent of the Eskdalemuir Observatory in southern Scotland but, at the outbreak of the First World War, he left this position to join the Friends' Ambulance Unit in France.

During the period 1903–1929, Richardson formulated new theoretical concepts and ideas that changed our understanding of the physics of turbulence.

His work was not at the time much appreciated by the scientific community, although many distinguished scientists were well aware of his creativity and his achievements. There are perhaps several reasons why Richardson remained outside the academic mainstream. As previously remarked, his ideas were simply too new for most of his contemporaries. In the period 1900–1930, physics underwent two dramatic and exciting revolutions triggered by the discoveries of quantum mechanics and relativity. An equally exciting revolution occurred much later in the 1970s through the discovery of chaotic behavior in dynamical systems, the statistical mechanics of complex systems, and the universality properties in second-order phase transitions. Many new discoveries were triggered by a combination of these new theoretical ideas coupled with computer simulations based on accurate and efficient numerical schemes for solving complex nonlinear dynamics. Richardson was an early pioneer in this field. However, without computers, his numerical schemes could not be implemented, and it was perhaps for this reason that his ideas and pioneering work were not fully appreciated at the time.

In order to appreciate Richardson's contribution to the study of turbulence, one should read his original papers. It is an interesting experience from which one may recognize that Richardson had a profound respect for any question that is scientifically meaningful no matter how strange or naive the question may appear to be. For Richardson, science consisted in investigating any such question using observation or experimental data and in building a mathematical framework capable of explaining such data. In all his papers, curiosity was the driving force, coupled with the motive of providing useful results for real applications. With these initial conditions satisfied, science (for Richardson) reaches its highest level and many extraordinary tasks can be accomplished.

After a short period at the Meteorological Office in Benson, Richardson became a Lecturer at Westminster Training College (then near Westminster Abbey) and in 1929 he moved north to the Technical College in Paisley, an industrial town near Glasgow in Scotland. Although Richardson was elected Fellow of the Royal Society in 1926, there was "at that time ... no academic position available for someone wishing to pursue research in meteorology ..." (Hunt (1987)).

Richardson was, like his parents, a Quaker and a committed pacifist, and this had a profound influence on his whole life: as a Conscientious Objector during the First World War, he chose as mentioned above to work in France as an ambulance driver; after the war, he abandoned his studies on dispersion (even going so far as to burn his notes on the subject) when he discovered that the military were using his ideas to aid their understanding of the dispersion of poisonous gas; he resigned from the Meteorological Office in 1926 when it was put under military control. Most significantly, after 1929, Richardson's interest switched to studying, from a scientific point of view, how to prevent warfare, a field of research that he virtually created. (This may appear remote from his research on turbulence; however, in attempting to analyze a great variety of war-related data, he introduced a new concept with far-reaching consequences, namely the concept of fractal dimension.) He retired in 1943 and lived in Hill House at Kilmun, about 25 miles from Glasgow, where he died on 30 September 1953, having spent the last year of his life working on a mathematical theory of war and conflict.

Richardson's achievements are now well-established and he is recognized as one of the leading scientists of the 20th century. In the following discussion, we shall review his three most important results for turbulence:

- The diffusion law in turbulent flows
- The energy cascade in turbulence and numerical weather forecasting
- The introduction of fractal dimension

In reviewing the ideas underlying these results, we shall outline some basic physical laws concerning turbulence. It is notable that some of the problems discussed in Richardson's original papers are still to this day matters of active current research.

5.2 The 4/3 law

Everyone knows that in order to mix two fluids thoroughly, it is usual to generate turbulence in some way to accelerate the process. All who enjoy preparing cocktails know that shaking is a good way to improve mixing! It is not so well-known that without shaking, mixing between two fluids (through molecular diffusion) may require hours, or even days. These elementary facts provide a starting point for many courses on turbulent flow. It is not easy however to reach a satisfactory explanation of just why and how mixing is enhanced by turbulence.

Richardson's (1926) paper made important progress in this respect. It is important to understand his result within the theoretical and experimental framework of the period. In 1905, Einstein had derived an important theoretical relation between the diffusivity of a small particle in a fluid at rest and its mass and radius. Einstein's relation is based on the atomic theory of matter, which in 1905 was still not yet fully accepted. Using Einstein's relation, it is possible to derive Avogadro's number, and in 1908 Jean Perrin found that the result of this derivation agreed with experiment, thus providing strong support for the atomic theory of matter.

Einstein's basic idea was to recognise that diffusivity is the best quantity to observe in an experiment. Diffusivity is defined as

$$K = \lim_{t \to \infty} \frac{|X(t) - X(0)|^2}{2t} \tag{5.1}$$

where $X(t)$ denotes the observed displacement of the particle in a fixed direction at time t. By introducing the concentration density $C(\mathbf{x}, t)$ as proportional to the probability of finding a particle at position $\mathbf{x}$ at time t (given the initial position to be $\mathbf{x} = 0$), the diffusivity enters as a constant in the macroscopic evolution of $C(\mathbf{x}, t)$:

$$\frac{\partial C}{\partial t} = K \nabla^2 C, \tag{5.2}$$

i.e. the well-known diffusion equation. The question raised by Richardson (1926) starts with the experimental observation that if we assume equation (5.2) to be valid for turbulent flows, then the so-called 'constant' K can vary by a factor of anything from two to a billion, according to experimental data. Thus we cannot blindly apply the concept of diffusivity for turbulent flows: something new and non-trivial is happening which fundamentally modifies the diffusive process.

Richardson argued that in a turbulent flow, diffusivity is enhanced because two Lagrangian particles separated at $t = 0$ by some initial distance $R(0)$ may behave completely differently as the result of advection by two different 'wind gusts' or 'eddies'. It is not therefore possible to apply the theory of Brownian motion as developed by Einstein, without explicitly considering the small-scale turbulent fluctuations. Today we speak of 'eddy diffusivity' whenever the diffusion constant K emerges by some kind of averaging over turbulent fluctuations. Richardson's brilliant idea was to discuss diffusion in a turbulent flows in terms of the separation between different particles. It is an important conceptual step which is worth discussing in detail. Figure 5.2, from Richardson's original paper, illustrates the qualitative ideas that he introduced. In his own words:

> a small dense cluster of marked molecules, represented by the dots of fig. 1, which, by molecular diffusion alone, would spread through successive spherical clusters, shown in fig. 2 and fig. 3, actually seldom passes through the large spherical stage 3, because it is first sheared in two detached clusters as suggested in fig. 4. These are carried far away from one another, and are likely to be again torn into smaller pieces as in fig. 5.

Given two particles whose initial separation is $R(0)$, molecular diffusion and turbulent advection conspire to increase their separation in time. The separation $R(t)$ at time t is a suitable variable for describing both effects. If we consider a large number, N, of particles, diffusion in turbulent flows can be

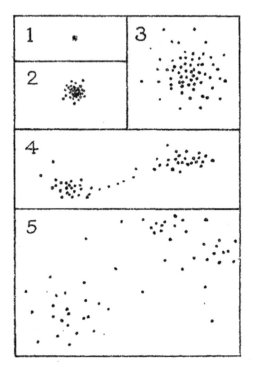

Figure 5.2 The figure shows the qualitative behaviour of particle dispersion in a turbulent flow. Richardson showed this figure in his original 1926 paper in order to illustrate the effect of wind gusts on particle dispersion.

described in terms of the root-mean-square (RMS) separations between these particles. This point of view is radically different from that describing diffusion in terms of the single-particle probability distribution $C(\mathbf{x}, t)$. Richardson pointed out that the information to look for can be obtained through the quantity $P_2(\mathbf{l}, t)$ defined as the probability of finding a particle at the point $\mathbf{x} + \mathbf{l}$ given that there exists a particle at point $\mathbf{x}$. In the language of physics, we call P_2 the 'two-point correlation function'. Richardson in fact showed that P_2 can be computed as

$$P_2(\mathbf{l}, t) = \frac{\int d\mathbf{x}_l C(\mathbf{x}_l, t) C(\mathbf{x}_l + \mathbf{l}, t)}{\int d\mathbf{x}_l C(\mathbf{x}_l, t)}, \tag{5.3}$$

thus relating P_2 to the two-point correlation function of the concentration field. If there is no turbulence, the function P_2 satisfies (5.2) with a diffusivity constant equal to $2K$. Richardson argued that in general, due to turbulence, the

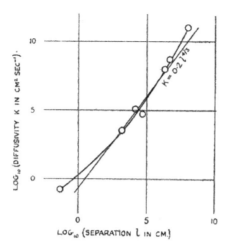

Figure 5.3 In his 1926 paper, Richardson computed the diffusivity $K(l)$ as a function of l using different experimental results available at that time. The 'scale' l in the figure refers to the RMS displacement of the particles. The figure, taken from the original 1926 Richardson's paper, is a log–log plot of K versus l and the straight line is Richardson's fit to the data leading to the famous 4/3 law.

correct expression should be

$$\frac{\partial P_2}{\partial t} = \sum_i \frac{\partial}{\partial l_i}\left(K(l)\frac{\partial}{\partial l_i}P_2\right), \tag{5.4}$$

i.e. diffusivity in a turbulent flow must depend on the 'scale' l at which we look at the phenomenon. The function $K(l)$ must depend on l in order to properly take account of the effect of turbulence at small scales.

Richardson's next step was to analyse observations of diffusion in the atmosphere in order to understand the functional form of $K(l)$. In his paper, Richardson quoted different observational results and presented them all in a log–log plot reproduced in Figure 5.3. With reference to this figure, Richardson wrote

> The straight line on the logarithmic diagram which corresponds to $K = 0.2\, l^{4/3}$ also fits the observation as well as the curve in the limited range between $l = a$ metre and $l = 10$ kilometres. For mathematical simplicity this formula will be used in the illustrations which follow.

Reading Richardson's papers, one may see that he was in general reluctant to express numbers derived from observations as 'vulgar' fractions like 4/3. Indeed, in his book, *Weather Prediction by Numerical Process*, we find many places where vulgar fractions are written by preference in the decimal form, e.g. 0.2857 rather than 2/7. Actually this is one point that makes Richardson's papers difficult to read. Even in his 1926 paper, Richardson does not use any vulgar fraction in fitting the data of Figure 5.3. According to his previous work, one would expect to find the fit given by $l^{1.31}$ or with some other exponent in decimal form *close* to 4/3. The data in the figure do not even indicate that a power law will provide the best fit! It is therefore all the more remarkable that Richardson made his first and perhaps only 'simplification' by choosing $l^{4/3}$, a significant choice because it is the only one compatible with the fundamental ideas leading to the Kolmogorov (1941) theory on homogeneous isotropic turbulence! Kolmogorov conjectured that in such turbulence the probability distribution of velocity fluctuations (in the 'inertial range' of scales) should depend only on the average rate of energy dissipation ϵ_K (the suffix K here chosen to denote Kolmogorov) and the scale l. Kolmogorov further assumed that ϵ_K is independent of viscosity (or equivalently of Reynolds number). Since the physical dimension of ϵ_K is length^2/time^3 and K has dimensions length^2/time, one immediately obtains for K the estimate $K = g\epsilon^{1/3}l^{4/3}$, where g is a numerical constant of order unity. So Richardson's simplification in fitting the data of Figure 5.3 is quite special, and we may conjecture that he was aware, at least intuitively, of possible links between the diffusion constant and the rate of energy dissipation. It is clear that by assuming the exponent 4/3 as a good fit for the observational data of Figure 5.3, the numerical prefactor, i.e. the number 0.2, actually has the physical dimensions length$^{2/3}$/time. Knowing how insistent Richardson was in understanding the physics behind any numerical results, we may conjecture that he must have thought about the physical meaning of the constant 0.2 and how it is related to turbulence. From an historical perspective, it is also interesting to remark that the fitting law for Figure 5.2 was written by Richardson as "$\epsilon l^{4/3}$ where $\epsilon = 0.2 \text{cm}^{2/3}/\text{sec}$". The choice of the Greek letter ϵ, anticipating Kolmogorov's choice of symbol for the rate of energy dissipation, is also intriguing.

In 1949, together with H. Stommel, Richardson published a short note in the *Journal of Meteorology* entitled *Note on Eddy Diffusion in the Sea* (Richardson & Stommel (1949)). The idea was to apply the results of the 1926 paper to the ocean, using a new data set and new information. In their note, Stommel and Richardson observed that the 1926 results are roughly validated by the new data, and that "any power between $l^{1.3}$ and $l^{1.5}$ would be a tolerable fit to the observations". Thus, the requirement for a fit $\sim l^{4/3}$ was no longer regarded as mandatory, nor was it made for simplicity. At the end of the paper, the authors wrote:

> After this manuscript was submitted the writers have read two unpublished manuscripts by C.L. von Weizsäcker and W. Heisenberg in which the problem of turbulence for large Reynolds number is treated deductively with the results that

they arrive at the 4/3 law. The agreement between von Weizsäcker and Heisenberg's deduction and our quite independent induction is a confirmation of both.

These results of von Weizsäcker and Heisenberg come from their version of the 'Kolmogorov' (1941) theory, developed independently but leading to essentially the same results.

Concerning the work with Stommel, Hunt (1987) has related a nice story: while at Kilmun, Richardson

> received the famous visit by H. Stommel, from Woods Hole, which was the opportunity to return to the question of how pairs, or a cloud, of particles separate in a turbulent flow, such as he saw from his house every day on the waters of the Holy Loch. They threw parsnips from a small pier into Loch Long, and using a remarkable measuring instrument they confirmed that the rate of spreading increased as the distance between the pairs of parsnips increased, consistent with the four-thirds law.

Whatever ideas and intuitions Richardson had in fitting the observational data, he made an important note on the constant ϵ in front of the $l^{4/3}$ factor:

> A more detailed study will reveal variations of 10 times or more in ϵ according to the up-gradient of temperature and mean-wind or other circumstances. Even so ϵ will be remarkably more constant than the diffusivity K for the concentration, which, as we have seen, varies with l a billion of times.

Since $\epsilon^3 \sim \epsilon_K$, the rate of energy dissipation, Richardson's remark perhaps reveals the first experimental evidence for the result that the rate of energy dissipation in fully developed turbulence is almost independent of the Reynolds number, a fundamental cornerstone of our understanding of turbulence.

One consequence of Richardson's result is highly nontrivial. Consider the RMS separation $l \equiv \langle R^2(t) \rangle^{1/2}$ between two Lagrangian particles, where the symbol $\langle \cdots \rangle$ stands for the average over many realizations. Then Richardson's diffusion implies

$$\frac{d}{dt}\langle R^2(t) \rangle = \epsilon \langle R^2(t) \rangle^{2/3} \qquad (5.5)$$

(note that the right-hand side of (5.5) has the dimension of a diffusivity). Equation (5.5) can be easily solved in terms of $R(0)$, the distance between the two particles at time $t = 0$:

$$\langle R^2(t) \rangle = [R^2(0)^{1/3} + 3\epsilon t]^3. \qquad (5.6)$$

By equation (5.6), the average distance between two Lagrangian particles in a fully developed turbulent flow increases algebraically in time even if its initial value is extremely small. According to Richardson's picture in Figure 5.2, the initial growth of $R(t)$ is controlled by molecular diffusion (panels 1 and 2 in the figure), while at later time R grows much faster (compared to molecular diffusion alone), namely as t^3. This picture implies that there exists some characteristic time scale τ_η such that for $t \gg \tau_\eta$ the effect of molecular motion becomes negligible and diffusion is dominated by turbulence. The notion of τ_η is not introduced in Richardson's paper but is an outcome of Kolmogorov's theory. In particular, τ_η can be identified as $(v/\epsilon_K)^{1/2}$ where v is the kinematic viscosity of the fluid. Notice that in the limit of large Reynolds number, i.e. in the limit $v \to 0$, $\tau_\eta \to 0$, the result $\langle R^2(t) \rangle \sim t^3$ may be expected to hold even for extremely small $R(0)$.

One way to appreciate the relevance of the above results is to consider the velocity difference $\delta\mathbf{v}(\mathbf{R}) \equiv \mathbf{v}(\mathbf{x}(t) + \mathbf{R}(t)) - \mathbf{v}(\mathbf{x}(t))$, where $\mathbf{x}$ and $\mathbf{x} + \mathbf{R}$ are the positions of the two particles and $\mathbf{v}(\mathbf{x}(t))$ and $\mathbf{v}(\mathbf{x}(t) + \mathbf{R}(t))$ the respective velocities. Since

$$\frac{d}{dt}\mathbf{x} = \mathbf{v}(\mathbf{x}, t), \tag{5.7}$$

$$\frac{d}{dt}(\mathbf{x} + \mathbf{R}) = \mathbf{v}(\mathbf{x} + \mathbf{R}, t), \tag{5.8}$$

by subtracting (5.7) from (5.8) we obtain

$$\frac{d\mathbf{R}}{dt} = \delta\mathbf{v}(\mathbf{R}). \tag{5.9}$$

Finally, multiplying (5.9) by R and averaging, we obtain

$$\frac{d\langle R^2 \rangle}{dt} = 2\langle \delta\mathbf{v}(\mathbf{R}) \cdot \mathbf{R} \rangle. \tag{5.10}$$

Comparing (5.9) with (5.5), we obtain $\epsilon\langle R^2 \rangle^{2/3} = 2\langle \delta v(R)R \rangle$. Therefore, for Richardson's diffusion to be true, we require that

$$\delta v(R) \sim \epsilon R^{1/3}, \tag{5.11}$$

which is the fundamental result derived by Kolmogorov (1941). The prediction of (5.6), reinterpreted by using (5.11), implies that in the limit of very large Reynolds number, the velocity difference is not differentiable, so that even for very small R (but still larger than the scale where molecular dissipation is important) the velocity difference cannot be simply proportional to R. This leads one to consider how to define the velocity field at very large Reynolds number, a consideration that probably led Richardson to open his 1926 paper with a section entitled *Does the wind possess a velocity?*. He wrote "The question, at first sight foolish, improves on acquaintance" and he gave an interesting and nontrivial example to illustrate his meaning, which is worth repeating here.

Consider the Weierstrass function

$$x(t) = kt + \sum_n \frac{1}{2^n} \cos(5^n \pi t).$$ (5.12)

For any t, the value of x is well defined since $\sum_n 2^{-n}$ is convergent. However, the *velocity dx/dt* is not defined since $\sum_n (5/2)^n$ is divergent. In this way, Richardson suggested that if we look at a turbulent velocity field, it may happen that the velocity differences at very small separations are similarly non-differentiable. This is precisely the physical meaning of (5.11). This observation, which is the starting point for a discussion of diffusion as a *scale-dependent quantity*, is crucial and it points directly at the heart of what we may call the turbulence problem. At very small scale, arbitrarily small for infinite Reynolds number, the velocity difference of two Lagrangian particles divided by their separation can be very large. Thus we are back to the notion, introduced and clarified later by Kolmogorov, that fully developed turbulence has a finite rate of energy dissipation in the limit of infinite Reynolds number. In modern language, we now refer to this important feature of fully developed turbulence as the *dissipation anomaly*. Actually, in seeking to provide a clear, precise answer to Richardson's question, a major breakthrough in the theoretical understanding of Lagrangian dynamics has been recently achieved (see the review by Falcovich et al. (2001)). This work has shown that Richardson's intuition concerning the nature of the turbulent velocity field was by no means *foolish*, and indeed has led to deep and important developments.

Particle dispersion in turbulent flow continues to be an active field of research, as reviewed by Salazar & Collins (2009). Great progress has been made in experiments that follow single-particle dynamics in turbulent flow, while large-scale direct numerical simulations of homogeneous isotropic turbulence are currently performed with several million Lagrangian particles, as reviewed by Toschi & Bodenschatz (2009). According to Salazar & Collins (2009), "Nothing has effectively challenged the existence of the Richardson regime. Similarly, there has not been an experiment that has unequivocally confirmed the Richardson scaling over a broad-enough range of time and with sufficient accuracy". The major difficulty in measuring Richardson scaling can be understood by the following argument. This scaling implies that the dispersion of two particles grows in time as t^3. Thus, the simplest way to verify this is by measuring $\langle R^2(t) \rangle$, where $\langle \cdot \rangle$ is the average over many particles and many different initial separations. Following our previous discussion, we may argue that Richardson scaling should hold within the time interval $[\tau_\eta, T_L]$ where τ_η is the Kolmogorov dissipation time-scale (i.e. the time-scale at which molecular effects are important) while $T_{\rm L}$ is the longest time-scale of the turbulent flow. Similarly, the range of spatial scales relevant for Richardson scaling are within the interval $[\eta, L]$ where η is the Kolmogorov dissipation scale while L is the largest spatial scale of the turbulent flows. For a turbulent flow at some Reynolds number Re, the ratio $L/\eta \sim Re^{3/4}$ while $T_{\rm L}/\tau_\eta \sim Re^{1/2}$, which implies $T_{\rm L}/\tau_\eta \sim (L/\eta)^{2/3} \sim Re_\lambda$, where Re_λ is the Reynolds number based on the Taylor scale. A reasonably good Kolmogorov scaling (5.11) is observed in numerical simulations and in laboratory experiments for $Re_\lambda \sim 500$ which implies $L/\eta \sim 10^4$. However, in order to achieve $T/\tau_\eta \sim 10^4$, one needs to perform experiments or numerical simulations at $Re_\lambda \sim 10^4$ or equivalently with $L/\eta \sim 10^6$, which cannot be achieved at present. The same result is obtained by following the analysis due to Batchelor (1950). Let r_0 be the characteristic scale of the initial separation of two Lagrangian particles, with r_0 in the inertial range, i.e. $\eta \ll r_0 \ll L$. Then, by using Kolmogorov's theory, Batchelor (1950) showed that Richardson scaling (5.6) should be observed for $t \geq t_0 \equiv r_0^{2/3}\epsilon^{-1/3}$, while for $t \leq t_0$ one should observe $\langle R^2(t)\rangle \sim t^2$, usually referred to as the Batchelor regime. In most laboratory experiments – see for instance Ouellette et al. (2006) and Salazar & Collins (2009) – t_0 lies in the range 50–$100\tau_\eta$ which implies that a reasonable Richardson scaling can be observed for $T_{\rm L} \sim 100 t_0 \sim 10^4 \tau_\eta$.

Using direct numerical simulations of isotropic turbulence, the situation can be improved as follows: let T_n be the time needed for two particles to separate from distance R_n to distance $R_{n+1} = AR_n$, where $A > 1$ is a fixed number. Then, upon averaging over many particles, the quantity $\langle T_n(R_n)\rangle$ should behave as $R_n^{2/3}$. This method has been successfully applied by Boffetta & Sokolov (2002) and Biferale et al. (2005) whose numerical results are consistent with Richardson scaling. In the light of the above discussion, it is all the more remarkable that Richardson was able to achieve so much, more than 80 years ago.

We may safely conclude by saying that Richardson's 1926 paper on diffusion is one of the key starting points in our understanding of the physics of turbulence. It introduces the concept of eddy diffusivity and its scale-dependent behaviour, and it poses some fundamental questions about turbulence. The concept of scaling is also introduced in this paper within the context of turbulent flow. Richardson's findings are still considered to be broadly valid, and major computational and experimental efforts have been devoted to verifying his early 1926 results. As we have discussed, he was close to discovering the role of the rate of energy dissipation in defining the statistical properties of turbulence, as later developed by Kolmogorov. We infer that Richardson had a good intuitive understanding of the consequences of his findings and their wide implications.

5.3 Richardson cascade and numerical weather prediction

One of Richardson's major works is without doubt his 1922 book *Numerical Weather Prediction by Numerical Processes*, which provides a comprehensive treatment of the methods and tools that can be used for weather forecasting, starting from the basic dynamic and thermodynamic equations of the atmosphere. This book may be regarded as the very first attempt at computational fluid dynamics (CFD), now a well-defined and specialized scientific field. It is no trivial task to read the book, which requires prior knowledge of dynamic meteorology, turbulence and numerical analysis. It contains many new concepts and ideas taken as starting points for the construction of operational weather services based on numerical forecasting. Here we shall discuss some of the questions and results highlighted by Richardson. Among them, one important concept that he introduced: the energy cascade in turbulent flows.

Citations on Richardson's cascade often quote his humorous rhyme:

> Big whirls have little whirls
> that feed on their velocity.
> Little whirls have lesser whirls
> and so on to viscosity
> – in the molecular sense.[1]

This rhyme appears in a paragraph of the book dealing with the effect of eddy motion and it conveys the idea that the energy contained at some scale is transferred to progressively smaller scales until ultimately dissipated by viscosity. In order to obtain the relevant equation of motion for the large-scale flows of the atmosphere, one needs to parametrize the effect of turbulence in some way. The concept of the energy cascade indicates that this parametrization must take account of how energy is dissipated in a turbulent flow.

We note that Richardson's book was published well before his 1926 paper; thus, he had already conceived the energy cascade before his famous $l^{4/3}$ law.

What then is this energy cascade in turbulent flow? The qualitative picture due to Richardson suggests that the characteristic scale of energy input by external forces (boundaries or pressure gradients) is much larger than the scale at which viscous effects, due to molecular motion, become relevant. Therefore, there must be a flux of energy from large to small scales which is due to the

---

[1] Richardson's oft-quoted rhyme is a parody of the verse of the Irish satirist Jonathan Swift:
So, naturalists observe, a flea
Hath smaller fleas that on him prey;
And these have smaller fleas to bite 'em
And so proceed ad infinitum.
Thus every poet in his kind,
is bit by him that comes behind.

strong nonlinearity of the Navier–Stokes equations. One can imagine a large-scale flow which, due to nonlinear instability, is broken down into smaller scale flows. These undergo a similar fate leading to yet smaller scale dynamics and so on until dissipative effects become dominant. The Richardson cascade has been quoted by Kolmogorov (1962) and it is likely that Kolmogorov was aware of the concept when working on his 1941 theory. It is a central concept of turbulence which has been questioned and discussed from time to time in many papers.

Another issue was of key relevance at the time. When Richardson was writing his book, the Norwegian school of meteorology, led by J. Bjerknes, had developed the concept of fronts and frontal development in mid-latitude circulation. The starting point for Bjerknes was the dynamics of sharp temperature gradients observed in front formation associated with convergence of low-level circulations. In the introduction to his book, Richardson wrote:

> V. Bjerknes and his collaborators J. Bjerknes, H. Solberg and T. Bergeron at Bergen, have enunciated the view, based on detailed observation, that discontinuities are vital organs supplying energy to cyclones.

Note that Richardson here drew attention to the 'supply of energy' which, in the Bjerknes theory, comes from the discontinuities (i.e. fronts) to the synoptic scale (cyclone). If such a theory is correct, any attempt to provide weather forecasts based on finite differences (as is the case for numerical forecasting) seems doomed to failure since the discontinuities will spoil the precision of the method. As reported by Hunt (1987), Richardson wrote in 1919 to Napier Shaw of the Meteorological Office asking whether he could return there in order to work on upper air 'sounding' experiments, with a view to making "weather predictions by a numerical process... a practical system" (Ashford (1985)). Shaw replied, saying that the 'graphical scheme' of V. and J. Bjerknes had been selected as the operational tool to be developed in the future. So the point raised by Richardson in his introduction was of crucial relevance.

The scientific controversy, which at that time was not perceived as such, between Bjerknes' *graphical* scheme and Richardson's *numerical* scheme is the first example of a long-standing controversy in dynamical meteorology and, more generally, in turbulence. To put the controversy in simple words, Bjerknes' approach was focused on well-defined structures (fronts), while Richardson's approach was based on the equations. This marks the beginning of the long debate on the importance of coherent structures (such as vortices and vortex filaments) as opposed to statistical scaling laws underlying the fundamental aspects of turbulence. In describing turbulent flows, perhaps the most important feature to understand is the dynamics of vortices, whose complex behaviour is governed by the Navier–Stokes equations. Vortex filaments can shrink in radius through the effect of large-scale straining flow, so that strong velocity differences can occur at relatively small scales. From this point of view, there is no need to introduce the energy cascade to explain why large velocity fluctuations are often observed at small scales in turbulent flows. On the other hand, if we think in terms of energy transfer from large to small scales, then the statistical properties at very small scales and at large Reynolds numbers should exhibit features which are independent of the detailed dynamics of turbulent generation and vortex interaction. If this is the case, then the statistical properties of turbulent flows should exhibit universal properties at large Reynolds number, which is one of the key results of the Kolmogorov theory. Recall that, as discussed in the previous section, Richardson's 4/3 law for eddy diffusion in turbulent flows is the first experimental evidence for the universality of the statistical properties of turbulent flow. It is fair to say that both points of view described above are crucial in understanding turbulence, although many experts still argue about which of these is the more relevant.

Returning to meteorology, it is a matter of fact that, as regards weather forecasting, the controversy was settled due to the work of Charney et al. (1950) and Phillips (1956). Using the ENIAC computer at the US army base in Maryland, Charney et al. (1950) showed that, by using a suitable version of the equations of motion describing weather dynamics, a successful 24-hour weather forecast could be achieved. The results were considered preliminary although, as reported by Smagorinsky (1981), it was "the beginning of a new era in weather forecasting – an era that will be based on the use of high-speed automatic computers". A few years later, Phillips (1956) simulated hemispheric motion on the high-speed computer at the Institute for Advanced Study, Princeton, using a simple energetically consistent model which was integrated for a simulated time of approximately one month. According to Charney (Lewis (1998)):

> the deformation field in the developing baroclinic wave produces frontogenesis in the form of the frontal wave, so that the primary cyclone wave does not form on a pre-existing front, rather it forms at the same time as the front and appears as the surface manifestation of the upper wave ... once the front has formed, it may permit frontal instabilities ... It would seem that the latter type is the 'cyclone wave' of Bjerknes & Solberg (1922), ...

Due to the scientific authority and impact of von Neumann, the brilliant theoretical insight of Charney and Phillips, and the availability (for the time) of powerful computers, numerical weather prediction was becoming a practical proposition, as foreseen by Richardson. As a side remark, Phillips' work was recognized through the award of the first Napier Shaw Memorial Prize, commemorating the centenary of the birth of England's venerated leader in meteorology, Sir Napier Shaw (1854–1945) (Lewis (1998)).

Another point reflecting Richardson's insight and character deserves mention. In his book, he applied his method to some relatively simple cases of forecasting. Thus for example, for a point in Central Europe he computed a surface pressure 'tendency' (i.e. the change of pressure over 6 hours) of 145 mb, whereas the observed change was only 0.2 mb, and −1.0 mb after 12 hours. He wrote that the reason for this huge discrepancy "is to be found in the accuracy of the observations" (Phillips (1973)). Phillips analysed Richardson' result, and demonstrated that initial errors in observations may lead to transient growth of internal gravity waves affecting the weather forecast just as Richardson had found. Filtering of gravity waves is necessary in order to eliminate this effect of initial errors. It is interesting to note that Richardson was so confident in his methods and tools that he was bold enough to include the clearly *wrong* prediction in his book. This was the right decision because the subsequent scrutiny of his results in fact opened the way to a better strategy in numerical weather forecasting. Charney sent Richardson a copy of his paper (Charney et al. (1950)), to which Richardson replied with the comment that "this was an enormous scientific advance on the single, and quite wrong result, ... in which the calculations of Richardson (1922) ended" (Ashford (1985)).

The reviews of the book (Ashford (1985)) were very positive about "its originality, its coordinated treatment of the dynamical processes". A short note of Woolard (1922) discussed the methods and the results obtained by Richardson. Woolard wrote:

> The method is applied to a very complete set of observations over middle Europe at 1910 May 20d. 7h. GMT, for which Bjerknes has published detailed charts. The resulting forecast is quite satisfactory, considering the amount and distribution of initial data: for all the reasons that have determined the present distribution of meteorological stations, the nature of the atmosphere as it has been summarized in its chief differential equations is not one; and in the present case, errors seems to have been present in the initial wind data also.

Thus, Richardson knew that his method was opening a new strategy in operational weather forecasting although many questions and problems remained. However, "the forecasters within meteorological organizations did not believe that this provided a practical approach at that time" (Hunt (1987)). In his review paper, Smagorinsky (1981) wrote:

> Weather forecasting was quite subjective, ... based on the experience of having observed and classified many such evolutions ... in 1943 I came into contact with Professor Bernard Haurwitz. When I asked him why physical principles had not been applied to the practical problem of weather prediction, he quickly pointed

out the futility of using the tendency equation ... When queried further, Haurwitz did recall the work by L.F. Richardson ...

It is interesting to remark that the entire print run of Richardson's 1922 book was a mere 750 copies. When re-issued in 1965 by Dover, it sold 3000 copies.

As mentioned above, an important issue raised in this book was concerned with the effect of initial conditions and their relevance for the predictability of atmospheric flows. Due to work by E. Lorenz, we now understand the importance of Richardson's findings. Lorenz (1963) was able to provide clear examples of relatively simple and well-defined dynamical systems whose error forecast $E(t)$ grew exponentially in time from an initial error $E(0)$, no matter how small this might be. This feature is what we now refer to as 'sensitive dependence on initial conditions', a feature of any chaotic system. Although this concept was not available to Richardson, his recognition of the effect of initial errors, so well summarized in Woolard's note, was influential in the later development of operational weather forecasts and in predictability studies in atmospheric dynamics.

In conjunction with the conceptual steps in his book, Richardson defined a systematic way to use finite difference methods in dealing with the complex nonlinear equations describing atmospheric flow; some of his prescriptions are still used in numerical simulations. Richardson was the first to address the question of how to improve computational efficiency by suitable organization:

> It took me the best part of six weeks to draw up the computing forms and to work out the new distribution into two vertical columns for the first time ... With practice the work of an average computer might go perhaps ten times faster. ...

By 'computer', he meant a person doing manual computations! He went on to obtain an estimate of about 64 000 computers in order to perform weather prediction on the whole globe. This number is so large that a suitable organization must be found.

> Imagine a large hall like a theatre ... The walls of this chamber are painted to form a map of the globe ... A myriad computers are at work upon the weather of the part of the map where each sits, ... The work of each region is coordinated by an official of higher rank. Numerous little 'night signs' display the instantaneous values so that neighbouring computers can read them ... From the floor of the pit a tall pillar rises ... In this sits the man in charge of the whole theatre ... One of his duties is to maintain a uniform speed of progress in all parts of the globe.

Next he analysed in detail how to pass messages and increase the efficiency of communication between the different areas. What Richardson was describing in this book was in effect the first architectural design of a massive parallel computer as we know it today. Coordination (the task of the man on the tall pillar) and message transmission are the core problems in massive parallel computation: organizing the data flow while parallelizing individual computations is the key to increased computational efficiency. In most reviews on the art of parallel computing, Richardson's ideas are still presented as a way of introducing the problem.

Richardson made many important contributions to our understanding of atmospheric physics. In particular, he obtained a general criterion for the suppression of turbulence by stable density stratification, (Richardson (1920)). While at Benson, he derived estimates for the relative contributions to the energy of the turbulence, on the one hand from the buoyancy forces caused by movement of eddies between levels of different temperatures, and on the other from acceleration of eddies moving between levels of different wind speed. The ratio of these two contributions is now called the Richardson number Ri. Richardson showed that if $Ri > 1$, then turbulence is suppressed. To quote Hunt (1987):

> this insight into the occurrence and strength of turbulence was soon acknowledged in the scientific world ... Prandtl (1931) incorporated Richardson's criterion into his textbook on fluid dynamics only a few years later.

5.4 Fractal dimension

After 1929, Richardson turned his attention from meteorology to quantitative sociology. His results are still highly cited and it is fascinating to read his papers and monographs on this subject. Among many topics that he studied, mostly outside the scope of this book, Richardson discussed the correlation between war occurrences and geographical boundaries. Again he started by collecting observational data in order to define a quantitative framework. During his research, while focusing on the definition of geographical boundaries, he discovered a new concept with important implications: the concept of fractional (or later 'fractal') dimension. Actually this concept had been earlier introduced in the mathematical literature by Hausdorff, Fatou and Julia before 1920, but it seems that Richardson was not aware of this. In his famous paper *How long is the coast of Britain?*, Mandelbrot (1967) wrote

> I propose to reexamine in this light, some empirical observations in Richardson (1961) and interpret them as implying, for example, that the dimension of the west coast of Great Britain is $D = 1.25$. Thus, the so far esoteric concept of a 'random figure of fractal dimension' is shown to have simple and concrete applications of great usefulness.

Mandelbrot here represented well the spirit of Richardson: science must be useful in some way and any scientific concepts must be applied to real-life situations. This was the motivation in all Richardson's work and his work on fractal dimension was no exception.

From a scientific point of view, his results in this area are closely related to the concepts of energy cascade and scale-dependent eddy diffusion that he had previously encountered in the turbulence context. Richardson explains that he "walked a pair of dividers along a map of the frontier so as to count the number of equal sides of a polygon, the corners of which lie on the frontier". Defining l as the separation of the divider points and $\sum l$ as the total 'length' of the boundary when viewed at the scale l, he found empirically that $\sum l \sim l^{-D_F}$, where D_F is a real number (the 'fractal' dimension of the boundary) with $1 < D_F < 2$. Through this discovery, he opened a new field of 'statistical topography'; see for instance Isichenko (1992). Fractal dimension, as later studied in detail by Mandelbrot, is a powerful concept as it characterizes the geometrical complexity of natural shapes. We can recognize a similar concept regarding the energy cascade in turbulence. Moreover, the concept of fractal dimension is naturally related to the concept of *scaling* and *scaling laws* in describing complexity. Scaling laws have played an important role in recent development of statistical mechanics; Richardson's $l^{4/3}$ law is one of the most important examples in fully developed turbulence.

Not surprisingly therefore, fractal dimension plays an important role in explaining some statistical features of homogeneous isotropic turbulence, namely the intermittent fluctuations of the velocity field. By introducing an object of fractal dimension D_F in a three-dimensional space, we can also introduce the probability for a box of length l to belong to the fractal object, which is simply given by l^{3-D_F}. Let us now assume that some positive-valued scale-dependent quantity $A(l)$ is defined on the same fractal object with the constraint $\langle A(l) \rangle = 1$. This implies that $A(l) \sim l^{D_F-3}$. We can also compute the moments of $A(l)$, which are given by

$$\langle A(l)^n \rangle \sim A(l)^{n(D_F-3)} l^{3-D_F} \sim l^{(n-1)(D_F-3)}. \tag{5.13}$$

Equation (5.13) implies that quantities like the kurtosis, defined as

$$K(l) \equiv \frac{\langle A(l)^4 \rangle}{\langle A^2(l) \rangle^2} \sim l^{D_F-3}, \tag{5.14}$$

increase with decreasing l. This implies that $A(l)$ is characterized by strong fluctuations at small scales. Similar behaviour is observed for many quantities in turbulence, like the rate of energy dissipation or the enstrophy (vorticity squared), which suggests that the complexity of turbulent flows may be characterized by non-integer fractal dimensions. Richardson was unaware of the potential application of the concept of fractal dimension in turbulence, as later introduced by Mandelbrot (1968). However, it is likely that he intuitively linked the complexity of a geometrical form (like the coast of Britain) to the concept of scale-dependent behaviour, a concept that he had developed so nicely in deriving the $l^{4/3}$ law.

Actually, Richardson's diffusion can be understood in a deeper way by looking at how mixing and entrainment are related to the concept of fractal dimension (Sreenivasan et al. (1989)). The basic qualitative argument can be understood by considering a drop of ink in a turbulent flow. Due to turbulence, the drop surface starts to change its shape in a rather complex way. More precisely, by defining l the scale at which we describe the surface (for instance the smallest scale of turbulent fluctuations), we may expect that the surface S behaves as a fractal object with $S \sim l^{-D}$ with $3 > D > 2$. Since the diffusion of particles is enhanced by increasing S, we also expect that, due to the fractal properties of S, the diffusion of ink will be much faster in a turbulent flows. Starting from this idea, more refined quantitative arguments can be developed to show exactly in what manner the concept of fractal dimension is relevant to the increased diffusion of turbulent flows.

5.5 Conclusions

As observed in the introduction, Richardson's scientific activity was characterized by his ability to pose deep questions concerning basic concepts. Some of these questions were answered by introducing new concepts or by developing new mathematical approaches. Richardson's results opened a new way of looking at the complex behaviour of physical systems, in particular turbulence and turbulent flows.

His character and life were unusual for a scientist of such outstanding creativity. According to Gold (1971):

> Research for Richardson was the inevitable consequence of the tendency of his mental machine to run almost, but not quite, by itself. So he was a bad listener, distracted by his thoughts, and a bad driver, seeing his dream instead of the traffic. The same tendency explains why he sometimes appeared abrupt in manner, otherwise inexplicable in one of his character.

Perhaps the most appropriate comment is that of Taylor (1958) who wrote "Richardson was a very interesting and original character who seldom thought on the same lines as his contemporaries and often was not understood by them".

References

Ashford, O. 1985. *Prophet or Professor: The Life and Work of L.F. Richardson*. Hilger.

Batchelor, G.K. 1950. The applicaton of the similarity theory of turbulence to atmospheric diffusion. *Quart. J. R. Meteorol. Soc.* **76** 133–146.

Biferale, L., Boffetta, G., Celani, A., Devenish, B.J., Lanotte, A. & Toschi, F. 2005. Lagrangian statistics of particle pairs in homogeneous and isotropic turbulence. *Phys. Fluids* **17** 115101.

Bjerkens, J. & Solberg, H. 1922. Life cycle of cyclones and the polar front theory of atmospheric circulation. *Geophys. Publ.* **3** 1–18.

Boffetta, G. & Sokolov, I.M. 2002. Relative dispersion in fully developed turbulence: the Richardson law and intermittency corrections. *Phys. Rev. Lett.* **88** 094501.

Charney, J.G., Fjørtoft, R. & von Neumann, J. 1950. Numerical integrations of the barotropic vorticity equations. *Tellus* **2** 237–254.

Falcovich, G., Gawedzki, K. & Vergassola, M. 2001. Particles and fields in fluid turbulence. *Rev. Mod. Phys.* **73** 914–973.

Gold, E. 1971. Lewis Fry Richardson. *Dictionary of National Biography*, Supp. (1951–60), London 1971, 837–839.

Hunt, J.C.R. 1987. Lewis Fry Richardson and his contribution to mathematics, meteorology, and models of conflict. *Ann. Rev. Fluid. Mech.* **30** xiii–xxxvi.

Isichenko, M.B. 1992. Percolation, statistical topography, and transport in random media. *Rev. Mod. Phys.* **64** 961–1043.

Kolmogorov, A.N. 1941. The local structure of turbulence in incompressible viscous fluid for very large Reynolds numbers. *C. R. Acad. Sci. USSR* **30** 301.

Kolmogorov, A.N. 1962. A refinement of previous hypotheses concerning the local structure of turbulence in a viscous incompressible fluid at large Reynolds number. *J. Fluid. Mech.* **13** 82–85.

Lewis, J.M. 1998. Clarifying the dynamics of the general circulation: Phillips's 1956 experiment. *Bull. Am. Met. Soc.* **79** 38–60.

Lorenz, E.N. 1963. Deterministic non-periodic flows. *J. Atmos. Sci.* **20** 130–141.

Mandelbrot, B. 1967. How long is the coast of Britain? Statistical self-similarity and fractional dimension. *Science* **156** 636–638.

Mandelbrot, B. 1969. On intermittent free turbulence. In *Turbulence of Fluids and Plasmas, 1968*, Brooklyn Polytechnic Institute, E. Weber, ed., Interscience, pp. 483–492.

Ouellette, N.T., Xu, H., Burgoin, M., & Bodenschatz, E. 2006. An experimental study of turbulent relative dispersion models. *New J. Phys.* **8** 109.

Phillips, N.A. 1956. The general circulation of the atmosphere: A numerical experiment. *Quart. J. Roy. Met. Soc.* **82** 123–164.

Phillips, N.A. 1973. Principles of large scale numerical weather prediction. In *Dynamic Meteorology*, P. Morel, ed., D. Reidel Publ. Co.

Prandtl, L. 1931. *Abriss de Stomunglere*. Vieweg.

Richardson, L.F. 1920. The supply of energy from and to atmospheric eddies. *Proc. Roy. Soc. London A* **97** 354–373.

Richardson, L.F. 1922. *Weather Prediction by Numerical Process*. Cambridge University Press.

Richardson, L.F. 1926 Atmospheric diffusion shown on a distance-neighbour graph. *Proc. R. Soc. London Ser. A* **110** 709–737.

Richardson, L.F. & Stommel, H. 1949. Note on eddy diffusion in the sea. *J. Met.* **5** 238–240.

Richardson L.F. 1961. The problem of contiguity: an appendix to statistics of deadly quarrels. *General Systems Yearbook* **6** 139–187.

Salazar, J.P.L.C. & Collins, L.R. 2009. Two-particle dispersion in isotropic turbulent flows. *Ann. Rev. Fluid Mech.* **41** 405–32.

Smagorinsky, J. 1981. The beginning of numerical weather prediction and general circulation modeling: early recollections. *Adv. Geophys.* **25** 3–37.

Sreenivasan, K.R., Ramshankar, R. & Meneveau, C. 1989. Mixing, entrainment, and fractal dimensions of surfaces in turbulent flows. *Proc. R. Soc. Lond. A* **421** 79–108.

Toschi, F. & Bodenschatz, E. 2009. Lagrangian properties of particles in turbulence. *Ann. Rev. Fluid. Mech.* **41** 375–404.

Taylor, G.I. 1958. The present position in the theory of turbulent diffusion. *Adv. Geophys.* **6** 101–111.

Woolard, E.W. 1922. L.F. Richardson on weather prediction by numerical processes. *Mon. Weather Rev.* **50** 72–74.

6

The Russian school

Gregory Falkovich

The towering figure of Kolmogorov and his very productive school is what was perceived in the twentieth century as the Russian school of turbulence. However, important Russian contributions neither start nor end with that school.

6.1 Physicist and pilot

> ...the bombs were falling almost the way the theory predicts. To have
> conclusive proof of the theory I'm going to fly again in a few days.
>
> A.A. Friedman, letter to V.A. Steklov, 1915

What seems to be the first major Russian contribution to the turbulence theory was made by Alexander Alexandrovich Friedman, famous for his work on non-stationary relativistic cosmology, which has revolutionized our view of the Universe. Friedman's biography reads like an adventure novel. Alexander Friedman was born in 1888 to a well-known St. Petersburg artistic family (Frenkel, 1988). His father, a ballet dancer and a composer, descended from a baptized Jew who had been given full civil rights after serving 25 years in the army (a so-called cantonist). His mother, also a conservatory graduate, was a daughter of the conductor of the Royal Mariinsky Theater. His parents divorced in 1897, their son staying with the father and becoming reconciled with his mother only after the 1917 revolution. While attending St. Petersburg's second gymnasium (the oldest in the city) Friedman befriended a fellow student Yakov Tamarkin, who later became a famous American mathematician and with whom he wrote their first scientific works (on number theory, received positively by David Hilbert). In 1906, Friedman and Tamarkin were admitted to the mathematical section of the Department of Physics and Mathematics of Petersburg University where they were strongly influenced by the great mathematician V.A. Steklov who taught them partial differential equations and regularly invited them to his home (with another fellow student V.I. Smirnov who later wrote the well-known *Course of Mathematics*, the first volume with Tamarkin). As his second, informal, teacher Alexander always mentioned Paul Ehrenfest who was in St. Petersburg in 1907–1912 and later corresponded with Friedman. Friedman and Tamarkin were among the few mathematicians invited to attend the regular seminar on theoretical physics in Ehrenfest's apartment. Apparently, Ehrenfest triggered Friedman's interest in physics and relativity, at first special and then general. During his graduate studies, Alexander Friedman worked on different mathematical subjects related to a wide set of natural and practical phenomena (among them on potential flow, corresponding with Joukovsky, who was in Moscow). Yet after getting his MSc degree, Alexander Friedman was firmly set to work on hydrodynamics and found employment in the Central Geophysical Laboratory. There, the former pure mathematician turned into a physicist, not only doing theory but also eagerly participating in atmospheric experiments, setting the measurements and flying on balloons. It is then less surprising to find Friedman flying a plane during World War I, when he was three times decorated for bravery. He flew bombing and reconnaissance raids, calculated the first bombardment tables, organized the first Russian air reconnaissance service and the factory of navigational devices (in Moscow, with Joukovsky's support), all the while publishing scientific papers on hydrodynamics and atmospheric physics. After the war ended in 1918, Alexander Alexandrovich was given a professorial position at Perm University (established in 1916 as a branch of St. Petersburg University), which boasted at that time Tamarkin, Besikovich and Vinogradov among the faculty. In 1920 Friedman returned to St. Petersburg. Steklov got him a junior position at the University (where George Gamov learnt relativity from him). Soon Friedman was teaching in the Polytechnic as well, where L.G. Loitsyansky was one of his students. In 1922 Friedman published his famous work *On the curvature of space* where the non-stationary Universe was born (Friedman, 1922). The conceptual novelty of this work is that it posed the task of describing the evolution of the Universe, not only its structure. The next year saw the dramatic exchange with Einstein, who at first published the paper that claimed that Friedman's work contained an error. Instead of public polemics, Friedman sent a personal letter to Einstein where he elaborated on the details of his derivations. After that, Einstein published the second paper admitting that the error was his. In 1924 Friedman published his work, described below, that laid down the foundations of the statistical theory of turbulence structure. In 1925 he made a record-breaking balloon flight to the height of 7400 meters to study atmospheric vortices and make medical self-observations. His personal life was quite turbulent at that time too: he was tearing himself between two women, a devoted wife since 1913 and another one pregnant with his child ("I do not have enough willpower at the moment to commit suicide" he wrote in a letter to the mother of his future son). On his way back from summer vacations by train in the Crimea, Alexander Friedman bought a nice-looking pear at a Ukrainian train station, did not wash it before eating and died from typhus two weeks later.

Friedman's work on turbulence theory was done in conjunction with his student Keller and was based on the works of Reynolds and Richardson, both cited extensively in Friedman and Keller (1925). Recall that Richardson derived the equations for the mean values which contained the averages of nonlinear terms that characterize turbulent fluctuations. Friedman and Keller cite Richardson's remark that such averaging would work only in the case of a so-called time separation when fast irregular motions are imposed on a slow-changing flow, so that the temporal window of averaging is in between the fast and slow timescales. For the first time, they then formulated the goal of writing down a closed set of equations for which an initial value problem for turbulent flow can be posed and solved. The evolutionary (then revolutionary) approach of Friedman to the description of the small-scale structure of turbulence parallels his approach to the description of the large-scale structure of the Universe. Achieving closure in the description of turbulence is nontrivial since the hydrodynamic equations are nonlinear. Indeed, if $\mathbf{v}$ is the velocity of the fluid, then Newton's second law gives the acceleration of the fluid particle:

$$\frac{d\mathbf{v}}{dt} = \frac{\partial \mathbf{v}}{\partial t} + (\mathbf{v}\nabla)\mathbf{v} = \text{force per unit mass.} \tag{6.1}$$

Whatever the forces, the acceleration already contains the second (inertial) term, which makes the equation nonlinear. Averaging the fluid dynamical equations, one expresses the time derivative of the mean velocity, $\partial\langle\mathbf{v}\rangle/\partial t$, via the quadratic mean $\langle(\mathbf{v}\nabla)\mathbf{v}\rangle$. Friedman and Keller realized that meaningful closure can only be achieved by introducing correlation functions between different points in space and different moments in time. Their approach was intended for the description of turbulence superimposed on a non-uniform mean flow. Writing the equation for the two-point function $\partial\langle\mathbf{v}_1\mathbf{v}_2\rangle/\partial t$, they then derived the closed system of equations by decoupling the third moment via the second moment and the mean: $\langle v_1^i v_2^j v_2^k \rangle = \langle v_{1i}\rangle\langle v_{2j}v_{2k}\rangle + \cdots$ (Friedman and Keller, 1925). It is interesting that Friedman called the correlation functions "moments of conservation" (Erhaltungsmomenten) as they express "the tendency to preserve deviations from the mean values" in a curious resemblance to the modern approach based on martingales or zero modes. The work was presented at the

First International Congress on Applied Mechanics in Delft in 1925. During the discussion after Friedman's talk he made it clear that he was aware that the approximation is crude and that time averages are not well-defined. He stressed that his goal was pragmatic (predictive meteorology) and that only a consistent theory of turbulence can pave the way for dynamical meteorology: "Instruments give us mean values while hydrodynamic equations are applied to the values at a given moment". The introduction of correlation functions was thus the main contribution to turbulence theory made by Alexander Friedman, a great physicist and a pilot.

One year after Friedman's death, the seminal paper of Richardson on atmospheric diffusion appeared. I cannot resist imagining what would have happened if Friedman saw this paper and made a natural next step: to incorporate the idea of cascade and the scaling law of Richardson's diffusion into the Friedman–Keller formalism of correlation functions and to realize that the third moment of velocity fluctuations, which they neglected, is crucial for the description of the turbulence structure. As it happened, this was done 15 years later by another great Russian scientist, mathematician Andrei Nikolaevich Kolmogorov.

6.2 Mathematician

> At any moment, there exists a narrow layer between trivial
> and impossible where mathematical discoveries are made.
> Therefore, an applied problem is either solved trivially or
> not solved at all. It is an altogether different story if an
> applied problem is found to fit (or made to fit!) the new
> formalism interesting for a mathematician.
>
> A.N. Kolmogorov, diary, 1943

Russians managed to continue, well into the twentieth century, the tradition of great mathematicians doing physics.

Andrei Kolmogorov was born in 1903. His parents weren't married. The mother, Maria Kolmogorova, died at birth. The boy was named according to her wish after Andrei Bolkonski, the protagonist from the novel *War and Peace* by Lev Tolstoy. Andrei was adopted by his aunt, Vera Kolmogorova, and grew up in the estate of his grandfather, district marshal of nobility, near Yaroslavl. The father, agronomist Nikolai Kataev, took no part in his son's upbringing: he perished in 1919, fighting in the Civil War. Vera and Andrei relocated to Moscow in 1910. In 1920, Andrei graduated from the Madame

Repman gymnasium (cheap but very good) and was admitted to Moscow University, with which he remained associated for the rest of his life. In a few months, he passed all the first-year exams and was transferred to the second year which "gave the right to 16 kg of bread and 1 kg of butter a month – full material prosperity by the standards of the day" (Kolmogorov, 2001). His thesis adviser was Nikolai Luzin who ran the famous research group 'Luzitania'. Apart from him, Kolmogorov was influenced by D. Egorov, V. Stepanov, M. Suslin, P. Urysohn and P. Aleksandrov, with whom Kolmogorov was close until the end of his life, sharing a small cottage in Komarovka village where they regularly invited colleagues and students, who described the unforgettable atmosphere of science, art, sport and friendship (Shiryaev, 2006). Kolmogorov completed his doctorate in 1929. In 1931, following a radical restructuring of the Moscow mathematical community, he was elected a professor. He spent nine months in 1930–31 in Germany and France, later citing important interactions with R. Courant, H. Weyl, E. Landau, C. Carathéodory, M. Frechet, P. Levy. Two years later he was appointed director of the Mathematical Research Institute at the university, a position he held until 1939 and again from 1951 to 1953. In 1938–1958 he was a head of the new Department of Probability and Statistics at the Steklov Mathematical Institute. Between 1946 and 1949 he was also the head of the Turbulence Laboratory in the Institute of Theoretical Geophysics.

Andrei Nikolaevich Kolmogorov was a Renaissance man: his first scientific work was on medieval Russian history; he then did research on metallurgy, ballistics, biology and statistics of rhythm violations in classical poetry, worked on educational reform, was the scientific head of a round-the-world oceanological expedition and used to make 40 km cross-country ski runs wearing only shorts. But first and foremost he was one of the greatest and most universal mathematicians of the twentieth century, if not of all time (Kendall et al., 1990). Kolmogorov put the notion of probability on a firm axiomatic foundation (Kolmogorov, 1933) and deeply influenced many branches of modern mathematics, especially the theory of functions, the theory of dynamical systems, information theory, logics and number theory. Seventy-one people obtained degrees under his supervision, among them several great and quite a few outstanding scientists. There is a certain grand design in the life work of Kolmogorov, to which one cannot give justice in this short essay. In his own words:

> I wish to stress the legitimacy and dignity of a mathematician, who understands the place and the role of his science in the development of natural sciences and technology, yet quietly continues to develop 'pure mathematics' according to its internal logics.

Kolmogorov used to claim that the mathematical abilities of a person are in inverse proportion to general human development: "Supreme mathematical genius has his development stopped at the age of five or six when kids like to tear off insect legs and wings". Kolmogorov estimated that he himself stopped at the age 13–14 when adult problems do not yet interfere with a boy's curiosity about everything in the world (Shiryaev, 2006, pp. 43, 171). Recall that Kolmogorov turned 14 in 1917 when the Revolution struck.

What brought this man to turbulence? Kolmogorov's interest in experimental aspects may have been triggered as early as 1930 when he met Prandtl, as did Friedman eight years earlier (Frisch, 1995). An impetus could have been the creation of the Institute of Theoretical Geophysics by academician O. Schmidt, who in 1939 made the newly elected academician Kolmogorov a secretary of the section of Physics and Mathematics (Ya.G. Sinai, private communications, 2009–2010). Kolmogorov's works on stochastic processes and random functions immediately predate his work on turbulence. Turbulence presents a natural step from stochastic processes (as functions of a single variable) to stochastic fields (as functions of several variables). His diary entry that starts this section may shed additional light; see also Yaglom (1994).

Kolmogorov later remarked that "it was important to find talented collaborators... who could combine theoretical studies with the analysis of experimental results. In this respect I was quite successful" (Yaglom, 1994). The first student of Kolmogorov to work on turbulence was the mechanical engineer M. Millionschikov who treated turbulence decay. In 1939, Loitsyansky used the Kármán–Howarth equation to infer the conservation of the squared angular momentum of turbulence: $\Lambda = \int r_{12}^2 \langle (\mathbf{v}_1 \cdot \mathbf{v}_2) \rangle d\mathbf{r}_{12}$. Considering the late (viscous) stage of turbulence decay when the size of the turbulence region grows as $l(t) \simeq \sqrt{\nu t}$, one can readily infer the law of the energy decay: $v^2(t) \simeq \Lambda l^{-5} \propto t^{-5/2}$. Also, neglecting the third moment (as had Friedman and Keller before), Millionschikov obtained a closed equation and solved it for the precise r, t dependencies of the second moment (Millionschikov, 1939). To describe turbulence at large Reynolds number Re, one needs to face eventually the third moment and account for the nonlinearity of hydrodynamics. Kolmogorov did that himself, estimating $dv^2/dt \simeq v^3/l$ and obtaining $l(t) \propto \Lambda^{1/7} t^{2/7}$ (Kolmogorov, 1941b). While Kolmogorov used his theory of small-scale turbulence (to be described below) to argue for these estimates, the relation $v \simeq l/t$ for integral quantities seems to be not very sensitive to the details of microscopic theories. The correction, unexpectedly, came from another direction: conservation of the Loitsyansky integral takes place not universally but depends on the type of large-scale correlations in the initial turbulent flow. In terms of

Fourier harmonics

$$\mathbf{v}(\mathbf{p}) = \int \mathbf{v}(\mathbf{r}) \exp(i\mathbf{p} \cdot \mathbf{r}) \, d\mathbf{r},$$

the energy spectral density $E(p) = p^2|\mathbf{v}(\mathbf{p})|^2$ is expected to go to zero at $p\,l \to 0$ as an even power of p. Only if the quadratic term is absent and $E(p) \propto \Lambda p^4$ is Λ then conserved. If, however, $E(p) \propto S p^2$, then it is the squared momentum (called the *Saffman invariant*), $S = \int \langle (\mathbf{v}_1 \cdot \mathbf{v}_2) \rangle d\mathbf{r}_{12}$, which is conserved and determines turbulence decay. This is treated in more detail in the chapters about Batchelor and Saffman. More interesting was the second paper of Millionschikov (1941), where the quasi-normal approximation was presented (apparently formulated by Kolmogorov who often attributed his results to students; Yaglom, 1994). This approximation consists in supplementing the equation for the second moment (which contains third moment) by the equation for the third moment (which contains the fourth moment, which is decoupled via the second moments, assuming Gaussianity) (Millionschikov, 1941). Such an approximation is valid only for weakly nonlinear systems such as the weak wave turbulence described in Section 6.4 below; for hydrodynamic turbulence it is a semi-empirical approximation which was extensively used for the next forty years.

In 1941 Kolmogorov was more occupied by the behavior of $E(p)$ for $p\,l \gg 1$ and by finding the third moment exactly. Already by the end of 1939, he had outlined the scheme of the mathematical description (of what we now call the Richardson cascade) based on self-similarity and predicted that $E(p)$ for $p\,l \gg 1$ will be a power law but did not obtain the exponent (see Yaglom (1994) and Obukhov (1988, p. 83)). Some time in 1940 (most likely in the Fall), Andrei Nikolaevich invited another student, mathematician Alexander Obukhov, and suggested thinking about the energy distribution in developed turbulence. At that time, Kolmogorov did not know about the Richardson cascade picture, while Obukhov did (Golitsyn, 2009). Obukhov later recalled that they met two weeks later, compared notes and found that the exponent was the same – the first Kolmogorov–Obukhov theory (KO41[1]) came into being (Golitsyn, 2009).

Alexander Mikhailovich Obukhov was born in 1918 into a middle-class family in Saratov. He finished school in 1934 and spent a year working on a weather observation station, which probably influenced his long-life fascination with atmospheric phenomena. There, he published his first scientific work *Atmospheric turbidity during the summer drought of 1934*. The following year

---

[1] I use the abbreviations KO41, KO62 deliberately (rather than the more usual K41, K62) in order to highlight the importance of Obukhov's contributions, which ran in parallel with those of his mentor Kolmogorov.

he was old enough to be accepted to Saratov University where he wrote in 1937 his first mathematical work *Theory of correlation of random vectors* which received first prize in the all-country student competition (on the occasion of the jubilee of the Revolution) and attracted Kolmogorov's attention. That was an extraordinary work on multivariate statistics where the young student proposed a new statistical technique which later became known as canonical correlation analysis (simultaneously proposed by the American statistician H. Hoteling). Kolmogorov invited Obukhov to transfer to the mathematical department of Moscow University in 1939. Obukhov graduated in 1940 and was allowed to stay in the University for research work, in particular, on spectral properties of sound scattered by a turbulent atmosphere. It was then natural that Obukhov took a spectral approach to turbulence.

After that fateful meeting, when Kolmogorov and Obukhov compared notes and found that their results agreed, they published separately. The first Kolmogorov paper was submitted on 28 December 1940 (Kolmogorov, 1941a). Kolmogorov considers velocities at two points, following Friedman and Keller, whose work he knew and valued (Yaglom, 1994), yet he was apparently the first to focus on the velocity differences $\mathbf{v}_{12} = \mathbf{v}_1 - \mathbf{v}_2$. Kolmogorov describes a multi-step energy cascade (without citing Richardson) as a "chaotic mechanism of momentum transfer" to pulsations of smaller scales. He then argues that the statistics of velocity differences for small distances (and small time differences) is determined by small-scale pulsations which must be homogeneous and isotropic (far from boundaries). That is, Kolmogorov introduces local homogeneity and turns Taylor's global isotropy (see Sreenivasan's chapter on Taylor) into local isotropy. Kolmogorov never invokes an accelerating nature of the cascade. He then makes a very strong assumption (later found to be incorrect) that the statistics of $\mathbf{v}_{12}$ at distances r_{12} much less than the excitation scale L is completely determined by the mean energy dissipation rate, defined as

$$\bar{\epsilon} = \left\langle \frac{\partial v^2}{2\partial t} \right\rangle = \frac{\nu}{2} \sum_{ij} \left\langle \left(\frac{\partial v^i}{\partial x_j} + \frac{\partial v^j}{\partial x_i} \right)^2 \right\rangle .$$

That allowed him to define the viscous (now called Kolmogorov) scale as $\eta = (\nu^3/\bar{\epsilon})^{1/4}$ and make the second (correct) assumption that for $r_{12} \gg \eta$ the statistics of velocity differences is independent of the kinematic viscosity ν. For $\eta \ll r_{12} \ll L$, one uses both assumptions and immediately finds from dimensional reasoning that $\langle v_{12}^2 \rangle = C(\bar{\epsilon} r_{12})^{2/3}$, where the dimensionless C is called the Kolmogorov constant (even though it is not, strictly speaking, a constant, as will be clear later).

What made a mathematician hypothesize so boldly?

> I soon understood that there was little hope of developing a pure, closed theory, and because of absence of such a theory the investigation must be based on hypotheses obtained on processing experimental data. While I didn't do experiments, I spent much energy on numerical and graphical representation of the experimental data obtained by others (Kolmogorov, 2001).

Sinai recalls Kolmogorov describing how he inferred the scaling laws after "half a year analyzing experimental data" on his knees on the apartment floor covered by papers (Shiryaev, 2006, p. 207). Some thirty years later, we find Andrei Nikolaevich again in this position on the ship's cabin floor catching mistakes in the oceanic data during a round-the-world expedition (Shiryaev, 2006, p. 54). In 1941, the data apparently were from the wind tunnel (Dryden et al., 1937); they were used in the third 1941 paper (Kolmogorov, 1941c) to estimate C.

More importantly, in this third paper, Kolmogorov uses the Kármán–Howarth equation, implicitly *assumes* that, although proportional to ν, the dissipation rate $\bar{\epsilon}$ has a finite limit at $\nu \to 0$, and derives the elusive third moment. Schematically, one takes the equation of motion (6.1) at some point 1, multiplies it by $\mathbf{v}_2$ and subtracts the result of the same procedure taken at point 2. All three forces acting on the fluid give no contribution in the interval $\eta \ll r_{12} \ll L$: viscous friction because $r_{12} \gg \eta$, external force because $r_{12} \ll L$ and the pressure term because of local isotropy. This is why that interval is called inertial, the term so suggestive as to be almost misleading, as we will see later. In this interval the cubic (inertial) term, which is the energy flux through the scale r_{12}, is equal to the time derivative term, which is a constant rate of energy dissipation: $\langle(\mathbf{v}_{12} \cdot \nabla)v_{12}^2\rangle = -2\langle\partial v^2/\partial t\rangle = -4\bar{\epsilon}$. Integrating this one gets

$$\langle(\mathbf{v}_{12} \cdot \mathbf{r}_{12}/r_{12})^3\rangle = -4\bar{\epsilon}r_{12}/5 . \tag{6.2}$$

For many years, the so-called 4/5-law (6.2) was the only exact result in the theory of incompressible turbulence. It is the first derivation of an 'anomaly' in physics in a sense that the effect of breaking the symmetry (time-reversibility) remains finite while the symmetry-breaking factor (viscosity) goes to zero; the next example, the axial anomaly in quantum electrodynamics, was derived by Schwinger ten years later (Schwinger, 1951).

Obukhov's approach is based on the equation for the energy spectral density written as $\partial E/\partial t + D = T$ where D is the viscous dissipation and T is the Fourier image of the nonlinear (inertial) term that describes the energy transfer over scales (Obukhov, 1941). Obukhov starts his paper by saying that for a given observation scale $l = 1/p$, larger-scale velocity fluctuations provide almost uniform transport while smaller-scale eddies provide diffusion. It is then natural to divide the velocity into two orthogonal components, containing respectively large-scale and small-scale harmonics: $\mathbf{v} = \bar{\mathbf{v}} + \mathbf{v}'$. Obukhov stresses that this is not an absolute Reynolds separation into the mean and fluctuations but decomposition conditional on a scale, as a harbinger of the renormalization-group approach which appeared later in high-energy physics and critical phenomena. In this very spirit, Obukhov averages his energy equation over small-scale fluctuations, shows that what contributes to T is the product $v'v'\nabla\bar{v}$ and then decouples it as $E(p)$ times the root-mean-square large-scale gradient $\Delta^{1/2}(p)$ defined by $\Delta(p) = \int_p^\infty k^2 E(k)d^3k$. That way of closure differs from that of Friedman and Keller since the focus is not on instantaneous values and solving the initial value problem for dynamic meteorology but on average values and finding a steady-state distribution. Obukhov then solves the resulting nonlinear (but closed!) equation and finds the spectrum $E(p) \propto p^{-5/3}$ which gives a Fourier transform $\langle v_{12}^2 \rangle = (Ar_{12})^{2/3}$. For the boundary of this spectrum he finds the Kolmogorov scale (which thus should be called Kolmogorov–Obukhov scale). He then derives the law of turbulent diffusion in such a velocity, $R^2(t) = At^3$, compares it with the Richardson diffusion law $R^2(t) = \bar{\epsilon}t^3$ and obtains $A \simeq \bar{\epsilon}$. Along the way, Obukhov gives a theoretical justification for the scaling of turbulent diffusivity $D(l) \propto l^{4/3}$ that was empirically established by Richardson. Obukhov ends by estimating the rate of atmospheric energy dissipation, assuming that 2 percent of the solar energy is transformed into winds[2], and obtains a factor comparable with that measured by Richardson. Magnificent work!

One can imagine the elation the authors felt upon discovering such beautiful simplicity in such a complicated phenomenon: the universality hypothesis was supported by the exact derivation of the third moment (6.2) and by the experimental data. One is tempted to conclude that the statistics of the velocity differences in the inertial interval is determined solely by the mean energy dissipation rate. What could possibly go wrong?

The answer came from a physicist. Lev Davidovich Landau was perhaps as great and universal a physicist as Kolmogorov was a mathematician. The fundamental contributions of Landau and his school and the monumental, unique Landau–Lifshits course of theoretical physics shaped, to a significant extent, the physics of the second half of the twentieth century. Landau was born in 1908 and grew up enthusiastic about the communist ideas. The years 1929–1931 he spent abroad, interacting with N. Bohr, W. Pauli, W. Heisenberg,

---

[2] Obukhov does not justify the estimate: my guess is that he took 2 percent as an estimate for the relative change of the Kelvin temperature between day and night; see Peixoto and Oort (1984) for the modern data.

R. Peierls and E. Teller among others. In the mid-1930s, Landau discovered that he could no longer travel abroad. Building his school and creating the course may be seen as an attempt to create a civilization in what he saw as a wilderness. In 1938, Landau co-authored an anti-Stalin leaflet, was arrested and spent a year in Stalin's jails; after Kapitza and Bohr wrote to Stalin, Landau was freed with his black hair turned gray.

Meanwhile the Second World War eventually came to the Soviet Union and a large part of the Academy was evacuated to Kazan. There, Kolmogorov gave a talk on their results on 26 January 1942. Landau was present. An official record of the talk contains a brief abstract by Kolmogorov and a short remark by Landau. In the abstract, Kolmogorov lucidly presents his results on the local structure and then adds something new: a closed system of three partial differential equations that describe large-scale flow and integrated properties of turbulence (the energy and the strain rate). That semi-empirical model is a significant step forward compared with the earlier models of Prandtl, von Kármán and Taylor, where the Reynolds equations for mean velocity were closed by hypothetical algebraic equations for Reynolds stresses. In 1945, Prandtl suggested a less sophisticated two-equation model. The models of the type suggested by Kolmogorov (later invented independently by Saffman and others), found, with the advent of computers, numerous engineering applications. Landau remarked:

> Kolmogorov was the first to provide a correct understanding of the local structure of turbulent flow. As to the equations of turbulent motion, it should be constantly born in mind ... that in a turbulent stream the vorticity is confined within a limited region; qualitatively correct equations should lead to just such a distribution of eddies.

It is reasonable to assume that the second part of Landau's remark is related to the second part of Kolmogorov's presentation, i.e. to the equations for the large-scale flows. In 1943 Landau derived his exact solution for a laminar jet from a point source inside a fluid (Landau and Lifshitz, 1987), so apparently he was thinking about flows of different shapes. Incidentally, I was unable to find a steady solution of Kolmogorov's equations that describe such a limited region. One may try to interpret Landau's remark as implicitly questioning the universality of small-scale motions: the further the probe from the axis of a turbulent jet, the less time it spends inside the turbulence region because of boundary fluctuations; therefore, the value of the Kolmogorov constant C must depend on the distance to the axis (Frisch, 1995). However, Kolmogorov explicitly postulated that his theory works away from any boundaries, so that the universal value of C is what he expects to be measured near the jet axis or deep inside other turbulent flows. It is likely that Landau started to have doubts about Kolmogorov's description of small-scale structure only later. In 1944, the sixth volume of the Landau–Lifshitz course, *Mechanics of Continuous Media*, appeared (Landau and Lifshitz, 1987). This book firmly set hydrodynamics as part of physics. The book contained a remark (attributed in later editions to Landau, 1944), which instantly killed the universality hypothesis:

> It might be thought that the possibility exists in principle of obtaining a universal formula, applicable to any turbulent flow, which should give $\langle v_{12}^2 \rangle$ for all distances r_{12} small compared with L. In fact, however, there can be no such formula, as follows from the following argument. The instantaneous value of v_{12}^2 might in principle be expressed in a universal way via the energy dissipation ϵ at that very moment. However, averaging these expressions is dependent on the variation of ϵ over times of large-scale motions (scale L), and this variation is different for different specific flows. Therefore, the result of the averaging cannot be universal.

Let us observe a moment of silence for this beautiful hypothesis.

To put it a bit differently: the third moment (6.2) is linearly proportional to the dissipation rate ϵ and is then related in a universal way to the mean dissipation rate $\bar{\epsilon}$. Yet other moments $\langle v_{12}^n \rangle$ are averages of nonlinear functions of the instantaneous value ϵ, so that their expressions via the mean value $\bar{\epsilon}$ depend on the statistics of the input rate determined by the motions at the scale L (that was more clearly formulated later by Kraichnan; see the chapter by Eyink and Frisch). The question now is whether such an influence of large scales changes factors that are of the order of unity (say making C non-universal) or changes the whole scale dependence of the moments, since now one cannot rule out the appearance of the factor (L/r_{12}) raised to some power. Kolmogorov and Obukhov themselves found the answers twenty years later, as will be described below.

During the war years the works of Kolmogorov and Obukhov were unknown to the rest of the world, becoming known after the war primarily through their discovery by Batchelor. Kolmogorov's 4/5-law was not independently derived but his 1/3-law and Obukhov's 5/3-law were rederived by Heisenberg, Weizsäcker and Onsager (Battimelli and Vulpiani, 1982; Frisch, 1995; Sreenivasan and Eyink, 2006). Apparently it is more difficult to get the factor than the scaling, all the more so because the factor is exact while the scaling is not. Note, however, that it is reasonable to expect that the moments depend in some regular way on the order n of the moment. If so, then the fact that $n = 2$ is not far from $n = 3$ means that KO41, which is exact for $n = 3$, must work reasonably well for $n = 2$, i.e. for the energy spectrum, which is indeed what measurements show. This is the reason that this flawed theory turned out to be very useful in numerous geophysical and astrophysical applications as long as one is interested in the energy spectrum and not high moments or strong fluctuations. For the next twenty years, Kolmogorov and Obukhov developed the applications and generalizations of KO41 instead of looking for a better theory. In retrospect, that seems to be a right decision. Its implementation involved the creation of a scientific school.

6.3 Applied mathematicians

One of my students rules the Earth atmosphere, another – the oceans.

A.N. Kolmogorov

Alexander Obukhov was soon joined by Andrei Monin and Akiva Yaglom, the two other key people that established the Kolmogorov school of turbulence. Andrei and Akiva were born the same year, 1921, and died the same year, 2007. They wrote the book (Monin and Yaglom, 1979) that for several decades was "the Bible of turbulence". The triple A of Alexander, Andrei and Akiva represented very different, and in some respects polar opposite, people. Alexander and Akiva were never Party members, with the latter even refusing to work on the nuclear project since he disliked the idea of developing a bomb for Stalin (Shiryaev, 2006, p. 440), while Andrei was a devoted communist who joined the Party during the war. That was a stark difference in the Soviet Union back then. Kolmogorov himself was not a Party member, yet allowed neither regime critique nor political conversations in his presence (Shiryaev, 2006, p. 442); descent from nobility and homosexuality (criminal under Soviet penal code) added extra vulnerability for Andrei Nikolaevich in Stalin's Russia.

Yaglom grew up in Moscow where his high-school friend was Andrei Sakharov (who later became a friend of Obukhov too). Akiva had a twin brother Isaak, with whom he shared a first prize at the Moscow Mathematical Olympiad in 1938. The prize was presented by Kolmogorov who never forgot good students (Yaglom, 1994) and in 1943 invited Yaglom to work on the theory of Brownian motion. Andrei Monin graduated in 1942 and the same year was also invited by Kolmogorov to work on probability distributions in functional spaces (where there is no volume element and thus no density). Both Akiva (in 1941) and Andrei (in 1942) volunteered for military service to fight in the war. Akiva was rejected because of poor eyesight. Andrei was drafted and spent the war as an officer-meteorologist serving at military airfields. He returned in 1946 ready to work on turbulence.

The first new result after 1941 was, however, obtained by Obukhov whose Kazan years were important and formative. In addition to Landau, he interacted there with the physicist M.A. Leontovich, a man of great integrity (who, among many other things, published with Kolmogorov the paper on Brownian motion in 1933). Landau and Obukhov were the first to suggest independently the Lagrangian analog of KO41. If $\mathbf{R}(t)$ describes the trajectory of a fluid particle, then the Lagrangian velocity is defined as $\mathbf{V}(t) = \mathbf{v}(\mathbf{R}, t)$. The relation $\mathbf{V}(t) - \mathbf{V}(0) \simeq (\epsilon t)^{1/2}$ first appeared in the Landau–Lifshitz textbook in 1944. Note however that the exact Lagrangian relation, which is a direct analog of the flux law (6.2), is not the (still hypothetical) two-time single-particle relation $|\mathbf{V}(t) - \mathbf{V}(0)|^2 \simeq \epsilon t$, but the Lagrangian time derivative of the *two-particle* velocity difference: $\langle d|\delta\mathbf{V}|^2/dt \rangle = -2\bar{\epsilon}$ (note that $\epsilon > 0$ in 3D and $\epsilon < 0$ in 2D) (Falkovich et al., 2001).

From 1946, Kolmogorov arranged a bi-weekly seminar on turbulence which was a springboard for the explosive development of KO41 and its applications. Obukhov started to work on the atmospheric boundary layer and dynamic meteorology. Already in 1943 he wrote a paper which because of the war was published in 1946 and yet was ahead of its time (Obukhov, 1988, p. 96; translated in Obukhov, 1971). Following Prandtl and Richardson, Obukhov considered the influence of stable stratification on turbulence. It is clear that turbulence disturbs stable stratification and increases the potential energy, thus decreasing the kinetic energy of the fluid. In other words, stratification suppresses turbulence. On the other hand, turbulence influences the vertical profile of the temperature. Obukhov developed a semi-empirical approach based on a systematic use of universal dimensionless functions. In addition to the dimensionless Richardson number that quantifies the relative role of stratification and wind shear, Obukhov measured the height in units of the sub-layer where the Richardson number is small and stratification is irrelevant. This defines what is now called the Obukhov–Monin scale, since the idea of the sub-layer was systematically exploited by Obukhov and Monin in 1954. In the paper which is a sequel to that of Obukhov (1988, p. 135), they showed that the profiles of the wind and the temperature are determined by the vertical fluxes of the momentum and heat; see Yaglom (1988) for more details.

The year 1949 was exceptionally productive. Kolmogorov applied KO41 to the problem of deformation and break-up of droplets of one liquid in a turbulent flow of another fluid: flow can break the droplet of the size a if the pressure difference due to flow $\rho(\delta v)^2 \simeq \rho(\bar{\epsilon}a)^{2/3}$ exceeds the surface tension stress σ/a (Kolmogorov, 1949). Obukhov established the basis of dynamic meteorology by his famous work on a geostrophic wind, derived what is now called the Charney–Obukhov equation for the rotating shallow water, known as Hasegawa–Mima for magnetized plasma (though I've heard Obukhov remarking that the plasma version was known to Leontovich before). Turbulence theory was significantly advanced when Obukhov published a pioneering work on the statistics of a passive scalar θ mixed by a turbulence flow (Obukhov, 1949a). Obukhov correctly describes the common action of turbulent mixing and molecular diffusion as a mechanism of relaxation. He then focuses on θ^2 (assuming $\langle\theta\rangle = 0$) which is a nontrivial step, missed by several people who got wrong answers; see Monin and Yaglom (1979). Obukhov identifies θ^2 as an analog of the energy density, arguing that when θ is the temperature then $\int \theta^2(\mathbf{r})\,d\mathbf{r}$ is the maximal work one can extract from an inhomogenously heated body. That opens the way to considering the cascade of this quantity in a direct analogy with the energy cascade. Obukhov's work then follows Kolmogorov's approach of his first 1941 paper, that is it considers the statistics of the differences $\theta_{12} = \theta_1 - \theta_2$. Obukhov assumes that there exists an interval of scales between the scales of production and dissipation where the statistics of θ_{12} is completely determined by the dissipation rates ϵ and $N = \langle\partial\theta^2/\partial t\rangle$. Dimensional reasoning then gives $\langle\theta_{12}^2\rangle \simeq N(r_{12}^2/\epsilon)^{1/3}$. Of course, the right-hand side here is the mean dissipation rate of θ^2 multiplied by the typical turnover time on the scale r_{12}. This 2/3-law was independently established by Corrsin in 1951 and is called the Obukhov–Corrsin law (see also the Corrsin chapter). The second exact relation in turbulence theory, the flux expression for a passive scalar analogous to (6.2) for energy, was derived by Yaglom the same year (Yaglom, 1949a). That same year Obukhov dispelled an erroneous belief (expressed in Millionschikov, 1941) that pressure fluctuations are zero in incompressible turbulence (Obukhov, 1949b). By taking the divergence of the Navier–Stokes equation, Obukhov obtained the incompressibility condition $\Delta p = -\nabla_i\nabla_j(v^i v^j)$, which allows one to express the second moment of pressure via the fourth moment of velocity, which is then decoupled via the product of the second moments, again assuming Gaussianity: $\langle p_{12}^2\rangle \propto r^{4/3}$. That 4/3-law together with 5/3, 2/3 and others was the basis for the joke that Obukhov discovered the fundamental 'all-thirds law'. There is a truth in every joke since the number 3 in the denominator of these scaling exponents arises because of two fundamental reasons: (i) the nonlinearity of the equation of motion is quadratic and (ii) the fluxes considered are of the quadratic integrals of motion. Immediately, Yaglom used Obukhov's approach to derive the mean pressure gradient and the mean squared fluid acceleration (Yaglom, 1949b). Remarkably, Yaglom's estimate for atmosphere showed that typical winds can make for accelerations exceeding that of gravity. Obukhov, Monin and Yaglom had a chance to experience that, flying on balloons in turns, thus continuing Friedman's tradition; in 1951 the wind data were obtained confirming KO41 scaling (Obukhov, 1951) (later, they also observed a layered structure of turbulence, the so-called turbulent 'pancakes', predicted by Kolmogorov in 1946 – Shiryaev, 2006, p. 181). In 1951, Obukhov and Yaglom published together a detailed paper that presented all the results on pressure and acceleration. Similar results were obtained independently by Heisenberg in 1948 and Batchelor in 1951.

The Kolmogorov turbulence seminar was attended by applied scientists and engineers as well, and discussions of applied problems went along with the focus on fundamental issues. In 1951, Kolmogorov accepted the next student, Gregory Barenblatt, whose name he remembered from the list of the students whose work won first prizes (following the familiar Obukhov–Yaglom pattern). Barenblatt was given the task of describing the transport of a suspended sediment by turbulent flows in rivers. Somewhat similarly to stably stratified flows, turbulence spends energy lifting sediments which, being small, then dissipate energy into heat when descending. Barenblatt built an elegant theory similar to that of Obukhov–Monin (Barenblatt, 1953).

Important insights into the advection mechanisms were obtained by eliminating global sweeping effects and describing the advected fields in a frame whose origin moves with the fluid. This picture of the hydrodynamic evolution, known under the name of quasi-Lagrangian description, was first introduced in Monin (1959). In a kind of a bridge between work on stratification and passive scalars, Obukhov considered unstable stratification, accounted for the buoyancy force and defined a new scale above which this force starts to be important (Obukhov, 1959). Bolgiano discovered this independently the same year and also suggested KO41-type scaling for turbulent convection at larger scales (Bolgiano, 1959).

In 1956 the Institute of Geophysics was divided into three parts and Obukhov was appointed director of the newly created Institute of Atmospheric Physics which now bears his name. That followed his long conversation with Leontovich which ended with the advice to "avoid administrative zeal" (Obukhov, 1990). In the Soviet Union, the Academy was a huge body that operated hundreds of scientific institutes with tens of thousands of researchers. Academic institutes worked under the strict Party control and a non-communist director was a rare bird. Obukhov flouted Party policy in another important respect: employing numerous Jewish scientists in his Institute. Since the late 1940s anti-Semitism as a Party policy was steadily gaining ground in Russian society and academia. Moscow University was particularly hostile: it was difficult for a Jew to be accepted as an undergraduate and next to impossible as a graduate student; this situation further deteriorated at the end of 1960s when undergraduate studies were closed as well (all the way to the 1970s when I avoided Moscow and went to Novosibirsk University). The mathematical students of Kolmogorov were particularly affected. For example, Sinai was not accepted for graduate studies after the committee failed him in Marxist philosophy: Kolmogorov was present at the exam but did not interfere (Ya.G. Sinai, private communications, 2009–2010). Kolmogorov then negotiated for Sinai a second attempt which succeeded. Turbulence researchers had it easier thanks to the Institute of Geophysics and later to Obukhov's Institute. Remarkably, that quite unusual director did not even fire refuseniks as was required by a direct Party order. Obukhov was universally admired by his co-workers despite his sometimes harsh style (its acceptance was softened by a common agreement that he was invariably the smartest person in the room, best equipped to "rule the Earth's atmosphere").

Andrei Monin was appointed "to rule the oceans" in 1965 when he was made director of the Institute of Oceanology. He had not only been a devout Party member since 1945, but a high-level if somewhat reluctant (Golitsyn, 2009) functionary in the Party hierarchy as an instructor and then the deputy chairman of the Science Department of the Party Central Committee. While the Academy kept some marginal degree of independence in electing (or rejecting) new members, the Department was the body which actually set the policy, appointed directors, issued permits for visits abroad etc. During the 1950s, the Department was particularly hostile towards "the group of non-communist scientists led by Tamm, Leontovich and Landau" (Monin, 1958).

Around 1960–61, Obukhov decided at last to address Landau's remark on dissipation rate fluctuations and initiated theoretical and experimental investigations into the subject (Golitsyn, 2009). Systematic measurements of wind velocity fluctuations were made by Gurvitz (1960). The calculations of the fluctuations of the energy dissipation rate ϵ, assuming quasi-normality, was done by Obukhov's student G.S. Golitsyn, who later extended the approach of KO41 to the analysis of the dynamics of planetary atmospheres (Golitsyn, 1973) and succeeded Obukhov as director of the Institute. Experimental data had shown that fluctuations were much stronger than the theoretical estimates. Strong non-Gaussianity of velocity derivatives was also observed before by Batchelor and Townsend. Looking for an appropriate model for the statistics of ϵ, Obukhov turned to another seminal paper of 1941 by Kolmogorov (1941d) on a seemingly different subject: ore pulverization. Breaking stones into smaller and smaller pieces presents a cascade of matter from large to small scales. A stone that appears after m steps is of size ϵ_m, which is a product of the size ϵ of an initial large stone and m random factors of fragmentation: $\epsilon_m = \epsilon e_1 \ldots e_m$, where $e_i < 1$. If those factors are assumed to be independent, then $\log \epsilon_m$ is a sum of independent random numbers. As m increases, the statistics of the sum tends to a normal distribution with the variance proportional to m. In other words, multiplicative randomness leads to log-normality.

Since the number of steps of the cascade from L to r is proportional to $\ln(L/r)$, Obukhov then assumed that the energy dissipation rate that is coarse-grained on a scale r has such a log-normal statistics with variance $\langle \ln^2(\epsilon_r/\bar{\epsilon}) \rangle = B + \mu \ln(L/r)$, where B is a non-universal constant determined by the statistics at large scales. Note that the variance grows when r decreases and so also do other (not very high) moments: $\langle \epsilon_r^q \rangle \propto (L/r)^{\mu q(q-1)/2}$. Obukhov then formulated the refined similarity hypothesis: KO41 is true locally; that is, the velocity difference at a distance r is determined by the dissipation rate coarse-grained on that scale: $\delta v(r) \simeq (\epsilon_r r)^{1/3}$. Averaging this expression over the log-normal statistics of ϵ_r one obtains new expressions for the structure functions that contains non-universal factors C_n and universal exponents: $\langle v_{12}^n \rangle = C_n r^{n/3}(L/r)^{\mu n(n-3)/18}$. This general formula was actually derived by Kolmogorov who was shown the draft of Obukhov (1962) (containing only $n = 2$) before boarding the train that took him to the Marseille conference, separately from rest who flew there. Kolmogorov arrived at Marseille with his own draft (Kolmogorov, 1962) and their two presentations were highlights of the conference. The Marseille gathering between 28 August and 3 September 1961 was a remarkable event that brought almost all the leading researchers together, many for the first time. Yaglom recalls:

> The USSR delegation included Kolmogorov..., his two pre-war students M.D. Millionschikov and A.M. Obukhov, and me – a war-years student. Such a composition had the flavor of Khrushchev's liberalization (for me it was the first time I was permitted to attend a meeting in a 'capitalist country').

Russians at last had a chance to meet turbulence's great scholars from all generations. Most of the heroes of this book were present: von Kármán, Taylor, Batchelor, Townsend, Corrsin, Saffman and Kraichnan. It is poignant to see Kolmogorov and Kraichnan (whose names are forever linked by the 2D–3D 5/3-scaling) in the same photograph.

The new theory KO62 gives the same linear scaling for the third moment. Attempts to estimate μ from experimental data on the variance of dissipation or velocity structure functions give $\mu \simeq 0.2$, so that KO62 only slightly deviates from KO41 for $n < 10 \div 12$. Its importance must be then mostly conceptual. The main point is understanding that the relative fluctuations of the dissipation rate grow unboundedly with the growth of the cascade extent, L/r (in his paper, Kolmogorov credits that to Landau even though the latter's 1944 remark did not mention any scale-dependence of the fluctuations – Frisch, 1995). That understanding opened the way to the description of dissipation concentrated on a measure (Novikov and Stewart, 1964), which was later suggested to be fractal (Mandelbrot, 1974), and shown to be actually multi-fractal (Parisi and Frisch, 1985; Meneveau and Sreenivasan, 1987). Let us stress another

Figure 6.1 Kolmogorov and others at the Marseille conference.

conceptual point: the 5/3-law for the energy spectrum is incorrect despite being (outside of the turbulence community) the most widely known statement on turbulence. Still, KO62 does not seem to be such a momentous achievement as KO41. First, it evidently does not make sense for sufficiently high n. Second and more important, it is still under the spell of two magic concepts of the Kolmogorov school: Gaussianity and self-similarity. Compared with KO41, the new version KO62 somehow pushes these two further down the road: the new (refined) self-similarity is local and Gaussianity is transfered to logarithms, replacing additivity with multiplicativity. Still, KO62 is based on the belief that a single conservation law (of energy) explains the physics of turbulence and that the (local) energy transfer rate completely determines local statistics. As we now believe, direct turbulence cascades (from large to small scales) have, at a fundamental level, nothing to do with either Gaussianity or self-similarity, even though these concepts can help to design useful semi-empirical models for applications. There is more to turbulence than just cascades. Energy conservation determines only a single moment (the third for incompressible turbulence). To understand the nature of turbulence statistics, one returns to the old remark of Friedman that the correlation functions are "moments of conservation". In this way, one discovers an infinite number of statistical conservation laws having a geometrical nature, each determining its own correlation function; to this must be added that the exponents are now measured with higher precision and they are neither KO41, nor KO62; see for example Falkovich et al. (2001) and Falkovich and Sreenivasan (2006).

Note in passing that the interaction between Landau and Kolmogorov was a two-way street. We described above how Landau's reaction to KO41 changed the theory of turbulence. No less fruitful was Kolmogorov's reaction to Landau's suggestion in 1943 that, as the Reynolds number Re grows, the sequence of instabilities leads to multi-periodic motion; that is, the attractor in the phase space of the Navier–Stokes equation is a torus whose dimensionality grows with Re. Superficially, this seems to be very much in the spirit of Kolmogorov's own 1941 argument that "at large Re, pulsations of the first order are unstable in their own turn so that the second-order pulsations appear" (Kolmogorov, 1941a). However, Kolmogorov developed deeper insights into the onset of turbulence and posed the question of whether it is possible that a continuous spectrum appears at finite Re. That was answered by work on dynamical systems theory, which he started in 1953 "because the hope appeared and my spirit uplifted" (Stalin died). The resulting KAM theory (after Kolmogorov, Arnold and Moser) describes which invariant tori survive under a slight change of Hamiltonian and forms the basis of understanding Hamiltonian chaos. Later, Kolmogorov initiated a great synthesis of the random and deterministic, based on the notions of entropy and complexity, magnificently carried out by his student Sinai and others. To overcome the natural prejudice of considering dynamic systems as deterministic, one needs to be profoundly aware of the finite precision of any measurement and of the exponential divergence of trajectories (Kendall et al., 1990). Kolmogorov–Sinai entropy and dynamical chaos are fundamental to our understanding of numerous phenomena; in particular, related ideas were used later for describing the statistics of turbulence below the Kolmogorov–Obukhov scale where the flow is spatially smooth but temporally random; see for example Falkovich et al. (2001). In addition, Kolmogorov's program for the 1958 seminar included the task of developing the theory of one-dimensional (Burgers) turbulence which was completed by Sinai and others some 40 years later.

I find it puzzling though that Kolmogorov himself never applied his powerful probabilistic thinking and understanding of stochastic processes and complexity to quantum mechanics and statistical physics (it was done by his students Gelfand and Sinai respectively). It seems that Kolmogorov's direct contact with physics was only via classical mechanics and hydrodynamics (Novikov, 2006).

Obukhov started a new chapter in Obukhov (1969) by introducing what he called systems of hydrodynamic type and what were later known as shell models. He was inspired by the 1966 work of Arnold on the analogy between the

Euler equation for incompressible flows and the Euler equation for solid body motion; see Arnold and Kesin (1998) for the detailed presentation[3]. Obukhov approximated fluid flow by a system of ordinary differential equations with quadratic nonlinearity and quadratic integrals of motion. Since there was no consistent way of determining the number of equations for this or that type of flow, Obukhov initiated laboratory experiments and their detailed comparison with computations. It is worth noting that Obukhov and his co-workers worked on few-mode dynamic models (apparently independently of E. Lorentz) as well as on chains intended to model turbulence cascades (Gledzer et al., 1981).

We conclude this section by referring the reader to the magnificent opus by Monin and Yaglom where much more can be found on KO41, KO62 and many other subjects including the field-theoretical approaches of Edwards, Kraichnan and others. "If ever a book on turbulence could be called definitive", declared *Science* in 1972, "it is this book by two of Russia's most eminent and productive scientists in turbulence, oceanography, and atmospheric physics." As does the presentation here, it stresses the physics of KO41 and KO62, but also makes it clear that the theory in its entirety is definitely that of mathematicians. The mathematical foundations were laid before and after 1941 in the works of Kolmogorov, Obukhov, Gelfand, Yaglom and others. A complete analysis of stationary processes using the Hilbert space formulation was done in 1941. Considerable work was done on spectral representations of random processes; subtle points of legitimacy and convergence were cleared for the Fourier transform and other orthogonal expansions for translation-invariant random functions, which physicists take for granted without much thought. Part 2 of the Monin–Yaglom book was finished in 1966 and published in 1967, in time to cite the first 1965 paper of Zakharov on wave turbulence. That is the subject of the next section.

6.4 Theoretical physicist

> Keep your hands off our light entertainment,
> Do not tempt us with crumbs of attainment,
> Do not teach us the right aspirations,
> Do not tease us with serving the nation.
> V. Zakharov (2009b)

Another stream in Russian work on turbulence originates from the Landau school. Apart from his cameo appearance in the Kolmogorov–Obukhov

---

[3] I believe that exploration of this analogy will bear even more fruit in the future.

part of the story, Landau himself didn't work on the theory of developed turbulence, despite his firm belief that the problem belongs in physics. In the
1950s, he was interested in plasma physics and steered in this direction the
young Roald Sagdeev, who went to work in the theoretical division of the Russian project on controlled thermonuclear fusion. Plasmas are subject to various instabilities and practically always are turbulent. Inspired by the works
of David Bohm on an anomalous diffusion in plasmas (Bohm et al., 1949)
and the needs of thermonuclear fusion, the theory of plasma instabilities and
turbulence was intensely developed in Russia by B. Kadomtsev, A. Vedenov,
E. Velikhov and R. Sagdeev during the 1950s and 1960s. Sagdeev's uniform
approach to plasma hydrodynamics (extended then to other continuous media)
was a trademark of the Landau school: at first all dynamical equations of continuous media were supposed to be written in a canonical Hamiltonian form,
then particular solutions are found and their stability analyzed, then perturbation theory applied to the description of random fields.

To carry on this project, a most unlikely figure appeared: a student expelled
for a fistfight from the Moscow Energy Institute. Vladimir Zakharov was born
in Kazan in 1939 to the Russian family of an engineer. When at elementary
school, he did well and had a slight burr so was considered a Jew by his peers
– an experience conducive to an early formation of personal independence.
Zakharov knew Sagdeev first as a friend of his older brother and he met him
in the Energy Institute where Sagdeev was teaching physics part-time. After
expulsion, Sagdeev brought Zakharov to G. Budker who led parallel experimental projects in two fields (high energy and plasma physics) and two cities
(Moscow and Novosibirsk). In 1957, a new scientific center was created some
3500 kilometers east of Moscow. In 1961 Budker convinced Sagdeev and Zakharov to leave Moscow and come to that new center in Novosibirsk. Sagdeev
was to lead the plasma physics department in the newly established Nuclear
Physics Institute (now the Budker Institute) while Zakharov was admitted to
Novosibirsk University, leaving all his troubles behind and starting a new life
in the brave new world of hastily built barrack-style buildings in the middle of
the taiga.

Note in passing that Zakharov's poetry was published by the main Russian
literary magazines, included in anthologies etc.: there exists a bilingual book
with English translations (Zakharov, 2009a). As a scientist, he grew up inside a
strongly interacting community of physicists and mathematicians, particularly
influenced by M. Vishik, V. Pokrovsky and G. Budker. Zakharov succeeded in
making important advances in the directions usually considered far apart: in-
tegrability and exact solutions on the one hand, and turbulence on the other. In
particular, he was able to find turbulence spectra as exact solutions.

Following Sagdeev's program, Zakharov reformulated the equations for plasma and water waves in terms of Hamiltonian variables. Written as the amplitudes of plane waves, all such equations have the form $\dot{a}_k = -i\omega_k a_k +$ nonlinear terms (quadratic, cubic, etc.). Now, if the wave amplitudes are small while the frequencies ω_k are large and different for different k, one can treat nonlinear terms as small perturbations. Considering a set of random small-amplitude waves in a random-phase approximation, one expresses the time derivative of the second moment, $\langle a_k a_{k'}^* \rangle = n_k \delta(\mathbf{k} - \mathbf{k'})$, via the third moment, which is relevant if three-wave resonances are possible; i.e. one can find triads of wave vectors such that $\omega_{k+k'} = \omega_k + \omega_{k'}$. Exactly as in a quasi-normal approximation, one then writes the equation for the third moment, decouples the fourth moment and obtains the kinetic equation for waves:

$$\dot{n}_k = \int W(\mathbf{k}, \mathbf{p}, \mathbf{q})\delta(\omega_k - \omega_p - \omega_q)\delta(\mathbf{k} - \mathbf{p} - \mathbf{q})(n_p n_q - n_k n_p - n_k n_q)d\mathbf{p}d\mathbf{q}$$

$$+ \text{ cyclic permutations } k \rightarrow p \rightarrow q \rightarrow k.$$

The right-hand side is a collision term very much like in the Boltzmann kinetic equation. The idea of phonon collisions was introduced in Peierls (1929); the collision term was used by Landau and Rumer in 1937 to calculate sound absorption in solids (Landau and Rumer, 1937). In the article (Zakharov, 1965) submitted on 28 October 1964, Zakharov took this equation (which he learnt from Camac et al., 1962) and asked if it has a stationary solution different from the equilibrium Rayleigh–Jeans distribution $n_k = T/\omega_k$. Inspired by the Kolmogorov–Obukhov spectrum he set to look for a power-law solution $n_k \propto k^{-s}$. Taking first the case of acoustic waves when the coefficients are relatively simple, $\omega_k \propto k$ and $W \propto kpq$, Zakharov first checked that the collision integral tends to $-\infty$ when $s \rightarrow 4$ and to $+\infty$ when $s \rightarrow 5$, so it has to pass through zero at some intermediate s. He then bravely substituted $s = 4.5$ and obtained for the collision integral 18 gamma-functions that promptly canceled each other. The first Kolmogorov–Zakharov spectrum was born. Still, it took some time for Zakharov to appreciate that the spectrum indeed describes a cascade of energy local in k-space and is an exact realization of KO41 ideas: by checking the convergence of the integrals in the kinetic equation at $p \rightarrow 0$ and $p \rightarrow \infty$, one can directly establish that the ends of the inertial interval really do not matter, in contrast with hypotheses about turbulence of incompressible fluids. Interestingly, the position of the Kolmogorov–Zakharov exponent exactly in the middle of the convergence interval is a general property now called counterbalanced locality: the contributions of larger and smaller scales are balanced on the steady spectrum (Zakharov et al., 1992). In 1966, Zakharov submitted his PhD thesis (under the supervision of Sagdeev) which was devoted to waves on a water surface (Zakharov, 1966; Zakharov and Filonenko, 1966). There, one finds a complete description for the case of capillary waves: obtaining the spectrum from the flux constancy condition, checking locality as integral convergence and showing that this is indeed an exact solution by using conformal transforms that were independently invented by Kraichnan for his direct interaction approximation at about the same time (see Chapter 10). Then Zakharov takes on the turbulence of gravity waves whose dispersion relation, $\omega_k \propto \sqrt{k}$, does not permit three-wave resonances. In this case, the lowest possible resonance corresponds to four-wave scattering. Every act of scattering conserves not only the energy, $E = \int \omega_k n_k d\mathbf{k}$, but also the wave action, $Q = \int n_k d\mathbf{k}$, which can be also called 'number of waves'. The situation is thus similar to the two-dimensional Euler equation which conserves both the energy and squared vorticity. In his thesis, Zakharov derives two exact steady turbulent solutions of the four-wave kinetic equation, one with the flux of E and another with the flux of Q. He then argues that the energy cascade is direct, i.e. towards small scales. While Zakharov derived an exact solution that describes an inverse cascade, he didn't explicitly interpret it as such (he also gave some arguments in the spirit of Onsager about transport of Q to large scales in a decaying turbulence). After Kraichnan's 1967 paper was published and brought to his attention by B. Kadomtsev, Zakharov realized the analogy and interpreted the spectra he derived as a double-cascade picture. In 1967, he published the direct-cascade spectrum for Langmuir plasma turbulence (Zakharov, 1967), where the inverse-cascade spectrum was obtained in 1970 by E. Kaner and V. Yakovenko from the Kharkov branch of the Landau school (Kaner and Yakovenko, 1970). Note that the hypotheses Kolmogorov formulated in 1941 are true for Zakharov's direct and inverse cascades of weak wave turbulence and are probably true for Kraichnan's inverse cascade in incompressible two-dimensional turbulence as well. In 2006, Kraichnan and Zakharov were together awarded the Dirac medal for discovering inverse cascades.

The early years of the Novosibirsk scientific center were also the years of Khrushchev's brief thaw. At that time, there was probably no other place in the country where academicians, professors and young students lived in such a close proximity and had so few barriers for scientific and social interaction. A small town in the forest, "Siberia's little Athens", was for a while allowed some extra degrees of freedom. That was about to end in 1968 when Zakharov became one of the initiators and signatories of the open letter to the Party Central Committee protesting the arrests of dissidents. But Brezhnev's time was vegetarian compared with that of Stalin: Zakharov's only punishment was a ban on foreign travel, then thought to be forever.

Hard is Athenian mien,
harder still 'midst feasting vultures.
He who will get on the wing,
sees half the world as his home.
Zakharov (2009b)

6.5 Epilogue

> In twenty years no one will know what
> actually happened in our country.
> A.N. Kolmogorov, 1943 (Nikol'skii, 2006)

Our story ends (somewhat arbitrarily) in 1970. What followed – study of shell models by Obukhov's school, development of the weak turbulence theory by Zakharov's school, works on Lagrangian formalism and zero modes – deserves a separate essay which may be too early to write.

In his old age Kolmogorov suffered from Parkinson's disease and from an eye illness that made him almost blind. Nevertheless, he tried to work practically until the end, always surrounded by his former students, who also took turns in providing necessary help. Landau was seriously injured in an automobile accident in 1962; he was 59 days in a coma and survived with the help of his students and colleagues in the country and abroad; he lived for six more years but was unable to work. Obukhov and Yaglom worked until their last days, monuments of unaging intellect.

Kolmogorov died in 1987 and Obukhov in 1989. That year, the Berlin Wall fell, the Soviet Union opened the gates and disintegrated within two years. An exodus of scientists brought substantial parts of the Kolmogorov, Obukhov and Landau schools to the West. These schools then turned international but also weakened their links to Russia and started to lose their distinct Russian spirit.

Under one of the most oppressive regimes in the twentieth century, in the country which lost most of its educated class to emigration, civil war and terror, and which was often plagued by war, diseases, poverty and hunger, great mathematical and physical schools flourished. Scholars raised in these schools had a specific code of behavior. Long corridor chats were the most effective forums of exchanging the latest ideas. Most seminars had no sharply defined ends, some even had no clear beginning, as the people came before to discuss related subjects (Ya.G. Sinai, private communications, 2009–2010). Everyone worked inside a coherent group of people familiar with the details of each other's work (a downside was that some people never had much incentive to learn how to present their results to the outside world). Much has been said about the aggressive style and interruptions at Russian seminars. One must however understand the context: in a life which was a sea of official lies, doing science was perceived as building a small solid island of truth; even unintentional errors risked decreasing the solid ground on which we stand. Landau used to say: "An error is not a misfortune, it is a shame". One is reminded of monastic orders that preserved and advanced knowledge during the dark ages (though in other respects, most Soviet scientists weren't monks). A more prosaic reason that bonded people within a school was an impaired mobility of scientists – recall that both Kolmogorov and Landau had a postdoctoral period abroad, a possibility denied to most of their students[4]. Still, the main attraction of the schools was the personalities of the leaders.

By radically restricting creative activities, a tyrannical society channeled the creative energy into the narrow sector of natural sciences and mathematics. Russian society is more open now, and the choice of science as one's occupation is rarely placed in the context of morality. Will we ever again be blessed with universalist geniuses of the caliber of Kolmogorov and Landau?

Acknowledgements In preparing this essay, I have benefitted from conversations with U. Frisch, G. Golitsyn, K. Khanin, B. Khesin, D. Khmelnitskii, S. Medvedev, R. Sagdeev, Ya. Sinai, K. Sreenivasan, V. Tseitlin and V. Zakharov.

References

Arnold, V.I. and B.A. Khesin. 1998. *Topological Methods in Hydrodynamics*, Appl. Math. Sci. Series **125**, Springer-Verlag.

Battimelli, G. and A. Vulpiani. 1982. Kolmogorov, Heisenberg, von Weizsäcker, Onsager: un caso di scoperta simultanea. In *Atto III Congresso Nazionale di Storia della Fisica (Palermo 11–16.10.1982)*, 169–175.

Barenblatt, G.I. 1953. Motion of suspended particles in a turbulent flow. *Prikl. Mat. Mekh.*, **17**, 261–74; 1955. Motion of suspended particles in a turbulent flow occupying a half-space or plane channel of finite depth. *ibid.*, **19**, 61–68.

Bohm, D. et al. 1949. In *The Characteristics of Electrical Discharges in Magnetic Fields*, A. Guthrie and R.K. Wakerling (eds), McGraw–Hill.

Bolgiano, R. 1959. Turbulence spectra in a stably-stratified atmosphere, *J. Geophys. Res.* **64**, 2226–9; 1962. Structure of turbulence in stratified media, *ibid.*, **67**, 3015–23.

Camac, M. et al. 1962. Shock waves in collision-free plasmas. *Nuclear Fusion Supplement* **2**, 423.

Dryden, H.L. et al. 1937. Nat. Adv. Com. Aeronaut., Rep 581.

---

[4] I believe that impaired mobility was the main reason why Soviet science as a whole never lived up to our expectations.

Falkovich, G., K. Gawędzki, and M. Vergassola. 2001. Particles and fields in fluid turbulence, *Rev. Mod. Phys.*, **73**, 913.

Falkovich, G. and K.R. Sreenivasan. 2006. Lessons from hydrodynamic turbulence, *Physics Today*, **59**(4), 43.

Frenkel, V. 1988. Alexander Alexandrovich Friedman, Biography, *Sov. Phys. Uspekhi.*, **155**, 481.

Friedman, A.A. 1922. Über die Krümmung des Raumes, *Z. Physik*, **10**, H6, 377–387.

Friedman, A.A. and L.V. Keller. 1925. Differentialgleichungen für die turbulente Bewegung einer compressibelen Flüssigkeit. In *Proc. First. Internat. Congress Appl. Mech. Delft*, C. Beizano and J. Burgers (eds), 395–405.

Frisch, U. 1995. *Turbulence: The Legacy of A.N. Kolmogorov*, Cambridge University Press.

Gledzer, E.B., F.V. Dolzhanskii and A.M. Obukhov. 1981. In *Systems of Hydrodynamic Type and Their Application*, Nauka, Moscow (in Russian).

Golitsyn, G.S. 1973. *An Introduction to Dynamics of Planetary Atmospheres*, Leningrad, Gidrometeoizdat (in Russian).

Golitsyn, G.S. 2009, 2010. Private communications.

Gurvitz, A.S. 1960. Experimental study of the frequency spectra of the vertical wind in the atmospheric boundary layer, *Dokl. Akad. Nauk*, **132**, 806–809.

Kaner, E.A. and V.M. Yakovenko. 1970. Weak turbulence spectrum and second sound in a plasma, *Sov. Phys. JETP*, **31**, 316–30.

Kendall, D. et al. 1990. Andrei Nikolaevich Kolmogorov (1903–1987), *Bull. London Math. Soc.*, **22** (1), 31–100.

Kolmogorov, A.N. 1933. *Grundbegriffe der Wahrscheinlichkeitsrechnung*, Springer-Verlag.

Kolmogorov, A.N. 1941a. The local structure of turbulence in incompressible viscous fluid for very large Reynolds number, *C. R. Acad. Sci. URSS*, **30**, 301–305.

Kolmogorov, A.N. 1941b. On decay of isotropic turbulence in an incompressible viscous fluid, *C. R. Acad. Sci. URSS*, **31**, 538–540.

Kolmogorov, A.N. 1941c. Energy scattering in locally isotropic turbulence in incompressible viscous fluid for very large Reynolds number, *C. R. Acad. Sci. URSS*, **32**, 19–21.

Kolmogorov, A.N. 1941d. Logarithmically normal distribution of the size of particles under fragmentation, *Dokl. Akad. Nauk SSSR*, **31**, 99–101.

Kolmogorov, A.N. 1949. On the disintegration of drops in a turbulent flow, *Dokl. Akad. Nauk SSSR*, **66**, 825–828.

Kolmogorov, A.N. 1962. Precisions sur la structure locale de la turbulence dans un fluide visqueux aux nombres de Reynolds élevés (in French and Russian). In *Mécanique de la Turbulence* (Coll. Int. du CNRS a Marseille), 447–58. Paris: CNRS; A refinement of previous hypotheses concerning the local structure of turbulence in a viscous incompressible fluid at high Reynolds numbers. *J. Fluid Mech.*, **13**, 82–85.

Kolmogorov, A.N. 2001. *Private Recollections*, translated from Russian, http://www.kolmogorov.info/curriculum-vitae.html.

Landau, L.D. and E. Lifshitz. 1987. *Fluid Mechanics*, Pergamon Press.

Landau, L.D. and Yu.B. Rumer. 1937. On sound absorption in solids, *Phys. Zs. Sovjet.*, **11**, 8–15.

Mandelbrot, B. 1974. Intermittent turbulence in self-similar cascades: divergence of high moments and dimension of the carrier, *J. Fluid Mech.*, **62**, 331–58.

Millionschikov, M.D. 1939. Decay of homogeneous isotropic turbulence in a viscous incompressible fluid, *Dokl. Akad. Nauk SSSR.* **22**, 236.

Millionschikov, M.D. 1941. Theory of homogeneous isotropic turbulence. *Dokl. Akad. Nauk SSSR*, **22**, 241–42; *Izv. Akad. Nauk SSSR, Ser. Geogr. Geojiz.*, **5**, 433–46

Meneveau, C. and K.R. Sreenivasan. 1987. Simple multifractal cascade model for fully developed turbulence, *Phys. Rev. Lett.*, **59**, 1424–1427.

Monin, A.S. 1958. Report to the Party Central Committee, published in *Moskovskaya Pravda*, **137**, (735), 19 June 1994.

Monin, A.S. 1959. The theory of locally isotropic turbulence, *Sov. Phys. Doklady*, **4**, 271.

Monin, A.S., and A.M. Yaglom. 1979. *Statistical Fluid Mechanics*, MIT Press.

Nikol'skii, S.M. 2006. The Great Kolmogorov. In *Mathematical Events of the Twentieth Century*, Bolibruch, A.A et al. (eds), Springer-Verlag.

Novikov, S.P. 2006. Mathematics at the threshold of the XXI century (Historical Mathematical Researches) (in Russian). http://www.rsuh.ru/print.html?id=50768.

Novikov, E.A. and R.W. Stewart. 1964. The intermittency of turbulence and the spectrum of energy dissipation, *Izv. Akad. Nauk SSSR, Ser. Geogr. Geojiz.*, **3**, 408–413.

Obukhov, A.M. 1941. On the spectral energy distribution in a turbulent flow, *Izv. Akad. Nauk SSSR, Geogr. Geofiz.*, **5**, 453–466.

Obukhov, A.M. 1949a. The structure of the temperature field in a turbulent flow, *Izv. Akad. Nauk SSSR, Geogr. Geofiz.*, **13**, 58.

Obukhov, A.M. 1949b. Pulsations of pressure in a turbulent flow, *Dokl. Akad. Nauk.*, **66**, 17–20.

Obukhov, A.M. 1951. Properties of wind microstructure in the near-ground atmospheric layer, *Izv. Akad. Nauk SSSR, Geofiz.*, **3**, 49–68.

Obukhov, A.M. 1959. Influence of Archimedes force on the temperature field in a turbulent flow, *Dokl. Akad. Nauk.*, **125**, 1246–1248.

Obukhov, A.M. 1962. Some specific features of atmospheric turbulence. *J. Geophys. Res.*, **67**, 311–314; *J. Fluid Mech.*, **13**, 77–81.

Obukhov, A.M. 1969. Integral invariants in systems of hydrodynamic type, *Dokl. Akad. Nauk.*, **184**, 309–312.

Obukhov, A.M. 1971. Turbulence in an atmosphere with a non-uniform temperature, *Boundary-layer Meteorology*, **2**, 7–29.

Obukhov, A.M. 1988. *Selected Works*, Gydrometeoizdat (in Russian).

Obukhov, A.M. 1990. Interest in Geophysics. In *Recollections about Academician M.A. Leontovich*, Nauka, Moscow (in Russian).

Parisi, G. and U. Frisch. 1985. On the singularity structure of fully developed turbulence. In *Turbulence and Predictability in Geophysical Fluid Dynamics*, Proc. Int. School 'E. Fermi', 1983, Varenna, Italy, 84–87, M. Ghil, R. Benzi and G. Parisi (eds), North-Holland, Amsterdam.

Peierls, R. 1929. Zur kinetischen Theorie der Würmeleitung in Kristallen, *Annalen der Physik*, **3**, 1055–1101; On kinetic theory of thermal conduction in crystals. In *Selected Scientific Papers by Sir Rudolf Peierls with Commentary*, World Scientific, 1997.

Peixoto J. and A. Oort. 1984. Physics of climate, *Rev. Mod. Phys.*, **56**, 365–429.

Schwinger, J. 1951. On gauge invariance and vacuum polarization, *Phys. Rev.*, **82**, 664–679.

Shiryaev A.N. (ed.) 2006. *Kolmogorov in Memoirs of Students* (in Russian). Nauka, Moscow.

Sreenivasan, K.R. and G. Eyink. 2006. Onsager and the theory of hydrodynamic turbulence, *Rev. Mod. Phys.*, **78**, 87–135.

Yaglom, A.M., 1949a. On a local structure of the temperature field in a turbulent flow, *Dokl. Akad. Nauk. SSSR*, **69**, 743.

Yaglom, A.M. 1949b. On acceleration field in a turbulent flow, *Dokl. Akad. Nauk. SSSR*, **67**, 795.

Yaglom, A.M. 1988. Obukhov's works on turbulence. In Obukhov (1988).

Yaglom, A.M. 1994. A.N. Kolmogorov as a fluid mechanician and founder of a school in turbulence research, *Ann. Rev. Fluid Mech.*, **26**, 1–22.

Zakharov, V.E. 1965. Weak turbulence in a media with a decay spectrum, *J. Appl. Mech. Tech. Phys.*, **4**, 22–24.

Zakharov, V.E. 1966. On a nonlinear theory of surface waves, PhD Thesis, *Ac. Sci. USSR Siberian Branch* (in Russian).

Zakharov, V.E. 1967. Weak-turbulence spectrum in a plasma without a magnetic field, *Sov. Phys. JETP*, **24**, 455; 1972. **35**, 908.

Zakharov V. 2009a. *The Paradise for Clouds*, Ancient Purple Translations.

Zakharov's poem (fragment), from Zakharov (2009a). Translated by A. Shafarenko.

Zakharov, V.E. and N.N. Filonenko. 1966. The energy spectrum for stochastic oscillations of a fluid surface, *Doklady Akad. Nauk SSSR*, **170**, 1292–1295; English: *Sov. Phys. Dokl.*, **11** (1967), 881–884.

Zakharov, V., V. Lvov and G. Falkovich. 1992. *Kolmogorov Spectra of Turbulence*, Springer-Verlag.

7

Stanley Corrsin

Charles Meneveau and James J. Riley

7.1 Early years

On 3 April 1920, a few years after G.I. Taylor's far-reaching observations of turbulent diffusion aboard the SS *Scotia* (Taylor, 1921), and at the time Lewis Fry Richardson was imagining vast weather simulations of atmospheric flow by human 'computers' (Richardson, 1922), across the Atlantic in the city of Philadelphia, Stanley Corrsin was born. His parents, Anna Corrsin (née Schorr) and Herman Corrsin had both emigrated to the United States only 13 years before. They came from Romania, where many Russian Jews had settled after leaving Russia in the late 19th and early 20th century. Following further hostilities in Romania, many emigrated again, this time to America. Anna and Herman Corrsin arrived separately at Ellis Island in 1907, Anna in July, and Herman in October. After brief stays in the New York and New Jersey area, where they met and married in 1912, they settled in the city of Philadelphia, in a mixed middle-class neighborhood, not far from the University of Philadelphia. They went into business in the clothing industry and raised their children. Their first son Eugene died young and their second, Lester, was born in 1918. Stan was their third and youngest son.

As a child, Stan Corrsin attended school in Philadelphia and, showing early signs of a highly gifted analytical mind, went on to skip two grades. He enjoyed following the ups and downs of his favorite baseball team, the Philadelphia Athletics. An appreciation for the game would accompany him throughout his life, including a keen interest in the subtle aerodynamic effects that can determine how balls fly through the air. Young Stanley enjoyed frequent visits to Philadelphia's Leary, a large used books store, and to Wanamaker's, Philadelphia's main department store. On these outings to downtown, he would have witnessed rapid developments thanks to mechanization and engineering. Like many American cities, Philadelphia in the 1920s saw ambitious projects of modernization with erections of steel and concrete skyscrapers, electrification of old buildings, widening of streets and construction of bridges, such as the Benjamin Franklin bridge over the Delaware river. The 1930s brought the Great Depression, and Philadelphia only began to recover with the massive further industrialization triggered by World War II and the expansion of giant shipyards that would – in time – supply the war effort.

After graduating from West Philadelphia High School in 1936, a bright and ambitious 16-year-old Stan Corrsin decided to study mechanical engineering. He had always been interested in how things work, and study in a technical field would allow an ambitious son of immigrants to make his mark. A 'mayor's scholarship' from the city of Philadelphia enabled his enrollment at the prestigious University of Pennsylvania located nearby in downtown Philadelphia. As a student, he distinguished himself with outstanding marks. And, presaging an unusual facility with the pen, it is said that he was the first engineering student at the University of Pennsylvania to receive the prize for a freshman essay in English. At some point he even considered becoming a professional writer. In his last year at U. Penn, he participated in a technical lecture competition organized by the American Society of Mechanical Engineers. His presentation entitled *An Optical Method for Visualizing Low Velocity Air Flow* earned him first prize. A proud headline from the 2 May 1940 university newspaper *The Daily Newsletter* proclaims "Corrsin dethrones Princeton as Pennsylvania takes first prize in engineer's Tourney".

In 1940, at the age of 20, Stan Corrsin graduated from U. Penn with a bachelors degree in mechanical engineering. Being drawn to the fundamentals underlying the engineering practice, he chose to continue his education and apply to graduate school in engineering. At the time, however, many top institutions still placed limits and quotas on students of Jewish heritage. It is said that Clark B. Millikan at Caltech heard of young Corrsin's promise and decided on the spot to offer him admission to Caltech. Thus Corrsin was accepted into the Caltech graduate program in aeronautics and he traveled to Southern California to begin graduate studies, just as the US war effort was ramping up.

7.2 First contributions at Caltech

Corrsin arrived at Caltech during what has come to be known as the golden age of aeronautics. Caltech was the US epicenter of aerodynamics at the time. The United States had found itself challenged to manufacture, on very short order, a massive fleet of war planes and needed to develop a basic understanding of the fluid dynamics of flight to aid in the design of modern aircraft. A vigorous

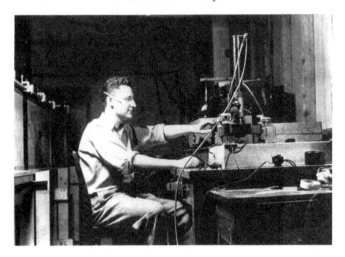

Figure 7.1 A 22-year-old Stan Corrsin in 1942 at the Graduate Aeronautical Laboratories, Caltech (GALCIT). The verso of the photograph states, in Corrsin's handwriting "Man doing research". The note is addressed to his mother, and adds for reassurance: "I'm really not so thin, it's just the light". Photograph courtesy of Dr. Stephen D. Corrsin.

program in fundamental and applied fluid dynamics research had been developing at Caltech's Guggenheim Aeronautical Laboratory (GALCIT), under the direction of Theodore von Kármán and Clark B. Millikan. Hans Liepmann, who in 1939 had just been hired by von Kármán after finishing his PhD in Zürich, became Corrsin's main academic adviser. At the time, Liepmann had begun experimental studies on boundary layers, transition to turbulence, and various turbulent shear flows. Corrsin began working in Liepmann's laboratory and distinguished himself for his dexterity in experimental science.

His first project at Caltech, which became his thesis in partial fulfillment of the requirements of Aeronautical Engineer, dealt with measurements of the decay of turbulence behind various grids. The subject of isotropic turbulence was in the air: on a visit to Caltech in 1936 and 1937, Leslie Howarth had collaborated with von Kármán and developed the equation for two-point correlation functions in decaying isotropic turbulence (von Kármán and Howarth, 1938). Corrsin's initial experiments provided data on the decay of standard deviations of two of the three turbulent velocity components behind three types of grids. More will be said later about ingenious measurement techniques of the time. In a photograph taken in the laboratory (Figure 7.1), he is seen reading a manometer, pencil tucked behind his ear. The results of the experiments, it turns out, were rather inconclusive. It was unclear whether turbulence was, or was not, observed to be sufficiently isotropic, or what the decay rate was. At the end of the thesis, which was never published in journal format, Corrsin writes: "The [...] conclusions are rather tentative; it is hoped that more certain results, and in particular the reasons for them, will come out of further investigation". Thus were laid the early seeds for Corrsin's work elucidating the fundamentals of isotropic turbulence. He completed the Masters thesis in 1942 (Corrsin, 1942) but, as further described in §7.6, his definitive experiments on decaying isotropic turbulence would have to await over two decades to become reality.

Corrsin then began to work in earnest towards his doctoral research and this work led to important, and no longer tentative or uncertain, results. The 1930s had seen initial developments in documenting basic properties of what are now known as the 'canonical' turbulent shear flows. By applying Prandtl's boundary layer concept to turbulent shear layers, thus assuming that they become asymptotically thin (although many never do), simplified parabolic equations had been developed describing the mean velocity in plane and round wakes and jets, in mixing layers, and in turbulent boundary layers along walls. The use of similarity variables and the eddy-viscosity assumption with local velocity and length-scales led to further simplifications. A series of experiments, most notably the measurements of mean velocity profiles in wakes by Townsend, had already begun to establish the validity and limitations of this approach. The popular textbooks by Townsend (1956), Hinze (1959) and Tennekes and Lumley (1972) provide excellent accounts of the accomplishments of that era.

By the early 1940s, after several of the canonical shear flows had been measured and documented in terms of mean velocity and Reynolds stresses, attention began to turn to the distribution of scalar fields. Examples of scalar fields include the temperature or the concentration associated with species being transported by turbulence. They are termed 'passive scalars' if they do not affect the motion, which therefore excludes cases with buoyancy effects that often occur in geophysical flows, or with strong volumetric expansion that accompany combustion. In the early 1940s, not much was known about distributions of passive scalars in turbulent shear flows. Of natural importance to propulsion and mixing, the turbulent jet was of great interest to von Kármán, Millikan, and the National Advisory Committee for Aeronautics (NACA). Thus, Corrsin's doctoral research project was on detailed measurements of the velocity and temperature fields in round jets. The work was closely followed by von Kármán, Millikan and supervised by Liepmann. Financial support was provided by NACA.

Corrsin went to work and designed and, with helpful laboratory technicians, built the experiment out of an existing, open-return 6 1/2 feet diameter wind tunnel. It was retrofitted with a contracting nozzle unit near its exit, thus creating a jet. Electrical units upstream would provide heating for the air, and warm air was also ducted outside the nozzle to improve uniformity of the temperature profile exiting the jet. In characteristic style, he writes:

> That this scheme was not completely successful can be seen from the temperature distribution measured at the mouth. It did represent, however, a distinct improvement over the wooden nozzle first tried.

The mean velocity and temperature readings were photographically recorded on automatically traversing photo-sensitive paper illuminated by a light beam. The latter was continually being deflected by a mirror mechanically connected to a Pitot pressure line for mean velocity measurements, and by a galvanometer connected to a thermocouple for mean temperature. Fluctuating velocity was measured using a platinum hot wire. In the second part of the report, an oscilloscope was used and the screen photographically recorded.

The results were reported in a NACA Wartime Report (Corrsin, 1943). The report contains 43 figures with profiles of mean velocity and temperature at various downstream distances, profiles of standard deviation of velocity fluctuations, log-log plots for downstream scaling, calibration curves, etc. Rather than simply showing experimental results, much of the effort was spent in detailed comparisons with profiles predicted using several variants of eddy-viscosity models. Based on the measurements, Corrsin reached conclusions about limits of validity of the similarity assumption and commented on differences between the scalar and momentum diffusivities (he confirmed in his measurements that the turbulent Prandtl number is less than unity). Notably, his first conclusion was "In a fully developed turbulent jet with axial symmetry, a completely turbulent flow exists only in the core region ...". The conclusion was based on his observations of hot-wire signals on the 'oscillograms', with fully turbulent signals when the probe is located in the centerline of the jet, but showing spotty turbulent regions interspersed with smooth, quasi-laminar portions of the signal when the probe was located off-center of the jet axis. This observation and conclusion already point to his keen interest in the detailed fundamental structure of turbulence. Corrsin would maintain interest in the phenomenon of what became known as 'outer intermittency' for several decades to come; some of this work addressing the geometry and intermittency of turbulence will be described in §7.7. A second experiment, on an array of plane jets, and a second NACA Wartime Report (Corrsin, 1944) soon followed. All the while, he contributed to important developments of the hot-wire anemometer.

According to Liepmann (1989), by 1943 Corrsin had completed the bulk of his PhD research. With the Second World War in full swing, however, he was charged with the instruction of Navy pilots and other military personnel on the basics of aerodynamics, and thus he remained actively involved in teaching at Caltech even after the end of the war, all the while writing his doctoral dissertation. His doctoral degree was awarded in May 1947, for his two-part dissertation entitled *I. Extended Applications of the Hot-Wire Anemometer; II. Investigations of the Flow in Round Turbulent Jets* (Corrsin, 1947).

At Caltech Corrsin had met a young woman, Barbara Daggett, who would become his wife. She was originally from the Los Angeles area, and worked as part of the Caltech administrative staff. They were soon married. Then came the call from Johns Hopkins University to join its faculty as Assistant Professor.

7.3 Arrival in Baltimore

The end of the Second World War and the transformative GI bill that provided college support for returning servicemen brought a renewed sense of direction that was felt on many campuses across the United States. At the Johns Hopkins University, located in the east coast city of Baltimore, there was talk of creating a department that would focus on the new science of aeronautics. This would be a new and forward-looking department, part of the university's School of Engineering. Johns Hopkins had been founded in 1876, at first occupying temporary spaces downtown, and only between 1914 and 1916 did classes move to the university's definitive seat on the Homewood campus – in what not too long before had been farmland but was fast becoming a leafy suburb of the city. The School of Engineering had existed as part of the university almost since its inception, and counted the traditional departments of Mechanical, Electrical, Civil, Chemical, and Sanitary Engineering. The addition of Aeronautical Engineering would develop synergies with laboratories in the area such as the Applied Physics Laboratory, Aberdeen Proving Ground, and the Naval Ordnance Laboratory that all pursued research and development in the rapidly developing field of aerodynamics.

The department began in 1946, under the direction of Francis H. Clauser, who was brought in as the department chair. Clauser, an earlier Caltech graduate who had worked under von Kármán, became well known as the developer of the 'Clauser plot' method to determine skin friction coefficients from measurements of mean velocity in turbulent boundary layers (Clauser, 1954). He began to hire faculty and among the first two was recently graduated Stan Corrsin who, together with his wife Barbara, thus moved across the country back to the East Coast. He began his work at the Johns Hopkins University in 1948, and would remain at the same institution for the rest of his life. The Corrsins lived several miles north of the Johns Hopkins campus in the suburban, almost rural, Towson area, in a house they bought in 1953. They had two children, Nancy E. Corrsin and Stephen D. Corrsin. Meanwhile, his parents Anna and Herman would retire to the state of Florida.

The Johns Hopkins Aeronautics Department continued to grow in the following years. In 1950 Leslie G. Kovasznay was appointed to the faculty, followed by Mark Morkovin and Robert Betchov who were appointed as research scientists (Hamburger Archives, JHU, 2009). In one of the most visible early contributions of the department, the faculty participated in a number of episodes of the critically acclaimed television series *The Hopkins Science Review*. Episodes included *Flight at Supersonic Speeds*, which aired on 2 February 1949, and a series *Man Will Conquer Space*, that aired in October 1952 and featured Wernher von Braun as the guest.

With him from Caltech, Corrsin brought Mahinder Uberoi, who had received his Masters degree there in 1946. Uberoi moved to Baltimore to become Corrsin's first doctoral student. They also brought along a hot-air jet unit that they had built at Caltech (Corrsin and Uberoi, 1950). It was more compact than the original wind tunnel add-on facility Corrsin had used earlier (Corrsin, 1947). The unit consisted of a centrifugal blower pushing air through a horizontal chamber with heating coils. A 90 degree elbow then turned the flow upward and, after passing through further screens and a smooth contraction, the 1 inch diameter heated jet emanated up into the laboratory. They also brought hot-wire anemometry from California. The experiment was set up in a laboratory in the Aeronautics Building (later known as Merryman Hall), an unassuming, grey concrete block building next to a wooded hillside at the edge of campus.

The velocity and temperature measurements from this experiment are described in some detail in a new NACA report (Corrsin and Uberoi, 1951). While this report was in press, Corrsin had performed initial analysis of the velocity data and published a rather remarkable brief communication in the *Journal of the Aeronautical Sciences* (Corrsin, 1949). This would be his third publication in 1949 and since joining Johns Hopkins. [He had written two other short notes published earlier in 1949, with Kovasznay on a hot-wire length correction (Corrsin and Kovasznay, 1949) and on transformation formulae between one and three-dimensional scalar spectra (Kovasznay et al., 1949).] He begins the short *Journal of Aeronautical Sciences* note with the statement that "The most significant idea contributed to the problem of turbulent shear flow in many years is the hypothesis of local isotropy due to Kolmogorov" (a statement of remarkable longevity still valid, some would say, to this day). He goes on to present a plot of the correlation coefficient between band-pass filtered signals of stream and cross-stream components (the normalized cross-spectrum). The data were taken at the maximum shear region in the jet using X-hot wires and the voltage readings from both wires were band-pass filtered using analog filter banks. The difference of their mean-square voltages, evaluated using vacuum thermocouple units, are proportional to the co-spectrum. The correlation coefficient as function of frequency decays rapidly to zero at the high frequencies characteristic of small-scale motions. Statistical isotropy demands that the two fluctuating components be uncorrelated at high frequencies. Corrsin's observation, therefore, gave significant and direct support to the notion that small scales in turbulence are isotropic, in a flow where the large scales clearly are not isotropic. This would be his first of many direct experimental examinations of theories pertaining to the small-scale structure of turbulence.

7.4 Structure of scalar fields in isotropic turbulence

Having begun to ponder the fine-scale structure of turbulence and having now temperature and velocity data available from the experiments with Uberoi, Corrsin turned his attention to the expected forms of the temperature two-point correlations and spectra in isotropic turbulence. In Corrsin (1951a), he applied the methodology of von Kármán and Howarth (1938) to derive the equation for scalar correlations $\langle \theta(\mathbf{x} + \mathbf{r})\theta(\mathbf{x})\rangle = \langle \theta^2\rangle m(r)$. He also used the von Kármán and Howarth (1938) argument about the vanishing pressure–velocity correlations in isotropic turbulence to reason that the temperature–velocity correlation vanishes, and established the dimensionless third-order scalar-variance velocity correlation and its cubic behavior with distance at small displacements. He went on to define integral and Taylor micro-scales appropriate for the scalar field, namely

$$L_\theta = \int_0^\infty m(r)\,dr, \quad \text{and} \quad \lambda_\theta^2 = -2/m''(0,t),\qquad(7.1)$$

respectively. He also examined various possible consequences of assuming self-preserving solutions and discussed the role of the invariant,

$$N = \langle\theta^2\rangle \int_0^\infty r^2 m(r)\,dr,\qquad(7.2)$$

during the decay of scalar fluctuations.

In parallel, he examined the spectral structure of the scalar fluctuations by repeating the arguments presented by Kolmogorov (1941) using Fourier space.

Taking the Fourier transform of the equation for scalar correlations, Corrsin (1951b) derived the spectral equation for the temperature spectrum, $G(k)$. He derived the solution in the case of small Péclet number, where the nonlinear transfer terms are negligible. In this case the problem reduces to the heat equation with its characteristic exponential decay as $\sim \exp(-2\gamma t k^2)$, where γ is the scalar diffusion coefficient and k is the magnitude of the wavenumber. At low wavenumbers, Corrsin showed that the spectrum grows as k^2, a result intimately related to the existence of the invariant N mentioned above. For the intermediate range of wavenumbers, he went on to generalize the Kolmogorov (1941) approach using dimensional arguments.

Of crucial relevance are the rates of dissipation of kinetic energy

$$\epsilon = 2\nu \int_0^\infty k^2 E(k) dk$$

and of scalar variance

$$\epsilon_\theta = 2\gamma \int_0^\infty k^2 G(k) dk.$$

Quoting directly from Corrsin (1951b):

> The dimensions of the pertinent quantities are:
>
> $$k = L^{-1}$$
> $$G = L\mathcal{T}^2$$
> $$\epsilon_\theta = \mathcal{T}^2 T^{-1}$$
> $$\epsilon = L^2 T^{-3},$$
>
> where L=Length, $\mathcal{T}$=temperature, and T=time. Hence, the only possible arrangement is
>
> $$G(k) = A\epsilon_\theta \epsilon^{-1/3} k^{-5/3}, \tag{7.3}$$
>
> where A is a dimensionless constant.

Thus Corrsin arrived at the $-5/3$ spectral scaling in the inertial-convective range of wavenumbers. Unbeknownst to him, on the other side of the iron curtain Obukhov (1949) had undertaken very similar steps and arrived at the same form for the spectrum of scalar fluctuations. Consequently, the dimensionless constant A in (7.3) is now called the 'Obukhov–Corrsin constant' C_θ.

Having produced this well-known prediction for the power-law decay of the scalar spectrum, the experimental evidence for it in Corrsin's papers from that time is, however, not overwhelming. The NACA report (Corrsin and Uberoi,

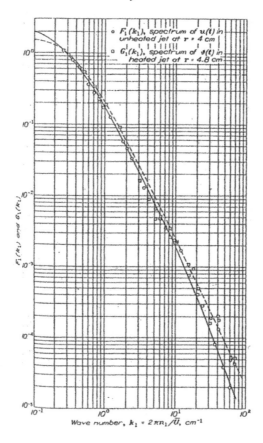

Figure 7.2 One-dimensional power spectra of velocity (circles) and temperature (squares) measured in a jet near the peak shear off-center position, adapted from Corrsin and Uberoi (1951).

1951) is only one of the very few publications where measured scalar spectra are reported. Figure 7.2 shows a reproduction of the measured spectra in the maximum shear region in the heated round jet. The velocity and scalar spectra display similar decay, not inconsistent with $-5/3$; but with the scatter of the data, as well as slightly different results obtained on the jet centerline where the scalar spectra were a bit flatter, Corrsin never argued that the data really supported his predictions of $-5/3$ scaling. The subject of the scaling of power spectrum continued to elicit many further studies over the subsequent decades, including several by his students and junior collaborators (Kistler et al., 1954; Mills et al., 1958; Sreenivasan et al., 1980; Sreenivasan, 1996).

7.5 Scalar transport and diffusion

Have you ever stopped to watch smoke billowing out of a chimney? On a calm day it moves generally upward; but there is much irregular meandering in its small-scale motions. On a windy day the wind overshadows this slow buoyant rising and washes the smoke along. Again, the smoke-cloud motion, now chiefly due to the wind, shows not only a gross pattern but also a thoroughly chaotic motion of various parts relative to each other.

Thus opens Corrsin's general interest article 'Patterns of Chaos' published in the *Johns Hopkins Magazine* (Corrsin, 1952). He wrote the article to introduce his research to the university community just a few years after arriving on campus. Ever since his PhD thesis in his study of the heated turbulent jet, a central theme of his research was the turbulent transport of scalars, e.g. smoke, temperature, and chemical species. He continued work on this topic throughout his career, making a number of major contributions.

In his thesis research (Corrsin, 1947), in addition to his important observations of the turbulent velocity field, he was – for the first time – able to measure the turbulent heat flux and, therefore, to present data to test some of the existing theories for turbulent heat transfer. This was followed, in particular, by a detailed study of diffusion of heat from a line source in isotropic turbulence (Uberoi and Corrsin, 1953), another piece of work he completed with Uberoi after they had moved to Johns Hopkins. The line source experiment was meant to address Taylor's theory of 'diffusion by continuous movements' (Taylor, 1921). In the paper were listed the following important statistical measures of the diffusive powers of turbulence:

(1) average rate of dispersion of particles from a fixed point;
(2) average rate of increase of the spacing between different fluid particles;
(3) the average rate of transport of particle concentration under a given mean concentration gradient;
(4) the average rate of increase of the length of a fluid line;
(5) the average rate of increase of the area of a fluid surface.

These and closely related topics were to consume the attention of much of Corrsin's future research.

In addressing the first of these topics, Uberoi and Corrsin realized the approximate correspondence between the average temperature, $\overline{\Theta}$, downstream from a heated wire, stretched perpendicular to the flow direction, and the probability density of Y, the fluid particle displacement in the direction normal to both the wire and the flow direction. Using this correspondence the mean-square displacement $\overline{Y'^2}$ versus downstream distance was measured, and compared with Taylor's theory. The data offered one of the first consistency checks of Taylor's theory, and an estimate for the turbulent heat transfer coefficient (turbulent diffusivity) from the Lagrangian analysis. In addition Eulerian and Lagrangian micro-scales were measured, and a correction and generalization of a theoretical expression of Heisenberg (1948) relating them was made.

After these ground-breaking results obtained in the late 1940s and early 1950s, institutional challenges called on Corrsin to lead the Mechanical Engineering Department. Thus in 1954 he moved from the Aeronautics Department in Merryman Hall on the campus periphery, to the more centrally located Maryland Hall where Mechanical Engineering was housed. As chairman of ME, he would be expected to devote some part of his time to administrative duties such as dealing with faculty hirings, teaching assignments, and managing the departmental infrastructure. There were teaching laboratories and halls containing large machinery, steam engines and, as remarked by John Lumley, other "examples of man's ingenuity" (Lumley and Davis, 2003). Lumley had arrived at Johns Hopkins in 1952, and would become Corrsin's third PhD student after Uberoi and Kistler. He recalls that one of Corrsin's major efforts was to modernize the department by removing the old machines and replacing them with wind tunnels. This did not occur without some resistance by more tradition-bound alumni and administrators in the Dean's office at the time.

A move away from engineering towards engineering science was to become one of the hallmarks of mechanics at Johns Hopkins. This move culminated with the closure of the engineering school altogether and the creation of the Mechanics Department in 1960. As recalled by Phillips (1986), Corrsin happily relinquished the chairmanship to George Benton and had a rubber stamp made that said "let George do it". He would use it with gusto on the incessant paperwork that could now proceed to be dealt with somewhere else.

With his family, he would continue to live in their house in Towson, the calm suburban area north of the city. Every morning he would drive the children to their school along tree-lined Charles Street, on his way to the Homewood campus. He never left Baltimore for extended periods of time. He did not absent himself for sabbatical leaves, preferring to host extended visitors rather than being a visitor himself. More will be said in §7.9 about the extraordinary environment at Johns Hopkins at that time, and about Corrsin's role in shaping it.

He and his students continued to work on the diffusive properties of turbulence through the next three decades. Taylor's theory, which gives a prediction of the Eulerian heat transfer coefficient, is expressed only in terms of Lagrangian quantities, namely, the Lagrangian mean-square displacement $\overline{Y'^2}$ and the Lagrangian velocity time autocorrelation $R_L(\tau) = \overline{V(t)V(t + \tau)}$, where V is a Lagrangian velocity and τ the time separation. Almost all data are taken, however, in an Eulerian frame. This led Corrsin to carefully define and then address the so-called Euler–Lagrange problem. That is, given the statistical properties of the Eulerian velocity field, say $\mathbf{u}(\mathbf{x}, t)$, what are the statistical properties of interest of the Lagrangian field, in particular the mean-square displacement and velocity time autocorrelation. Corrsin (1959; and later in more detail, 1962b) pointed out the exact relationship between the Lagrangian velocity time correlation, $R_{Lij}(\tau)$, the joint Lagrangian displacement/Eulerian velocity probability density, and the Eulerian space-time velocity autocorrelation, $R_{Eij}(\zeta, \tau) = \overline{u_i(\mathbf{x}, t) u_j(\mathbf{x} + \zeta, t + \tau)}$. Assuming that, for large time separations, the Eulerian velocity field becomes independent of the Lagrangian displacements ζ field, he then obtained

$$R_{Lij}(\tau) = \iiint p_Y(\zeta, \tau) R_{Eij}(\zeta, \tau)\, d\zeta, \tag{7.4}$$

where $p_Y(\zeta, \tau)$ is the Lagrangian displacement probability density. This result has been utilized by many turbulent dispersion model developers, and is widely known as the 'Corrsin independence hypothesis'.

Corrsin (1963) made some of the first estimates of the relationship between various Eulerian and Lagrangian length and time scales. Assuming high Reynolds numbers and the existence of inertial ranges for both the Eulerian velocity spatial spectrum and the Lagrangian velocity frequency spectrum, he concluded that, for homogeneous turbulence, $v_1' T_{11}/L_{11} \sim 1$ where v_1' is the Lagrangian root-mean square velocity in a flow direction of interest (which equals the Eulerian rms velocity u_1' for a homogeneous flow; Lumley, 1962), T_{11} is the Lagrangian integral time scale for the velocity v_1, and L_{11} is the Eulerian integral spatial scale for u_1. He found this result surprising, given the complex relationship between the Eulerian and Lagrangian autocorrelations; see, for example, equation (7.4) above. Continuing with this same reasoning, he concluded that $T_{11}/\Theta_{11} \sim 1$, where Θ_{11} is the integral time scale of u_1. Finally, he also concluded, using similar arguments, that $\theta_{11}/\alpha_{11} \sim 1$, where θ_{11} and α_{11} are temporal Taylor microscales corresponding to u_1 and v_1, respectively. Corrsin expressed doubts about this last relationship, reasoning that this estimate disagrees

> with a plausible intuitive expectation: since the Eulerian time autocorrelation involves new fluid continuously wandering past the observation point, while the Lagrangian one follows along a material point, we might expect the latter to be more persistent.

The testing of these predictions awaited accurate measurements of both the Eulerian and the corresponding Lagrangian quantities. The appropriate measurements of the Eulerian quantities were completed by Comte-Bellot and

Corrsin (1966, 1971) in their landmark measurements of homogeneous, isotropic turbulence in a frame of reference moving with the mean flow. These experiments will be described in more detail in the next section. Shlien and Corrsin (1974) repeated the heated wire experiments of Uberoi and Corrsin (1953), but with greater precision. They found, in particular, that $1.25\,T_{11} \simeq \Theta_{11}$, verifying the first estimate. On the other hand, they found that $\alpha_{11} \simeq 12\,\theta_{11}$, contradicting the second estimate, but consistent with his speculation.

Turbulent particle dispersion had often been modeled in a way similar to random walks in Brownian motion (see, for example, Einstein, 1905 and Goldstein, 1951). An innovative extension of this idea, addressing specifically the Euler–Lagrange problem, was Corrsin's work with John Lumley (Lumley and Corrsin, 1959) on random walks with both Eulerian and Lagrangian statistics. Limiting the problem to one dimension, they defined an Eulerian grid in space and time with specific rules of motion at each point, which defined the Eulerian space/time statistics. Then particles were allowed to 'walk' on this grid, determining the Lagrangian statistics. Analytical expressions were obtained for the relationships between various Eulerian and Lagrangian quantities, although the results could not be immediately applied to turbulence. This work was followed up by that of Patterson and Corrsin (1966), where more complex Eulerian fields and rules were defined, and computer simulations were used to obtain the various statistics. Although

> ... it was hoped that some empirical connection might be discovered between these two kinds of functions, the results show that no single Eulerian two-point correlation function is a good approximation to the Lagrangian function, ...

a result that is probably true also for the turbulence case.

As mentioned earlier, in dealing with turbulent dispersion, Corrsin realized the importance of understanding the bending and folding of iso-surfaces of transported scalars, and hence the growth of lines embedded in these surfaces, and the growth of the surfaces themselves. In first addressing this problem (Corrsin, 1955), he realized that the length of a scalar iso-line in two dimensions could be related to the number of crossings of that surface with a straight line through the fluid. Then using a theorem from Rice (1944, 1945) and assuming the scalar, say $\phi(\mathbf{x})$, is normally distributed, the line length $\mathcal{L}_\phi$ can be related to the autocorrelation of the scalar as

$$\mathcal{L}_\phi = \frac{1}{\pi}\left\{-\frac{\Psi''(0)}{\Psi(0)}\right\}^{1/2} \exp(-\phi_c^2/2\Psi(0)), \qquad (7.5)$$

where $\Psi(\sigma)$ is the auto-correlation $\overline{\phi(s)\phi(s+\sigma)}$, and ϕ_c is the constant value of interest of ϕ. This result can easily be extended to an iso-surface in three

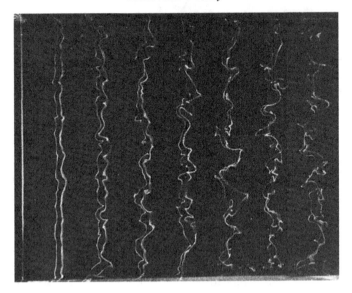

Figure 7.3 Evolution of lines formed with tiny hydrogen bubbles in a turbulent
water channel flow (reprinted from Corrsin and Karweit, 1969).

dimensions. The work was further extended by Corrsin and Phillips (1961) to
include contour lengths and surface areas of multiple-valued random variables.
These results have proven very useful, in particular, in theories of turbulent
combustion, where the area of the flame surface is often directly modeled (see,
for example, Poinsot and Veynante, 2001).

Corrsin and Karweit (1969) were the first to measure the fluid line growth
in turbulence. Michael Karweit was a graduate student pursuing his Masters
degree and would remain at Johns Hopkins as a long-time junior collaborator of Corrsin. Their experiment utilized a water tunnel with a test section of
dimension 8 in. square by 48 in., and approximately homogeneous turbulence
generated by a bi-plane grid of mesh size $\frac{1}{2}$ in. The grid Reynolds number was
1360. They used the 'hydrogen bubble' electrolysis method, with a platinum
wire stretched normal to the flow to generate the hydrogen bubble lines. These
lines were photographed at various distances downstream from the wire, and
their lengths were determined using an analysis relating length to the number of cuts and the angle of the line with respect to a straight reference line
(Corrsin and Phillips, 1961). Photographs and movies from this experiment
are now used in many classes in turbulence throughout the world, and a photograph is shown in Figure 7.3. Unfortunately, because of the limitation of
the length of the water tunnel, only short-time growth of the lines could be
observed. This growth was consistent with short-time growth estimates, but

the measurements could not confirm the conjectured long-time growth of the lines.

Batchelor (1952) had conjectured that, for long times in stationary, homogeneous turbulence, the number of eddies of each size acting to stretch the line is proportional to the line length. This leads immediately to the conclusion that the line will grow exponentially. This result was proven more rigorously by Cocke (1969) and Orszag (1970). Corrsin (1972) offered a simpler geometric proof of this result. His conclusions were weaker than the previous ones, but without restrictions to isotropy or constant density being required.

While pointing out the problems of a fundamental nature in the use of a turbulent scalar diffusivity (see below), Corrsin made theoretical and experimental estimates of this quantity. He extended Taylor (1921)'s theory to include a homogeneous, isotropic, stationary shear flow (Corrsin, 1953) and found, for example, for long times, the cubic dependence on time of the streamwise dispersion, i.e.

$$\overline{X_1'^2} \sim \frac{2}{3}\left(\frac{d\overline{u}_1}{dx_2}\right)^2 \overline{v_2'^2} T_{22} t^3 , \tag{7.6}$$

where $d\overline{u}_1/dx_2$ is the uniform mean shearing of the u_1 component of the velocity in the x_2 direction. This result was extended by Riley and Corrsin (1974) for the non-isotropic case. In particular, they computed the turbulent diffusivity tensor $\mathcal{K}_{ij}$ and found it to be non-diagonal, and to depend on the mean shear and the correlations of v_1' and v_2'. For example, the $\mathcal{K}_{11}$ component was found to be

$$\mathcal{K}_{11}(t) = \int_0^t R_{L11}(\tau)\,d\tau + \frac{d\overline{u}_1}{dx_2}\int_0^t \tau R_{L12}(\tau)\,d\tau . \tag{7.7}$$

Riley and Corrsin (1971) also performed computer simulations of fluid particle dispersion of homogeneous shear flows using an artificially constructed Eulerian flow field consisting of spatially and temporally varying Fourier modes, with amplitudes defined so that the statistics of the flow were similar to the laboratory measurements of Champagne et al. (1970). Their computed results were consistent with the analysis.

Almost all turbulence models employ, at some point, a linear gradient model, where the turbulent flux of a quantity (e.g. mass, heat, species concentration, momentum, kinetic energy) is assumed proportional to the linear gradient of that quantity. Corrsin would often express skepticism about closure models, in particular the eddy-diffusivity, gradient-based models and their motivating analogies to kinetic theory of gases. His arguments about the fundamental limitations of these models have served as motivating force to many researchers who in subsequent decades have attempted to develop more general and intricate closure models of turbulence.

In an influential paper entitled *Limitations of gradient transport models in random walks and in turbulence*, Corrsin (1974) presented a systematic analysis of closure models. Following ideas from continuum mechanics in deriving a relationship between the molecular flux of the quantity, the properties of the fluid, and the space and time gradients of the quantity, he assumed a general functional relationship between the turbulent flux of a quantity in the, say, z direction, $\bar{F}(z)$, and a functional of the average quantity $\bar{\Gamma}$ and the statistical properties of the velocity field. He then determined the assumptions required for the turbulent flux to be linearly related to the gradient of the average quantity. With ℓ, a length scale of the turbulence, τ, the time scale, and $V = \ell/\tau$ the corresponding turbulent velocity scale, the necessary conditions for the linear gradient model are found to be the following, where a subscript denotes a derivative with respect to that quantity:

(i) $|\bar{F}_{zzz}/\bar{F}_z|\ell^2 \ll 1$, i.e. the turbulent length scale should be much smaller than the distance over which the curvature of $\bar{\Gamma}$ changes appreciably;

(ii) $\tau|\bar{\Gamma}_{tz}/\bar{\Gamma}_z| \ll 1$, i.e. the turbulence time scale must be much smaller than the time over which $\bar{\Gamma}$ changes appreciably;

(iii) $|\ell_z/\ell + V_z/V| \ll |\bar{\Gamma}_z/\bar{\Gamma}|$, i.e. the changes in the turbulence properties must be very small over a distance for which $\bar{\Gamma}$ changes appreciably;

(iv) $|V_z/V| \ll |\ell_z/\ell|$, i.e. the turbulent velocity must be appreciably more uniform than ℓ; and

(v) the relative change in $\bar{\Gamma}$ must be very small over the turbulent time scale τ.

Corrsin then went on to compute these inequalities for several flows, pointing out that

> the archival literature is replete with data showing, either directly or indirectly, for both scalar and momentum transport, that the mean gradients vary considerably over distances comparable to the length scales characteristic of the 'eddies'.

He also argued that, for the turbulent flux of a scalar $\overline{\gamma u_i}$ (where $\gamma = \Gamma - \bar{\Gamma}$ is the scalar fluctuation and u_i is the fluctuating part of the velocity vector), the turbulent diffusivity must be considered as a second-order tensor, i.e.

$$\overline{\gamma u_i} = -\mathcal{K}_{ij}\frac{\partial \bar{\Gamma}}{\partial x_j}. \tag{7.8}$$

He pointed out that there is no reason to assume that the diffusivity matrix is diagonal, as assumed in many models. In fact, using estimates from various sets of data, he argued that the off-diagonal terms are often comparable to the diagonal terms.

Several years later, he and co-workers (Sreenivasan et al., 1981) revisited these conditions and analyzed experimentally obtained turbulent heat flux and temperature gradients across various types of homogeneous and inhomogeneous shear flows. They identified additional conditions and concluded that there was a need for models based on more than just the mean field properties of the flow.

7.6 Homogeneous turbulence: decay and shear

Ever since his early experiments as part of his Masters thesis at Caltech, Corrsin had been interested in quantifying the precise decay rate of kinetic energy in isotropic turbulence unconstrained by boundaries. Several statistical theories and models predicted different decay rate exponents n of kinetic energy with time, i.e. $\overline{u'^2} \sim t^n$ for a particular component of turbulence kinetic energy. Depending on what quantity (invariant) was assumed to be constant during the decay, different values of n were obtained. As discussed in Davidson (2004), most well-known are $n = -10/7$ (Kolmogorov, 1941), $n = -6/5$ (Saffman, 1967), or $n = -1$ for complete self-preservation which at the time had been discarded (Batchelor, 1948). Careful new experiments were needed to provide accurate data. Such data could be produced in a wind tunnel with a test section of large enough cross-section to prevent wall effects and long enough to enable turbulence to decay significantly. Also, the turbulence should be truly isotropic at the entrance of the test section. Many earlier attempts at generating isotropic turbulence, including Corrsin's own trials at Caltech, typically provided for larger velocity fluctuations in the streamwise direction than in the cross-stream directions. In order to take advantage of the unexpected availability of wood and of a team of carpenters who, it is said, had finished working on a Johns Hopkins building project earlier than planned, Corrsin designed a large closed-loop wind tunnel made almost entirely out of wood. The design called for a very large primary contraction with an area ratio of 25 to 1, in order to create a smooth, constant velocity air flow at the core of the test section.

Construction of the two-story facility occupying large areas of the basement and first floor of Maryland Hall proceeded quickly. Procurement and installation of a two-stage axial fan with adjustable pitch connected to a 150 horsepower electric motor on the first floor completed what would become a major facility for turbulence research. Cooling was needed to prevent excessive thermal contamination of the hot-wire probe readings. It was provided by a

Figure 7.4 Professor Stan Corrsin in later years explaining decaying isotropic turbulence behind a grid in a wind tunnel, including a secondary contraction.

cross-flow heat exchanger installed before the fan. Through the heat exchanger circulated cooling water siphoned off from a pond which, at the time, graced the east side of campus. For years, this arrangement would cause friction with the university's ground maintenance personnel.

The design also included a secondary contraction which would be located downstream of the turbulence producing grid. By forcing the initially some-what anisotropic turbulence to go through the secondary contraction, vorticity aligned in the streamwise direction would get amplified due to vortex stretch-ing, and the cross-stream turbulence variance would be increased relative to the streamwise turbulence component. Corrsin settled for a 1.27:1 secondary contraction which would greatly reduce the initial anisotropy of the turbulence. Years later, he would often explain the principle of the secondary contraction on a blackboard (Figure 7.4).

Geneviève Comte-Bellot arrived to Baltimore in 1963 as a Fulbright and postdoctoral fellow, having recently obtained her doctorate from the Univer-sity of Grenoble working with Antoine Craya. She went to work with Corrsin and implemented various improvements in hot-wire instrumentation and ana-log data acquisition. In two seminal papers on the decay of isotropic turbulence that arose from their collaboration, they presented what has become one of the most celebrated datasets of fluid mechanics. In the first paper (Comte-Bellot and Corrsin, 1966), they documented the performance of the secondary con-traction in promoting isotropy of the turbulence behind the grid. They also showed that the decay of turbulence variances proceeded according to a power law,

$$\frac{\overline{u'^2}}{U_0^2} \approx \frac{\overline{v'^2}}{U_0^2} \approx C \left(\frac{x - x_0}{M} \right)^n, \tag{7.9}$$

where $\overline{u'^2}$ and $\overline{v'^2}$ are the variances of streamwise and cross-stream velocity, respectively, U_0 is the mean velocity in the tunnel (that ranged between 10 and 20 m/s), M is the mesh-size of the turbulence-producing grid of bars (M ranged from 1 to 4 inches), and $x - x_0$ is the downstream distance to a virtual origin. In the presence of the secondary contraction, the isotropy requirement $\overline{u'^2} = \overline{v'^2}$ was met to a remarkable degree. Moreover, the data yielded decay exponents that fell mostly in the range between $n = -1.2$ and $n = -1.3$, over more than one decade of scaling. It was the most convincing experimental result showing that predictions from theories leading to either a $t^{-10/7}$ or a t^{-1} decay were not reproduced.

The second work was published sometime later (Comte-Bellot and Corrsin, 1971) and provided a detailed analysis of measured two-point correlation functions and spectra at various downstream distances from the grid. Hot-wire probes recorded velocity signals over a wide range of frequencies. Spectra for high frequencies were obtained using an HP wave analyzer. Since it was sensitive only down to 20 Hz, lower frequencies were captured by recording the signals to tape and replaying the tapes at higher speeds later on. Measuring correlation functions also involved playing back the tapes with varying time-delays. Additional analog signal processing included band-pass filters, multipliers and an electro-chemical integrator whose output finally corresponded to the time-converged correlation coefficients among narrow band-pass filtered signals.

The results show that correlation functions for band-pass filtered velocities decay at time-scales commensurate with the eddy-scale highlighted by the band-pass filtering. Also, all curves could be collapsed by an appropriate time scale, combining effects at various scales.

Comte-Bellot and Corrsin (1971) also report, in great detail, the precise energy spectra at various times (distances) during the decay. They used the measured one-dimensional energy spectrum $E_{11}(k, t)$ to deduce the radial three-dimensional energy spectrum using the assumption of isotropy. The resulting radial spectra $E(k, t)$, carefully tabulated, have been used by many researchers since to test and validate spectral closures such as eddy-damped quasi-normal theories and, in recent decades, subgrid-scale models for large eddy simulations (Moin et al., 1991). It has taken three decades for this ground-breaking experiment to be replicated using direct numerical simulations (de Bruyn Kops and Riley, 1998) as well as for a similar experiment to be remade at higher

Reynolds number in the same wind tunnel, this time using an active grid (Kang et al., 2003).

The question of the dynamics of narrow-band effects in turbulence continued to interest Corrsin for many years. He had been following the theoretical efforts of R.H. Kraichnan, who at the time lived in relative isolation, north in the New Hampshire woodlands. Once a year, Corrsin would travel to New Hampshire to visit with Kraichnan and discuss turbulence. One of the central quantities of the Kraichnan direct interaction approximation is the response function of turbulence to a spectrally local disturbance. Partly motivated by the discussions with Kraichnan, Kellogg and Corrsin (1980) performed an experiment in which the wake of a fine wire stretched across otherwise isotropic grid turbulence introduced a narrow-band disturbance. They recorded its decay and compared it to the linear perturbation response predicted by Kraichnan, noting 'fair agreement'. Interest in the dynamics of Fourier modes also led Corrsin to consider early uses of computer simulations. With J. Brasseur, a postdoctoral fellow at Hopkins in the early 1980s, they performed numerical experiments and followed the time-evolution of individual Fourier modes and observed their interactions within wave-number triads (Brasseur and Corrsin, 1987).

Towards the late 1970s and early 1980s Corrsin directed a concerted effort to study the most elemental non-isotropic turbulent flow, namely homogeneous shear flow in which the mean flow has a linear profile. Champagne et al. (1970) and Harris et al. (1977) produced such a mean velocity profile by forcing air flow through a set of parallel plates, each channel being associated with a screen of different solidity. The side with larger solidity corresponds to lower speeds due to the increased head losses suffered by the flow there. The evolution of turbulence, the growth of length-scales, and the resulting anisotropy were measured and to this day form a dataset used to calibrate turbulence models and compare to simulations.

Returning to the question of scalar transport, Tavoularis and Corrsin (1981a) made direct measurements of the turbulent diffusivity in a homogeneous shear flow. They used an experimental setup similar to that used in the homogeneous shear flow experiments of Harris et al. (1977), but with the exit turbulence-generating rods replaced with heating rods. This produced a uniform temperature gradient in the cross-stream (x_2) direction, to go along with their uniform velocity gradient across the same direction. Detailed measurements were made of the velocity field and temperature field statistics, including joint temperature/velocity statistics, spectra, autocorrelations, microscales and integral scales. In particular, with $d\bar{T}/dx_2 = $ constant, and $d\bar{T}/dx_1 = d\bar{T}/dx_3 = 0$, from measuring $\overline{u_1\theta}$ and $\overline{u_2\theta}$ they were able to determine $\mathcal{K}_{12} = -\overline{u_1\theta}/\frac{d\bar{T}}{dx_2}$ and $\mathcal{K}_{22} = -\overline{u_2\theta}/\frac{d\bar{T}}{dx_2}$. The result was that $\mathcal{K}_{12}/\mathcal{K}_{22} \simeq -2.2$. In re-examining existing data for heated turbulent boundary layers and heated pipe flows, they found approximate values of −2.4 and −2.1, respectively.

This work was extended by Tavoularis and Corrsin (1981b), using the same flow field, to the case with the mean temperature gradient transverse to the direction of the mean flow and the mean shear, i.e. $d\bar{T}/dx_3$ = constant, and $d\bar{T}/dx_1 = d\bar{T}/dx_2 = 0$. The only significant heat flux component was $\overline{u_3\theta}$ (the other two components were approximately zero by symmetry), and gave the results that $\mathcal{K}_{33} = -\overline{u_3\theta}/\frac{d\bar{T}}{dx_3} \simeq 1.6\mathcal{K}_{22}$.

7.7 The geometry and intermittency of turbulence

In his PhD thesis on the circular turbulent jet, Corrsin (1947) computed many of the statistical properties of the turbulent velocity and temperature field. But in observing oscillograms of the axial velocity signal he noticed that

> the 'turbulent' jet is completely turbulent only from the axis out to approximately $r = r_0$. For $r > r_0$, there exists first an annular transition region, in which the flow at a point alternates between the turbulent and laminar regimes.

(Here r_0 is the radial location where the mean axial velocity $\bar{U}$ drops to half of its peak value.) He went on to note that

> the general location of the transition region in the jet is about the same as the location of the $u'/\bar{U}$ maximum. This may mean that a part of the 'turbulence' is not due to the usual turbulent velocity fluctuations, but to actual differences in local mean velocity at a point, as the flow oscillates between the laminar and turbulent states.

Corrsin had discovered the intermittent layer between a laminar and a turbulent flow which is now known to be characteristic of any turbulent flow with a free-stream boundary (i.e. not a solid boundary) such as turbulent boundary layers, jets, wakes, shear layers, and other related flows.

The first definitive study of the intermittent regions between a laminar and a turbulent flow was by Corrsin and Kistler (1955), who addressed such regions for a turbulent boundary layer, a plane wake, and a circular jet. Although interesting experimental data were obtained in this study, one of its principal contributions was conceptual, in defining and clarifying the overall processes involved. The first issue is how to distinguish the turbulent and the non-turbulent regimes. Corrsin and Kistler realized that it was not the random motion that distinguished the turbulent region, since the flow in the laminar region was also quite random. They concluded that the characterizing feature of the turbulent region was its high vorticity, compared to the essentially irrotational flow of the non-turbulent region. Thus they concluded to apply "the word 'turbulent' to random rotational fields only".

They surmised that the rotational turbulent region must propagate into the non-turbulent region, much as "a flame front propagates through a combustible mixture". From the vorticity equation they reasoned that

> the random vorticity field . . . can propagate only by direct contact, as opposed to action at a distance, because rotation can be transmitted to irrotational flow only through direct viscous shearing action. This assures that . . . the turbulent front will always be a continuous surface; there will be no islands of turbulence out in the free stream disconnected from the main body of turbulent fluid.

Corrsin and Kistler reasoned that a very thin layer, which they called the laminar superlayer, separated the turbulent and non-turbulent regions. The turbulent side was characterized by strong vorticity amplification by vortex stretching, while the superlayer itself was characterized by viscous diffusion of vorticity across this layer. From simple physical/mathematical arguments, they concluded that the superlayer was very thin, with a width on the order of the Kolmogorov scale.

In order to address the intermittency of turbulence in the flow, Corrsin and Kistler followed Townsend (1948) and defined the intermittency γ as "the fractional time spent by the (fixed) probe in the turbulent fluid". Experimentally the intermittency γ was determined by electronically differentiating the hotwire signal for the axial component of the velocity, then rectifying, smoothing and clipping the resulting signal. A signal discriminator was used to determine whether the resulting signal was strong enough such that the region was turbulent; this signal discriminator was set by comparing the results of the signal output to a visual oscillogram output. In addition to the usual measurements of the velocity statistics, they were able to measure the intermittency γ and the position of the front Y as functions of time and downstream coordinate x. They were thus able to determine the intermittency γ, which is, in terms of Y, $\gamma(y) = \mathrm{prob}\{y \leq Y(t) \leq \infty\}$. In addition, they could determine the average position of the turbulent front, $\bar{Y}$, and its standard deviation $\sigma = \{\overline{(Y - \bar{Y})^2}\}^{1/2}$, which is a measure of the width of the intermittent zone, which they termed the wrinkle amplitude of the turbulent front.

From their data and using theoretical arguments, they found that the rate of increase of the wrinkle amplitude of the turbulent front was roughly predicted by Lagrangian analysis as $\sigma(x) \simeq \sqrt{2(v'/\bar{U})v'T_L}$, where v' and T_L are the local Lagrangian velocity fluctuation and velocity integral time scale, respectively. They also found that the downstream growth of the turbulent front, as measured by Y, was proportional to the growth of the shear-layer thickness.

Corrsin and Kistler drew two important additional conclusions from their study. First, they concluded "that the presence of the turbulent front with its attendant detailed statistical properties will have to be included in basic research on turbulent shear flows with free-stream boundaries". Secondly, they speculated that, in considering a scalar (e.g. heat, mass) in the flow for Prandtl and Schmidt numbers not much smaller than unity, "the front should apply equally well to heat or chemical composition. Oscillographic observations ... in a hot jet show a temperature fluctuation intermittency, presumably coincident with the vorticity intermittency".

Corrsin saw indications of outer intermittency in many other fluid dynamical systems. In a noteworthy interview in *Sports Illustrated* (Terrell, 1959), he was asked to explain the mechanism underlying the so-called 'knuckle ball'. It was the hallmark of Hoyt Wilhelm, a then famous pitcher for the Baltimore baseball team, the Orioles. Hoyt could throw a ball that would then move in unpredictable trajectories, thus confusing the opposing team's batter. A photograph in the article shows Corrsin in front of the blackboard with a sketch of the flow-field at the rear side of a baseball during flight. A jagged boundary line encloses the separated turbulent region. It is used to show that the unpredictable trajectories of the knuckle ball can be due to slight changes in lift and drag forces associated with the complicated geometry of the separated region. Quoting from the article:

> If the separation line was perfectly straight, the ball would go straight, for the pressure forces would be even. But since the separation line is highly irregular, so is the course of the ball. And since the separation line is constantly shifting and changing ... the course of the knuckle ball can change direction several times in flight.

Following Kolmogorov's (1941) important theory of local isotropy and similarity hypotheses regarding turbulent velocity fine-structure, and in fact his own along with Obukhov's (1948) theory for fine-scale scalar fields (Corrsin, 1951b), Corrsin became interested in the intermittent behavior and the geometric properties of fine-scale turbulence. Measurements by Batchelor and Townsend (1949) indicated that the fine-structures were strongly intermittent, localized in relatively small regions which were distributed somewhat randomly in space. This led many to question Kolmogorov's original theory (see, for example, Landau and Lifshitz, 1959), which led to a number of attempts to address the structure of the fine-scales as well as modifications of Kolmogorov's theory.

Corrsin's interest in understanding the spatial structure associated with the turbulent cascade of kinetic energy comes to light in a passage of his general-interest article published in *American Scientist* in 1961 (Corrsin, 1961b). Quoting from the article:

> From a geometrical viewpoint, the spectral transfer process in turbulence can be seen in the (empirical) fact that any blob of fluid momentarily having a fairly uniform local velocity, is stretched and twisted by its own motion (and that of neighboring fluid) into even longer, thinner and more convoluted 'strings' and 'sheets'. Since it is difficult to sketch such a locally coherent velocity field, we can illustrate this aspect via a similar phenomenon: turbulent mixing of a passive contaminant, like dye spots in a turbulent liquid.

He earned the 1961 American Scientist Prize for this article.

Corrsin (1962c) used a simple phenomenological model and some existing data to suggest that the fine-structure was distributed into thin sheets, with thickness on the order of the Kolmogorov scale, and separation distance of the order of the integral scale. This suggested that the velocity derivative flatness factor should scale linearly with the Taylor scale Reynolds number, $R_\lambda = u'\lambda/\nu$. On the other hand, Tennekes (1968) suggested the fine-structure was distributed as vortex tubes, with diameters of the order of the Taylor scale λ. This led to the prediction of the flatness factor scaling as $R_\lambda^{3/2}$. In addition, Obukhov (1962) and Kolmogorov (1962), attempting to take the fine-scale intermittency into account, assumed that the logarithm of the average energy dissipation rate over a very small volume had a normal distribution, and from this they were able to obtain modified expressions for the energy spectrum and structure functions. These hypotheses remained to be tested.

Working with A. Kuo (Kuo and Corrsin, 1971) in both grid-generated, nearly isotropic turbulence and on the axis of a round jet, Corrsin first addressed the size of the fine-scale regions, the dependence on Reynolds number, and the probability density of the locally averaged dissipation rate. Hot-wire anemometers were employed to make the velocity measurements, and three kinds of circuits were used to extract fine-scale signals from the outputs of the anemometers: differentiation circuits, band-pass filters, and high-pass filters. They found that there was a decrease in the relative fluid volume occupied by fine-structure of a given size as the turbulence Reynolds number R_λ increased. They also found that, for a fixed Reynolds number, the relative volume is smaller for smaller fine-structures. In addition, the average linear dimension of a volume of fine-structure (L_r) was found to be much larger than the size of the fine-structure r itself. For example, at $R_\lambda = 110$, they found that L_r/r varied from 15 to 30, decreasing with r. Finally, they found that $(\partial u/\partial t)^2$ was approximately log-normally distributed, at least when probabilities fall between about 0.3 and 0.95, in partial agreement with the assumptions of Obukhov and Kolmogorov.

Realizing that information about the shape of the fine-scale structures might eventually help understand the physical processes related to energy transfer to these scales, Kuo and Corrsin (1972) then attempted to determine the geometric character of the structures. Using the measurement technique of two-position coincidence functions for the presence of velocity fine-structure, they tried to distinguish the structures as being 'blobs', 'rods', or 'slabs'. Again hot-wire anemometer measurements were made in nearly isotropic turbulence. In order to determine the geometry of the structures, Kuo and Corrsin developed mathematical, geometric models for each structure; these models predicted, for each assumed structure, the simultaneous detection event rate as well as the simultaneous intermittency factor. Comparisons of the experimental results for these quantities to the predictions of the models then allowed the determination of the type of fine-scale structures.

Their tentative conclusion was that the fine-scale regions are more rod-like than blob-like or sheet-like. This implies a tendency for slightly 'stringy' structures, which may overlap with each other. Two other classes of structures were not eliminated by the measurements, ribbon-like structures, and a mixture of blobs and rods. Kuo and Corrsin suggested coincidence measurements using three or more probes to help determine among these alternatives. They also suggested using similar models for fine-scaled scalar fields to help distinguish the structures. These detailed results motivated many subsequent publications by other researchers on the intermittency statistics of turbulence, as well as an influential paper by Kraichnan (1974). He dealt with an analysis of the energy cascade along wavenumber bands arranged in octaves in an effort to provide a possible dynamical explanation for the spatial concentration of energy fluxes in smaller and smaller subregions of the flow during the cascade. A number of subsequent developments are recounted in some detail in the book by Frisch (1995).

7.8 Turbulence and chemical reactions

Corrsin was the first to apply statistical theory to turbulent, reacting flows. In a series of papers spanning the 1950s and 1960s he applied statistical analysis and results from turbulence and turbulence mixing developed over the past 20 years to determine the statistical properties of simple chemical reactions in turbulent flows. His research set the stage for much of the work on such flows that followed.

As was usually the case for him, Corrsin idealized the problem and considered homogeneous, isotropic turbulence, and assumed negligible effects of heat release on the fluid properties so that, in particular, the fluid density, the reaction-rate coefficients, and diffusion coefficients remain constant. Furthermore, he assumed that all of the reactant species concentrations but one were in great excess, so that the concentration of only the latter, say Γ, changes significantly in time. Finally, he assumed that the chemical reaction rates were of simple power-law form, i.e. proportional to Γ^n, and considered the cases where $n = 1$ or $n = 2$. Therefore, Γ satisfies the following convection–diffusion–reaction equation:

$$\frac{\partial \Gamma}{\partial t} + u_i \frac{\partial \Gamma}{\partial x_i} = \mathcal{D}\nabla^2\Gamma - \Phi(\Gamma),\tag{7.10}$$

where $\Phi(\Gamma) = k_n\Gamma^n$, k_n is the reaction-rate constant, $\mathcal{D}$ is the molecular diffusivity of Γ, and n is either 1 or 2.

First-order reactions ($n = 1$):   With the assumption of homogeneity, Corrsin (1958) averaged equation (7.10) and obtained the solution for the average as $\bar{\Gamma}(t) = \bar{\Gamma}_0 \exp(-k_1 t)$. He then obtained the equation for the mean-square fluctuation $\overline{\gamma^2}(t)$:

$$\frac{d\overline{\gamma^2}}{dt} = -2\mathcal{D}\overline{\frac{\partial \gamma}{\partial x_i}\left(\frac{\partial \gamma}{\partial x_i}\right)} - 2k_1\overline{\gamma^2}.\tag{7.11}$$

He introduced a microscale for γ, say λ_γ, and, based upon previous experiments, assumed $\lambda_\gamma/\lambda \simeq 2\mathcal{D}/\nu$ and obtained solutions for $\overline{\gamma^2}$ for different assumptions regarding λ. Here λ is the usual Taylor microscale. His solutions were of the form

$$\overline{\gamma^2}(t) = \overline{\gamma_m^2}(t)\exp(-2k_1 t),\tag{7.12}$$

where $\overline{\gamma_m^2}(t)$ is the solution for the nonreacting case ($k_1 = 0$). Therefore, he found that the effect of the first-order chemical reaction is to cause exponential decay in both $\bar{\Gamma}$ and $\overline{\gamma^2}$.

In a subsequent paper (Corrsin, 1961a), he focused on the spectral behavior of γ for first-order reactions. Using extensions of the spectral cascade arguments of Onsager (1949), and the mixing theories of Batchelor (1959) and Batchelor et al. (1959), he derived the following expressions for the energy spectrum of γ, say $G(k)$, for three different spectral subranges:

(i) inertial–convective subrange, $k \ll (\epsilon/\nu\mathcal{D}^2)^{1/4}$ if $\nu/\mathcal{D} \gg 1$, or $k \ll (\epsilon/\mathcal{D}^3)^{1/4}$ if $\nu/\mathcal{D} \ll 1$

$$G(k) \simeq Bk^{-5/3}\exp(3k_1\epsilon^{-1/3}k^{-2/3}).\tag{7.13}$$

(ii) viscous–convective subrange, $k \ll (\epsilon/\nu^3)^{1/4}$

$$G(k) \simeq Nk^{-(1+4k_1\nu^{1/2}\epsilon^{-1/2})}\exp\{-2(k/k_B)^2\}. \tag{7.14}$$

(iii) inertial–diffusive subrange, $(\epsilon/\mathcal{D}^3)^{1/4} \ll k \ll (\epsilon/\nu^3)^{1/4}$

$$G(k) \simeq \frac{1}{3}\frac{\epsilon_\theta^* \epsilon^{2/3}}{\mathcal{D}}\frac{1}{k^{5/3}[\mathcal{D}k^2 + k_1]^2}. \tag{7.15}$$

Here B and N are constants determined from the analysis, and $k_B = (\epsilon/\nu\mathcal{D}^2)^{1/4}$ is the Batchelor wave number. It is easy to see the effects of the reaction rate on the spectra. For example, in the inertial–convective subrange, where the spectrum is proportional to $\epsilon_\gamma\epsilon^{-1/3}k^{-5/3}$, the effect of chemical reaction is given by the factor $\exp(3k_1\epsilon^{-1/3}k^{-2/3})$. Corrsin points out that, for wave numbers above $k_c = k_1^{3/2}\epsilon^{-1/2}$, the effect of the chemical reaction on the spectral shape is negligible.

Having obtained results for the concentration of the reactant, Γ, Corrsin (1962a) then addressed the concentration of the product of the reaction, say P, for first-order reactions. The product concentration for this case satisfies the following equation:

$$\frac{\partial P}{\partial t} + u_i\frac{\partial P}{\partial x_i} = \mathcal{D}_P\nabla^2 P - k_1\Gamma. \tag{7.16}$$

Assuming equal diffusivities for the reactant and the product, i.e. $\mathcal{D} = \mathcal{D}_P$, the equation for $\bar{P}$ is closed and the solution is easily found to be $\bar{P} = \bar{P}_0 + \bar{\Gamma}_0\{1 - \exp(-k_1t)\}$. The equation for the mean-square fluctuations about $\bar{P}$, say $\overline{p^2}$, is similar to equation (7.11), except that the last term is now $+k_1\overline{p\gamma}$, introducing a new unknown. Arguing that the p and γ fields are perfectly correlated, Corrsin was then able to obtain a solution for $\overline{p^2}$ analogous to equation (7.12):

$$\overline{p^2}(t) = \overline{\gamma_m^2}(t)\{1 - \exp(-k_1t)\}^2. \tag{7.17}$$

He also obtained equations for the energy spectra of the product concentration for the inertial–convective, viscous–convective, and inertial–diffusive subranges, but these results will not be repeated here. These various predictions for both mean values and energy spectra are available to the community to guide experiments and modeling, and have been extensively utilized.

Second-order reactions ($n = 2$): Due to the nonlinearity of the reaction term, $k_2\Gamma^2$, the equation for mean reactant $\bar{\Gamma}$ is no longer closed, but contains the unknown $\overline{\gamma^2}$. In addition, the chemical reaction now causes a spectral flux in $\overline{\gamma^2}$. Corrsin (1958) again formed the equation for $\overline{\gamma^2}$, which now contained, in addition to the dissipation-rate term, the additional unknown $\overline{\gamma^3}$, for which he introduced an additional equation. Introducing microscales for the dissipation-rate terms in the equations for $\overline{\gamma^2}$ and $\overline{\gamma^3}$, he was left with equations for $\bar{\Gamma}$, $\overline{\gamma^2}$ and $\overline{\gamma^3}$ with a number of additional unknowns requiring additional assumptions. To simplify the problem further he then addressed three separate limiting problems:

(i) extremely low fluctuation levels, $\gamma'/\bar{\Gamma} \ll 1$;
(ii) very slow reactions;
(iii) very fast reactions;

and obtained solutions for each case. Corrsin (1958) also briefly addressed the equation for the spatial autocorrelation function $\overline{\gamma(\mathbf{x})\gamma(\mathbf{x}+\mathbf{r})}$.

Corrsin (1964) went on to consider the energy spectra for second-order reactions. Assuming small fluctuation levels ($\gamma' \ll \bar{\Gamma}$), extending the cascade method of Onsager (1949), and again following the approach of Batchelor (1959) and Batchelor et al. (1959), he developed expressions for the reactant energy spectra's inertial–convective, inertial–diffusive, and viscous–convective subranges.

7.9 The Johns Hopkins environment

During Corrsin's tenure at Johns Hopkins, the Mechanics Department was an exciting environment in which to work, and was considered one of the top centers for turbulence research in the world. Some of this is described in the *Annual Reviews of Fluid Mechanics* article by John Lumley and Steve Davis (Lumley and Davis, 2003). One of the authors of this chapter (J.J. Riley) was a graduate student there in the late 1960s and early 1970s, so this section relates mostly to that time period.

Owen Phillips has referred to the period in the late 1960s and early 1970s in the Mechanics Department at Johns Hopkins as the 'golden years', and much of the credit for creating and sustaining such an environment goes to Corrsin. The fluid mechanics faculty, which he had helped to build, was very active. Owen Phillips had completed his now classical work on surface wave and internal wave resonances, among many other things. Robert Long had finished his pioneering studies on stratified flow over complex terrain, and was now delving into stratified, turbulent flows. Leslie Kovasznay was continually improving experimental methods and studying structures in turbulent boundary layers. Francis Bretherton had just joined from Cambridge University and was embarking on several studies of geophysical flows which have now become famous; and Stephen Davis arrived from Imperial College London and immediately developed a very active program on various aspects of nonlinear instabilities. Clifford Truesdell and Jerald Eriksen were world-renowned for their work in theoretical continuum mechanics, and James Bell, Robert Pond and Robert Green had well-established programs in solid mechanics.

A cornerstone of the Hopkins environment was an unapologetic promotion of fundamental aspects of research in mechanics. According to Phillips (1986), Corrsin would say that science begins by asking simple questions about complex phenomena, but advances by asking more penetrating questions about simpler systems, whose solution could be obtained with rigor and explained with clarity. Again, quoting Phillips (1986), if an unwary colleague commented that some question was academic, Corrsin's inevitable response was, "This is an academic institution, where we consider academic questions". According to Michael Karweit the atmosphere was one of genuine pleasure in doing research on fundamentally important problems. Perhaps it was the realization of working on transcendental knowledge which resulted in the uniquely joyful atmosphere at that time.

Along similar lines, Corrsin had definite views about the importance of fundamental science in the context of the education of engineers. He did not favor those who would constantly clamor for an education mainly characterized by 'applied relevance'. His strong views come into focus in a 1968 letter he wrote to the editors of a publication that at the time reigned above all other illustrated magazines in the United States: *Life Magazine*. The letter was in response to the publication of an article advocating more "engineering relevance" in education.

> Sirs: I observe with wonder the demands of some college students and some faculty members (Professor Jerome in "The System Really Isn't Working," *Life*, 1 November) for an education characterized by 'relevance'. The primary weakness of the US engineering education from its inception until World War II was its wholehearted devotion to relevance: spending more time in shop and lab than in classroom, the students were well trained to cope with the design, operation and repair of the world's machinery, possibly to improve it a bit. Then along came new machines based on scientific principles no one guessed engineers would ever need to know. The result: many science-trained people had to be hastily recruited into doing engineering work. Their weakness in engineering principles led to mistakes – but at least they hadn't been given tunnel vision by a totally 'relevant' education.   Very truly yours, Stanley Corrsin, Professor.

The department seminar series was particularly interesting and lively. Both novice and established researchers would visit and present the results of their work, usually with much lively discussion from the faculty. Douglas Lilly from the National Center for Atmospheric Research discussed some new

numerical simulations of turbulence by him and his colleague, James Dear-
dorff; these happened to be the first large-eddy simulations of turbulence. Chris
Garrett discussed ideas about ocean internal waves which would later lead
to the now-famous Garrett–Munk energy spectrum for ocean internal waves.
There were always a number of visitors for extended visits. For example,
Geneviève Comte-Bellot had two long visits while engaged in her now-famous
collaboration with Corrsin on decaying grid turbulence (§7.6). Tim Pedley
was visiting from Cambridge and lectured on hydrodynamic stability. Keith
Moffatt, who was working on magnetohydrodynamic turbulence, was another
visitor from Cambridge. The Swedish oceanographer Pierre Welander was a
visiting Professor who lectured on ocean currents. James Serrin from the University of Minnesota was a visiting Professor lecturing on mathematical fundamentals of fluid mechanics. Frank Champagne arrived from Boeing to study
turbulent shear flows, while John Foss from Michigan State University worked
with Corrsin on turbulent diffusion experiments. Many postdoctoral fellows
working on various topics in fluid mechanics subsequently have had outstanding careers. They included, among others, John Allen, James Brasseur, John
Dugan, Fazle Hussain, Wolfgang Kollmann, Martin Maxey, and Katepalli R.
Sreenivasan. Of the students who received their PhD degrees from the Mechanics Department during the late 1960s and early 1970s, many are today
well known for their research, and occupy top positions in academia and research laboratories in industry and government.

There was considerable interaction with Hopkins researchers 'off-campus'
as well. For example, Akira Okubo, from the Chesapeake Bay Institute, at
the time located on the Homewood campus, often discussed his latest ideas
on turbulent dispersion. Vivian O'Brien of the JHU Applied Physics Laboratory, located between Baltimore and Washington, DC, would give seminars
on her latest research, often on bio-fluid mechanics. Besides turbulence, fluid
mechanics in and around living things was another emerging area to which
Corrsin made major contributions.

The morning coffee period was legendary, always drawing a large number
of faculty, students, and visitors. The late 1960s and early 1970s were the time
period of the Vietnam war, and feelings were very strong on all sides. Many
discussions were political. At the same time, Baltimore was rife with racial
and social problems. The downtown area of the city showed signs of considerable neglect. The harbor area in particular consisted of not much more than
a collection of abandoned and derelict warehouses. Racial tensions exploded
following the assassination of Martin Luther King Jr. and led to the Baltimore
riot of 1968, which lasted over a week. National Guard and federal troops had
to be called in to restore order. It was not until the early 1980s that, in a highly

successful example of American urban renewal, the entire inner harbor area was redeveloped.

Besides political discussions at the coffee period, the latest technical ideas were argued at length; if the ideas could survive a discussion at coffee somewhat intact, there was some hope for them. And, the coffee period was also a time of camaraderie and joking, especially if Corrsin were around. Many stories from this coffee period, perhaps often somewhat embellished, would be recalled at conference meetings for many years. And if it was an especially good, or bad, week, some students would go off and buy a supply of wine and cheese on Friday, and the coffee period would become an even livelier party.

By the late 1970s, however, internal disagreements in the department (by then called the Department of Mechanics and Materials Science) that had been developing for some time finally boiled to the surface. Moreover, the recognition that engineering required a separate administrative structure led to the closing of the department and the reestablishment of a distinct engineering school at Hopkins. It consisted of several traditional engineering departments which continued to nurse a distinctly strong science flavor.

7.10 Final years

In 1984, Corrsin became ill with cancer. He underwent an apparently successful operation followed by a none-too-aggressive therapy. Preparations for a conference in his honor, the 'Corrsin Birthday Symposium', were in full swing. It took place in Evanston, Illinois, early in 1985. The happy event celebrating his 65th birthday reunited many of his former students and postdocs. The contributions from the symposium are recorded as a collection of papers in a well-known book, *Frontiers in Fluid Mechanics*, edited by Davis and Lumley (1985).

Soon however the illness returned, this time much worse. There were months of treatments, extended hospital stays, and distress. Stanley Corrsin died on 2 June 1986. He was sixty-six. Another symposium that had been planned in his honor on occasion of the award of the American Society of Civil Engineers' Theodore von Kármán Medal took place in Minneapolis, Minnesota, on the day after his death (George and Arndt, 1988). The medal was awarded posthumously.

A memorial service was held on the Johns Hopkins campus in September at the beginning of the Fall semester. It was attended by many of Corrsin's students, postdocs and collaborators from around the world, his family, and university colleagues and staff.

Corrsin's impact on the field has been felt beyond his own scientific contributions, through those he instructed directly and through others he inspired, directly and indirectly. He had been honored by many professional awards, such as fellowship in the American Academy of Arts and Sciences, the American Physical Society and American Society of Mechanical Engineers, membership of the US National Academy of Engineering, and being named the Theophilus Halley Smoot Professor of Fluid Mechanics.

During his lifetime, Corrsin saw turbulence research progress from rudimentary single-probe hot-wire measurements in small bench-top shear flow experiments, all the way to large-scale turbulence measurement campaigns in large wind tunnels, and in the atmosphere and oceans. He saw turbulence theory develop from simple one-point and two-point closures to path-diagrammatic methods, and to the first several successful direct numerical and large-eddy simulations on supercomputers. His own contributions form the backbone of our present understanding of turbulent scalar transport, of fine-scale structure of passive scalars in turbulence, and of the phenomenon of outer intermittency. His contributions to homogeneous turbulence, decaying and sheared, as well as chemically reacting turbulence, are considered pivotal. Yet, what he called "the theoretical turbulence problem" (Corrsin, 1961b) remains to this day unsolved. The lack of systematic methodologies to make analytical predictions for even the simplest statistical objects continues to pose a serious challenge to the many fields where turbulence plays a crucial role. In the absence of a definitive theoretical framework to attack the problem, Corrsin's approach of joyful empiricism and fundamental analysis of canonical and carefully chosen example problems remains to this day the best approach to turbulence research.

Towards the end of his life, on occasion of Liepmann's 70th birthday in 1984, Corrsin penned the 'Sonnet to Turbulence'. It was read at the event by Anatol Roshko and loosely follows the form of William Shakespeare's Sonnet #18 and Elizabeth Browning's poem "How do I love thee? Let me count the ways ...". In the form received from SC by William K. George (1990), it is reproduced below as closing words about Corrsin's life. The sonnet evokes relevant turbulence phenomena and provides insights about his views on several new approaches that were being proposed at the time. For instance, in juxtaposing low-dimensional strange attractors versus supercomputing, he correctly predicted that the latter would be needed due to the very large number of degrees of freedom of turbulence (it is useful to recall that at the time the most powerful supercomputer was the Cray 2).

Sonnet to Turbulence (by S. Corrsin):
Shall we compare you to a laminar flow?
You are more lovely and more sinuous.
Rough winter winds shake branches free of snow,
And summer's plumes churn up in cumulus.
How do we perceive you? Let me count the ways.
A random vortex field with strain entwined.
Fractal? Big and small swirls in the maze
May give us paradigms of flows to find.
Orthonormal forms non-linearly renew
Intricate flows with many free degrees
Or, in the latest fashion, merely few –
As strange attractor. In fact, we need Cray 3's.
Experiment and theory, unforgiving;
For serious searcher, fun ... and it's a living!

Acknowledgements The authors thank the many friends and colleagues who have shared their memories. In particular they thank Stephen Davis, Michael Karweit, Mohamed Gad-El-Hak, K.R. Sreenivasan, and William K. George, for their comments on an early version of this chapter. They are especially grateful to Dr. Stephen D. Corrsin, SC's son, for his valuable recollections of family and other important events, as well as for comments on this text.

References

Batchelor, G.K. 1948. Energy decay and self-preserving correlation functions in isotropic turbulence. *Q. Appl. Maths.*, **6**, 97–116.

Batchelor, G.K. 1952. The effect of homogeneous turbulence on material lines and surfaces. *Proc. Roy. Soc. A*, **213**, 349–366.

Batchelor, G.K. 1959. Small-scale variation of convected quantities like temperature in turbulent fluid. 1. General discussion and the case of small conductivity. *J. Fluid Mech.*, **5**, 113–133.

Batchelor, G.K., and Townsend, A.A. 1949. The nature of turbulent motion at large wave-numbers. *Proc. Roy. Soc. A*, **199**, 238–255.

Batchelor, G.K., Howells, I.D., and Townsend, A.A. 1959. Small-scale variation of convected quantities like temperature in turbulent fluid. 2. The case of large conductivity. *J. Fluid Mech.*, **5**, 134–139.

Brasseur, J.G., and Corrsin, S. 1987. Spectral evolution of the Navier–Stokes equations for low order couplings of Fourier modes. In *Advances in Turbulence; Proceedings of the First European Turbulence Conference, Ecully, France, July 1–4, 1986*, Springer Verlag, 152–162.

Champagne, F.H., Harris, V.G., and Corrsin, S. 1970. Experiments on nearly homogeneous turbulent shear flow. *J. Fluid Mech.*, **41**, 81–139.

Clauser, F.H. 1954. Turbulent boundary layers in adverse pressure gradients. *J. Aeronautical Sciences*, **21**, 91–108.

Cocke, W.J. 1969. Turbulent hydrodynamic line stretching. Consequences of isotropy. *Phys. Fluids*, **12**, 2488–2492.

Comte-Bellot, G., and Corrsin, S. 1966. The use of a contraction to improve the isotropy of grid-generated turbulence. *J. Fluid Mech.*, **25**, 657–682.

Comte-Bellot, G., and Corrsin, S. 1971. Simple Eulerian time correlation of full- and narrow-band velocity signals in grid-generated, 'isotropic' turbulence. *J. Fluid Mech.*, **48**, 273–337.

Corrsin, S. 1942. *Decay of Turbulence behind Three Similar Grids*. Aeronautical Engineering Thesis, Caltech.

Corrsin, S. 1943. Investigation of flow in an axially symmetrical heated jet of air. NACA Wartime Report – ACR No. 3L23.

Corrsin, S. 1944. Investigation of the behavior of parallel two-dimensional air jets. NACA Wartime Report – ACR No. 4H24.

Corrsin, S. 1947. *I. Extended Applications of the Hot-Wire Anemometer – II. Investigations of the Flow in Round Turbulent Jets*. PhD Thesis, Caltech.

Corrsin, S. 1949. An experimental verification of local isotropy. *J. Aero Sci.*, **16**, 757–758.

Corrsin, S. 1951a. The decay of isotropic temperature fluctuations in an isotropic turbulence. *J. Aeronautical Sci.*, **18**(6), 417–423.

Corrsin, S. 1951b. On the spectrum of isotropic temperature fluctuations in an isotropic turbulence. *J. Appl. Phys.*, **22**, 469.

Corrsin, S. 1952. Patterns of chaos. *The Johns Hopkins Magazine*, **III**(4), 2–8.

Corrsin, S. 1953. Remarks on turbulent heat transfer. In *Proc. First Iowa Symp. Thermodynamics*, 5–30.

Corrsin, S. 1955. A measure of the area of a homogeneous random surface in space. *Quart. Appl. Math.*, **12**(4), 404–408.

Corrsin, S. 1958. Statistical behavior of a reacting mixture in isotropic turbulence. *Phys. Fluids*, **1**(1), 42–47.

Corrsin, S. 1959. Progress report on some turbulent diffusion research. *Adv. Geophysics*, **6**, 161–164.

Corrsin, S. 1961a. The reactant concentration spectrum in turbulent mixing with a first-order reaction. *J. Fluid Mech.*, **11**, 407–416.

Corrsin, S. 1961b. Turbulent flow. *American Scientist*, **49**, 300–325.

Corrsin, S. 1962a. Some statistical properties of the product of a turbulent first-order reaction. In *Fluid Dynamics and Applied Mathematics*. Edited by J.B. Diaz and S.I. Pai. Gordon and Breach, 105–124.

Corrsin, S. 1962b. Theories of turbulent dispersion. In *Mécanique de la Turbulence*, Editions du CNRS Paris, 27–52.

Corrsin, S. 1962c. Turbulent dissipation fluctuations. *Phys. Fluids*, **5**, 1301–1302.

Corrsin, S. 1963. Estimates of the relations between Eulerian and Lagrangian scales in large Reynolds number turbulence. *J. Atmos. Sci.*, **20**(2), 115–119.

Corrsin, S. 1964. Further generalizations of Onsager cascade model for turbulent spectra. *Phys. Fluids*, **7**(8), 1156–1159.

Corrsin, S. 1972. Simple proof of fluid line growth in stationary homogeneous turbulence. *Phys. Fluids*, **15**(8), 1370–1372.

Corrsin, S. 1974. Limitations of gradient transport models in random walks and in turbulence. *Adv. Geophysics*, **18A**, 25–71.

Corrsin, S., and Karweit, M. 1969. Fluid line growth in grid-generated isotropic turbulence. *J. Fluid Mech.*, **39**, 87–96.

Corrsin, S., and Kistler, A.L. 1955. Free-stream boundaries of turbulent flows. NACA Report, **1244**.

Corrsin, S., and Kovasznay, L.S.G. 1949. On the hot-wire length correction. *Phys. Rev.*, **75**, 1954.

Corrsin, S., and Phillips, O.M. 1961. Contour length and surface area of multiple-valued random variables. *J. Soc. Indust. Appl. Math.*, **9**(3), 395–404.

Corrsin, S., and Uberoi, M. 1950. Further experiments on the flow and heat transfer in a heated turbulent air jet. NACA Report 998 – formerly NACA TN 1865.

Corrsin, S., and Uberoi, M. 1951. Spectra and diffusion in a round turbulent jet. NACA Report, **1040**.

Davidson, P.A. 2004. *Turbulence: An Introduction for Scientists and Engineers*. Oxford University Press.

Davis, S.H., and Lumley, J.L. (eds.) 1985. *Frontiers in Fluid Mechanics: A Collection of Research Papers Written in Commemoration of the 65th Birthday of Stanley Corrsin*. Springer Verlag.

de Bruyn Kops, S. M., and Riley, J. J. 1998. Direct numerical simulation of laboratory experiments in isotropic turbulence. *Phys. Fluids*, **10**, 2125–2127.

Einstein, A. 1905. On the movement of small particles suspended in stationary liquids required by molecular-kinetic theory of hear. *Annalen der Physik*, **17**, 549–560.

Frisch, U. 1995. *Turbulence, the Legacy of A.N. Kolmogorov*. Cambridge University Press.

George, W.K. 1990. The nature of turbulence. In *FED-Forum on Turbulent Flows*. Edited by W.M. Bower, M.J. Morris and M. Samimy. Am. Soc. Mech. Eng. Book No H00599, **94**, 1–10.

George, W.K., and Arndt, R. (eds.) 1988. *Advances in Turbulence*. Taylor & Francis.

Goldstein, S. 1951. On diffusion by discontinuous movements, and on the telegraph equation. *Quart. J. Mech. Appl. Math.*, **4**(2), 129–156.

Hamburger Archives, JHU. 2009. The Ferdinand Hamburger Archives of The Johns Hopkins University. Department of Aeronautics (http://ead.library.jhu.edu/rg06-080.xml#id39664206), Record Group Number 06.080.

Harris, V.G., Graham, J.A.H., and Corrsin, S. 1977. Further measurements in nearly homogeneous turbulent shear flow. *J. Fluid Mech.*, **81**, 657–687.

Heisenberg, W. 1948. Zur statischen theorie der turbulenz. *Z. Physik*, **134**, 628–657.

Hinze, J.O. 1959. *Turbulence: An Introduction to its Mechanism and Theory*. McGraw-Hill.

Kang, H.S., Chester, S., and Meneveau, C. 2003. Decaying turbulence in an active-grid-generated flow and comparisons with large-eddy simulation. *J. Fluid Mech.*, **480**, 129–160.

Kármán, von T., and Howarth, L. 1938. On statistical theory of isotropic turbulence. *Proc. Roy. Soc. (London) A*, **164**, 192–215.

Kellogg, R.M., and Corrsin, S. 1980. Evolution of a spectrally local disturbance in grid-generated, nearly isotropic turbulence. *J. Fluid Mech.*, **96**, 641–669.

Kistler, A.L., O'Brien, V., and Corrsin, S. 1954. Preliminary measurements of turbulence and temperature fluctuations behind a heated grid. NACA Research Memorandum, **54D19**.

Kolmogorov, A.N. 1941. The local structure of turbulence in incompressible viscous fluid for very large Reynolds number. *C.R. Acad. Sci. USSR*, **30**, 301.

Kolmogorov, A.N. 1962. A refinement of previous hypotheses concerning the local structure of turbulence in a viscous incompressible fluid at high Reynolds number. *J. Fluid Mech.*, **13**, 82–85.

Kovasznay, L.S.G., Uberoi, M., and Corrsin, S. 1949. The transformation between one- and three-dimensional power spectra for an isotropic scalar fluctuation field. *Phys. Rev.*, **76**, 1263–1264.

Kraichnan, R.H. 1974. On Kolmogorov's inertial-range theories. *J. Fluid Mech.*, **62**, 305–330.

Kuo, A.Y-S., and Corrsin, S. 1971. Experiments on internal intermittency and fine-structure distribution functions in fully turbulent fluid. *J. Fluid Mech.*, **50**, 285–319.

Kuo, A.Y-S., and Corrsin, S. 1972. Experiment on the geometry of the fine-structure regions in fully turbulent fluid. *J. Fluid Mech.*, **56**, 447–479.

Landau, L.D., and Lifshitz, E. 1959. *Fluid Mechanics*. Addison-Wesley (1944 – 1st Russian edition, Moscow).

Liepmann, H.W. 1989. Stanley Corrsin: 1920–1986. *Memorial Tributes: National Academy of Engineering (The National Academies Press)*, **3**.

Lumley, J.L. 1962. Approach to Eulerian–Lagrangian problem. *J. Math. Phys.*, **3**, 309–312.

Lumley, J.L., and Corrsin, S. 1959. A random walk with both Lagrangian and Eulerian statistics. *Adv. Geophysics*, **6**, 179–183.

Lumley, J.L., and Davis, S.H. 2003. Stanley Corrsin: 1920–1986. *Ann. Rev. Fluid Mech.*, **35**, 1–10.

Mills, R.R., Kistler, A.L., O'Brien, V., and Corrsin, S. 1958. Turbulence and temperature fluctuations behind a heated grid. NACA Tech. Note, **4288**.

Moin, P., Squires, K., Cabot, W., and Lee, S. 1991. A dynamic subgrid-scale model for compressible turbulence and scalar transport. *Phys. Fluids A*, **3**, 2746–2757.

Obukhov, A.M. 1949. Structure of the temperature field in turbulent flows. *Izv. Akad. Nauk SSSR, Ser. Geofiz.*, **13**, 58–69.

Obukhov, A.M. 1962. Some specific features of atmospheric turbulence. *J. Fluid Mech.*, **13**, 77.

Onsager, L. 1949. Statistical hydrodynamics. *Nuovo Cim.*, **6**(2), 279–287.

Orszag, S.A. 1970. Comments on turbulent hydrodynamic line stretching – consequences of isotropy. *Phys. Fluids*, **13**(8), 2203–2204.

Patterson, G.S. Jr., and Corrsin, S. 1966. Computer experiments on random walks with both Eulerian and Lagrangian statistics. In *Dynamics of Fluids and Plasmas*, Proceedings of the Symposium held in honor of Professor Johannes M. Burgers, 7–9 October, 1965, at University of Maryland. Edited by S.I. Pai, A.J. Faller, T.L. Lincoln, D.A. Tidman, G.N. Trytton, and T.D. Wilkerson. Academic Press, 275–307.

Phillips, O.M. 1986. Book review of *Frontiers in Fluid Mechanics. J. Fluid Mech.*, **171**, 563–567.

Poinsot, T., and Veynante, D. 2001. *Theoretical and Numerical Combustion*. R.T. Edwards, Inc.

Rice, S.O. 1944. Mathematical analysis of random noise. *Bell Systems Tech. J.*, **23**(3), 282–332.

Rice, S.O. 1945. Mathematical analysis of random noise.. *Bell Systems Tech. J.*, **24**(1), 46–156.

Richardson, L.F. 1922. *Weather Prediction by Numerical Process*. Cambridge University Press.

Riley, J.J., and Corrsin, S. 1971. Simulation and computation of dispersion in turbulent shear flow. *Conf. on Air Pollution Met. AMS*.

Riley, J.J., and Corrsin, S. 1974. The relation of turbulent diffusivities to Lagrangian velocity statistics for the simplest shear flow. *J. Geophys. Res.*, **79**(12), 1768–1771.

Saffman, P.G. 1967. The large scale structure of homogeneous turbulence. *J. Fluid Mech.*, **27**, 581–594.

Shlien, D.J., and Corrsin, S. 1974. A measurement of Lagrangian velocity autocorrelation in approximately isotropic turbulence. *J. Fluid Mech.*, **62**, 255–271.

Sreenivasan. 1996. The passive scalar spectrum and the Obukhov-Corrsin constant. *Phys. Fluids*, **8**, 189–196.

Sreenivasan, K.R., Tavoularis, S., Henry, R., and Corrsin, S. 1980. Temperature fluctuations and scales in grid-generated turbulence. *J. Fluid Mech.*, **100**, 597–621.

Sreenivasan, K.R., Tavoularis, S., and Corrsin, S. 1981. A test of gradient transport and its generalizations. In *Turbulent Shear Flows 3*. Edited by L.J.S. Bradbury, F. Durst, B.E. Launder, F.W. Schmidt and J.H. Whitelaw. Springer Verlag, 96–112.

Tavoularis, S., and Corrsin, S. 1981a. Experiments in nearly homogeneous turbulent shear flow with a uniform mean temperature gradient. Part 1. *J. Fluid Mech.*, **104**, 311–347.

Tavoularis, S., and Corrsin, S. 1981b. Theoretical and experimental determination of the turbulent diffusivity tensor in homogeneous turbulent shear flow. In *3rd Symp. Turb. Shear Flows, Univ. California, Davis*, pp. 15.24–15.27.

Taylor, G.I. 1921. Diffusion by continuous movements. *Proc. London Math. Soc. Ser. A*, **20**, 196–211.

Tennekes, H. 1968. Simple model for small-scale structure of turbulence. *Phys. Fluids*, **11**, 669–670.

Tennekes, H., and Lumley, J.L. 1972. *A First Course in Turbulence*. MIT Press.

Terrell, R. 1959. Nobody hits it. *Sports Illustrated*, June 29 issue.

Townsend, A.A. 1948. Local isotropy in the turbulent wake of a cylinder. *Austral. J. Sci. Res. A*, **1**(2), 161–174.

Townsend, A.A. 1956. *The Structure of Turbulent Shear Flow*. Cambridge University Press.

Uberoi, M., and Corrsin, S. 1953. Diffusion of heat from a line source in isotropic turbulence. NACA Report **1142**.

8

George Batchelor: the post-war renaissance of research in turbulence

H.K. Moffatt

8.1 Introduction

George Batchelor (1920–2000), whose portrait (1984) by the artist Rupert Shephard is shown in Figure 8.1, was undoubtedly one of the great figures of fluid dynamics of the twentieth century. His contributions to two major areas of the subject, turbulence and low-Reynolds-number microhydrodynamics, were of seminal quality and have had a lasting impact. At the same time, he exerted great influence in his multiple roles as founder Editor of the *Journal of Fluid Mechanics*, co-Founder and first Chairman of EUROMECH, and Head of the Department of Applied Mathematics and Theoretical Physics (DAMTP) in Cambridge from its foundation in 1959 until his retirement in 1983.

I focus in this chapter on his contributions to the theory of turbulence, in which he was intensively involved over the period 1945 to 1960. His research monograph *The Theory of Homogeneous Turbulence*, published in 1953, appeared at a time when he was still optimistic that a complete solution to 'the problem of turbulence' might be found. During this period, he attracted an outstanding group of research students and post-docs, many from his native Australia, and Senior Visitors from all over the world, to work with him in Cambridge on turbulence. By 1960, however, it had become apparent to him that insurmountable mathematical difficulties in dealing adequately with the closure problem lay ahead. As he was to say later (Batchelor 1992):

> by 1960 ... I was running short of ideas; the difficulty of making any firm deductions about turbulence was beginning to be frustrating, and I could not see any real break-through in the current publications.

Over the next few years, Batchelor focused increasingly on the writing of his famous textbook *An Introduction to Fluid Dynamics* (Batchelor 1967), and in the process was drawn towards low-Reynolds-number fluid mechanics and

Figure 8.1 Portrait of George Batchelor by Rupert Shephard 1984; this portrait hangs in DAMTP, Cambridge, the Department founded under Batchelor's leadership in 1959.

suspension mechanics, the subject that was to give him a new lease of research life in the decades that followed. After 1960, he wrote few papers on turbulence, but among these few are some gems (Batchelor 1969, 1980; Batchelor, Canuto & Chasnov 1992) that show the hand of a great master of the subject.

I got to know George Batchelor myself from 1958, when he took me on as a new research student. George had just completed his work to be published in two papers the following year (Batchelor 1959; Batchelor, Howells & Townsend 1959) on the 'passive scalar problem', i.e. the problem of determining the statistical properties of the distribution of a scalar field which is convected and diffused within a field of turbulence of known statistical properties. There was at that time intense interest in the rapidly developing field of magnetohydrodynamics, partly fuelled by the publication in 1957 of Cowling's Interscience Tract *Magnetohydrodynamics*. Batchelor had himself written a famously controversial paper "On the spontaneous magnetic field in a conducting liquid in turbulent motion" (Batchelor 1950a; see also Batchelor 1952b), and it was natural that I should be drawn to what is now described as the 'passive vector problem', i.e. determination of the statistical evolution of a weak magnetic field, again under the dual influence of convection and diffusion by a 'known' field of turbulence. George gave me enormous encouragement and support during my early years of research in this area, for which I shall always be grateful.

My view of Batchelor's contributions to turbulence is obviously coloured by my personal interaction with him, and the following selection of what I regard as his outstanding contributions to the subject has a personal flavour. But I am influenced also by aspects of his work from the period 1945–1960 that still generate hot debate in the turbulence community today; among these, for example, the problem of intermittency which was first identified by Batchelor & Townsend (1949), and which perhaps contributed to that sense of frustration that afflicted George (and many others!) from 1960 onwards.

8.2 Marseille (1961): a watershed for turbulence

These frustrations came to the surface at the now legendary meeting held in Marseille in September 1961 on the occasion of the opening of the former Institut de Mécanique Statistique de la Turbulence (Favre 1962). This meeting, of which Batchelor was a key organizer, turned out to be a most remarkable event. Kolmogorov himself was there, together with Obukhov, Yaglom and Million-shchikov (who had first proposed the zero-fourth-cumulants closure scheme, in which so much work and hope had been invested during the 1950s), in which so much work and hope had been invested during the 1950s); von Kármán and G.I. Taylor were both there – the great father-figures of pre-war research in turbulence – and the place was humming with all the current stars of the subject – Stan Corrsin, John Lumley, Philip Saffman, Les Kovasznay, Bob Kraichnan, John Laufer, Hans Liepmann, Ian Proudman, Anatol Roshko, and George Batchelor himself among many others.

One of the highlights of the Marseille meeting was when Bob Stewart presented results of the measurement of ocean spectra in the tidal channel between Vancouver Island and mainland Canada (not mentioned in the Proceedings, but published soon after by Grant, Stewart & Moilliet 1962). These were the first convincing measurements showing several decades of a $k^{-5/3}$ spectrum, and providing convincing support for Kolmogorov's (1941a,b) theory which had been published 20 years earlier. But then Kolmogorov gave his lecture, which I recall was in the sort of 'Russian' French that was as incomprehensible to the French themselves as to the other participants. However the gist was clear: he said that quite soon after the publication of his 1941 papers, Landau had pointed out to him a defect in the theory, namely that wheresoever the local rate of dissipation of energy ϵ is larger than the mean, there the energy cascade will proceed more vigorously, and an increasingly intermittent distribution of $\epsilon(\mathbf{x}, t)$ is therefore to be expected. Arguing for a log-normal probability distribution for ϵ, a suggestion that he attributed to Obukhov, Kolmogorov showed that the exponent $(-5/3)$ should be changed slightly, and that higher-order statistical quantities would be more strongly affected by this intermittency.

This must in fact have been no real surprise to Batchelor, because as indicated above, it was he and Townsend who had remarked on the phenomenon of intermittency of the distribution of vorticity as the Kolmogorov scale $(\nu^3/\epsilon)^{1/4}$ is approached, in their 1949 paper. They had noticed the puzzling increase of flatness factor (or 'kurtosis') of velocity derivatives with increasing Reynolds number $Re = \langle \mathbf{u}^2 \rangle^{1/2} L/\nu$ in standard notation, a behaviour that is inconsistent with the original Kolmogorov theory (for on that theory, the flatness factor and similar dimensionless characteristics of the small-scale features of the turbulence, should be determined solely by ϵ and kinematic viscosity ν, and so necessarily universal constants, independent of Reynolds number). They interpreted this in terms of a tendency to form "isolated regions of concentrated vorticity", and it is interesting to note that much of the research on turbulence of the last two decades has been devoted to identifying such concentrated vorticity regions, both in experiments and in numerical simulations. Townsend himself thought in terms of a random distribution of vortex tubes and sheets (Townsend 1951b) in his theory for the dissipative structures of turbulence, a theory that is described in Batchelor's (1953) monograph.

I regard the 1961 Marseille meeting as a watershed for research in turbulence. The very foundations of the subject were shaken by Kolmogorov's presentation; and the new approaches, particularly Kraichnan's (1959) direct interaction approximation, were of such mathematical complexity that it was really difficult to retain that link between mathematical analysis and physical understanding that is so essential for real progress.

Given that Batchelor was already frustrated by the mathematical intractability of turbulence, it was perhaps the explicit revelation that all was not well with Kolmogorov's theory that finally led him to abandon turbulence in favour of other fields. He had invested huge effort himself in the elucidation and promotion of Kolmogorov's theory (see below) and regarded it as perhaps the one area of the subject on which reasonable confidence could be placed; to find this theory now undermined at a fundamental level by its originator, and that, ironically, just as experimental 'confirmation' of the flawed theory was becoming available, must have been deeply disconcerting, and one may well understand how it was that over the subsequent decade, Batchelor's energies were more or less totally deflected not only to his textbook, but also to the Editorship of the *Journal of Fluid Mechanics*, which he had founded in 1956 and which was now in a phase of rapid growth, and equally to the Headship of the Department of Applied Mathematics and Theoretical Physics (DAMTP) in Cambridge, which had been established largely under his visionary impetus in 1959.

8.3 Personal background

But first, some early background. George Batchelor was born on 8 March 1920 in Melbourne, Australia. He attended school in Melbourne and won a scholarship to Melbourne University where he studied mathematics and physics, graduating at the age of 19 in 1939, just as World War II broke out. He was guided to take up research in Aerodynamics with the CSIR Division of Aeronautics. Throughout the war, he worked on a succession of practical problems which were not of great fundamental interest, but which served to motivate him towards the study of turbulence, which he perceived not only as the most challenging aspect of fluid dynamics but also as the most important in relation to aerodynamic applications.

In the course of this work, he read the papers of G.I. Taylor, particularly those from the 1930s on the statistical theory of turbulence, and resolved that this was what he wanted to pursue as soon as the war ended. He wrote to G.I. offering his services as a research student, and G.I. agreed to take him on. At the same time, and most significantly, George persuaded his fellow-Australian Alan Townsend to join him in this voyage of discovery. I became aware in later years of George's powers of persuasion, of which this was perhaps a first manifestation. Alan described in a later essay (Townsend 1990) his initial encounter with George in Melbourne, and how it was that he was induced to switch from research in nuclear physics to experimental work on turbulence. In his last published paper "Research as a life style", Batchelor (1997) relates that on his suggesting to Townsend that they should join forces in working on turbulence under G.I. Taylor, Townsend responded that he would be glad to do so, but he first wanted to ask two questions: Who is G.I. Taylor? and What is turbulence? The former question was obviously the easier to answer; the latter has provoked much philosophical debate over the years! [Batchelor (1953) himself described turbulent flow in simple terms as flow in which the velocity takes random values but whose average properties are uniquely determined by the controllable data (e.g. boundary conditions) of the flow.] In any event, George's answer must have been sufficient to convince Alan on what turned out to be an excellent career move; the partnership between George Batchelor and Alan Townsend, combining brilliant intuition in both theory and experiment, was to endure for the next 15 years during which the foundations of modern research in turbulence were to be established.

George married Wilma Raetz, also of Melbourne, in 1944, and in January 1945 they set off on an epic sea voyage to Cambridge, via New Zealand, Panama, Jamaica and New York, and then in a convoy of 90 ships across the Atlantic to Tilbury docks in London; and thence to Cambridge where George and Wilma were destined to spend the rest of their lives. George was then just 25 years old.

Alan Townsend came independently to Cambridge, and when he and George met G.I. Taylor and talked with him about the research that they would undertake, they were astonished to find that G.I. himself did not intend to resume work on turbulence, but rather to concentrate on a range of problems – for example the rise of large bubbles from underwater explosions, or the blast wave from a point release of energy – that he had encountered through war-related research activity. George and Alan were therefore left more-or-less free to determine their own programme of research, with guidance but minimal interference from G.I. – and they rose magnificently to this challenge!

8.4 Batchelor and the Kolmogorov theory of turbulence

Batchelor spent his first year in Cambridge searching the literature of turbulence in the library of the Cambridge Philosophical Society. There, he made an amazing discovery – he came upon the English language editions of the 1941 issues of *Doklady*, the *Comptes Rendus of the USSR Academy of Sciences*, in which the seminal papers of Kolmogorov had been published; amazing, because the Academy had been displaced from Moscow to Kazan in the foothills of the Ural Mountains, in the face of the German advance from the west; amazing, because it is hard to imagine how the Academy could have continued to produce an English language edition of *Doklady* in the crisis situation then prevailing; and even more amazing because it is hard to imagine how any mail, far less consignments of scientific journals, could have found their way from the USSR to England during those dreadful years. [Barenblatt 2001 relates that bound volumes of *Doklady* and other Soviet journals were used as ballast for supply ships making the dangerous return journey from Russia through Arctic waters to the West!]

But the fact is that Batchelor did indeed find these papers and immediately recognised their significance. In his lecture "Fifty Years with Fluid Mechanics" at the 11th Australasian Fluid Mechanics Conference (Batchelor 1992), he said: "Like a prospector systematically going through a load of crushed rock,

I suddenly came across two short articles, each of about four pages in length, whose quality was immediately clear". Four pages was the normal limit of length imposed by the USSR Academy for papers in *Doklady*, a limit that perhaps suited Kolmogorov's minimalist style of presentation, but at the same time made it exceptionally difficult for others to recognize the significance of his work. Batchelor did recognize this significance, and proceeded to a full and thorough discussion of the theory in a style that was to become his hallmark: the assumptions of the theory were set out with the utmost care, each hypothesis being subjected to critical discussion both as to its validity and its limitations; and the consequences were then derived, and illuminated with a penetrating physical interpretation at each stage of the argument.

The VIth International Congress of Applied Mechanics, which was held in Paris in September 1946, provided an early opportunity for Batchelor to announce his findings to a wide international audience. (This sequence of quadrennial Congresses had been established in the 1920s through the forceful initiative of von Kármán, Prandtl, G.I. Taylor and Jan Burgers, but had been interrupted by the war. It was a remarkable achievement to reinstate the sequence so soon after the war, in the chaotic and straitened conditions that must still have prevailed in Paris at that time. Paul Germain told me with a degree of chagrin that the Proceedings of the Congress were duly delivered to Gautiers-Villars following the Congress, but have never yet appeared, perhaps a record in publication delay!)

Batchelor wisely published his contribution to the Congress in a brief communication to *Nature* (Batchelor 1946b). In this, he described the Kolmogorov theory and he simultaneously drew attention to parallel lines of enquiry of Onsager, Heisenberg and von Weizsäcker. There is an interesting historical aspect to this: both Heisenberg and von Weizsäcker were taken, together with other German physicists, to Britain at the end of the war, and placed under house arrest in a large country house not far from Cambridge. At their own request, they were allowed to visit G.I. Taylor, no doubt under surveillance (this was probably in August 1945 – see Batchelor 1992) to discuss energy transfer in turbulent flow, and it was in subsequent discussion between Taylor and Batchelor that the link with the work of Kolmogorov was recognized. But as Batchelor said: "The clearest formulation of the ideas was that of Kolmogorov, and it was also more precise and more general".

Kolmogorov himself gave due credit to the prior work of Lewis Fry Richardson, who had conceived the 'energy cascade' mechanism immortalized in a rhyme reproduced elsewhere in this volume. Kolmogorov's signal achievement, on which Batchelor rightly focused, was to identify the mean rate of dissipation of energy per unit mass ϵ and the kinematic viscosity of the fluid ν

as the sole dimensional quantities on which all small-scale statistical properties of the turbulence should depend.

Batchelor's definitive paper on the subject appeared in 1947 in the *Proceedings of the Cambridge Philosophical Society*. There can be little doubt that it was this paper that effectively disseminated the Kolmogorov theory to the Western world. Barenblatt has claimed that in Russian translation it also served to make the theory comprehensible to turbulence researchers in the Soviet Union! The theory is described in Chapter 6 of this volume, and I need not labour the details here. There is however one point that does deserve mention: Batchelor draws particular attention to Kolmogorov's derivation of the 'four-fifths' law for the third-order structure function in isotropic turbulence:

$$B_{ddd}(r) = \langle (u'_d - u_d)^3 \rangle = -\tfrac{4}{5} \epsilon r,$$

where, in Kolmogorov's notation, the suffix d indicates components of velocity parallel to the separation between points $\mathbf{x}$ and $\mathbf{x}' = \mathbf{x} + \mathbf{r}$; r is in the inertial range, and ϵ is the rate of dissipation of energy per unit mass of fluid. Batchelor's careful re-derivation of this result (which strangely he chose not to reproduce in his 1953 monograph) still merits study. The result is of central importance because, as pointed out by Frisch (1995), "it is both exact and nontrivial"; indeed it is perhaps the only exact nontrivial result in the whole of the dynamic (as opposed to merely kinematic) theory of turbulence.

Over the following years, Batchelor was to be increasingly preoccupied with exploiting the new insights that the Kolmogorov theory provided in a range of problems (e.g. the problem of turbulent diffusion) for which the small-scale ingredients of the turbulence play a key role. Batchelor kept a research notebook during the late 1940s, in which is found a page significantly entitled "Suggestions for exploitation of Kolmog's [sic] theory of local isotropy"; on the other side of the page appear the following memoranda: "Apply theory to time-delay correlations; establish relation to space correlations; refer to diffusion analysis.... Apply to axisymmetric turbulence, e.g. to evaluate dissipation terms and establish tendency to isotropy". These were aspects of turbulence to which Batchelor devoted himself over the subsequent years.

Largely on the basis of his work on the elucidation of Kolmogorov's theory, Batchelor was elected in 1947 to a Fellowship at Trinity College, Cambridge, a position that enabled him to devote himself entirely to research over the next four years, during which he published some 15 papers (some with Townsend) on all aspects of turbulent flow. He took his PhD degree in 1948 (see Figure 8.2), and was by October 1949 installed as a lecturer in the Faculty of Mathematics, in succession to Leslie Howarth (of von Kármán–Howarth fame) who

Figure 8.2 George Batchelor with his mentor Sir Geoffrey (G.I.) Taylor, PhD graduation day, 1948.

had left Cambridge to take the Chair of Applied Mathematics at the University of Bristol.

Batchelor's PhD Examiners were in fact Leslie Howarth and G.I. Taylor himself (it being still accepted in those days that a Research Supervisor could also act as Examiner – defending counsel one day, prosecuting the next!). Batchelor recounted to me many years later that Howarth had asked him during the oral examination why he had dropped the terms that lack mirror symmetry in his discussion of the form of the spectrum tensor for isotropic turbulence, and what these terms might represent if they were retained. These were two questions to which Batchelor, on his own admission, was unable to give a satisfactory answer! They are in fact the terms that encapsulate the 'helicity' of turbulence, a concept that was to emerge two decades later and find important application in two contexts: the Euler equations (for which helicity is a topological invariant); and turbulent dynamo theory (see below).

The VIIth International Congress of Applied Mechanics was held at Imperial College, London, in September 1948; it was at this Congress that the International Union of Theoretical and Applied Mechanics (IUTAM) was formally established. In the group photograph of some 200 participants, which may be viewed on the IUTAM website, one may detect G.I. Taylor and Theodore von Kármán in the front row, and George Batchelor and Alan Townsend perched together on the plinth of a statue in the back row. I also recognize the unmistakable features of James Lighthill, Keith Stewartson and Michael Glauert, and some others, in this photograph. The Congresses of Paris (1946), London

(1948), Istanbul (1952), Brussels (1956), Stresa (1960) and Munich (1964) provided the opportunity for international contact and exchange of ideas during a period when international meetings in theoretical and applied mechanics were far fewer than they are today; Batchelor attended all of them, and encouraged his collaborators and students to do likewise. (I attended the Stresa Congress myself as a Research Student, and found it wonderfully stimulating and broadening; I have attended every ICTAM ever since, with just one exception; these Congresses are like family gatherings, where announcements of new results are keenly anticipated. G.I. Taylor used to say that he always saved his best results for these great international Congresses, and recommended that others should follow his example!)

8.5 Batchelor and the turbulent dynamo

Batchelor's involvement in magnetohydrodynamics stemmed from a symposium on 'Problems of motion of gaseous masses of cosmical dimensions' arranged jointly by the International Astronomical Union and the newly formed International Union of Theoretical and Applied Mechanics, and held in Paris in August 1949. It is interesting to go back to the Proceedings of that symposium (Batchelor 1951) to see what he said there. His paper starts:

> It is not a very enviable task to follow Dr von Kármán on this subject of turbulence. He explained things so very clearly and he has touched on so many matters that the list of things which I had to say is now torn to shreds by the crossings out I have had to make as his talk progressed. But there is one point on what ought to be called the pure turbulence theory, which I should like to make; this point concerns the spectrum and will be useful also for Dr von Weizsäcker, in his talk. After having made that point, I want to plunge straight into the subject that some of the speakers have lightly touched on and then hastily passed on from, namely the interaction between the magnetic field and the turbulence. That will perhaps give us something to talk about. I shall be thinking aloud so that everything may be questioned.

It takes some courage to "think aloud" in an international gathering of this kind; and here was Batchelor, at the age of 29, thinking aloud and indeed leading the debate, in the presence of such giants as von Kármán, von Neumann who was there also, and von Weizsäcker! It was in this setting and in his subsequent (1950a) paper that Batchelor developed the analogy between vorticity ω in a turbulent fluid and magnetic field $\mathbf{B}$ in a highly conducting fluid in turbulent motion. The analogy is one that has to be used with great care, because it is an imperfect one: vorticity is constrained by the relation $\omega = \nabla \times \mathbf{u}$ to the velocity field **u** that convects it, whereas **B** is free of any such constraint. Nevertheless, some valid results do follow from the analogy, despite their insecure foundation: in particular, the fact that in the ideal fluid limit, the magnetic flux through any material circuit is conserved, like the flux of vorticity in an inviscid non-conducting fluid.

Batchelor applied the analogy to the action of turbulence in a highly conducting cloud of ionized gas on a weak 'seed' magnetic field. He argued that, provided $\eta \lesssim \nu$ (where η is the magnetic diffusivity and ν the kinematic viscosity), stretching of the field, which is most efficient at the Kolmogorov scale $l_\nu = (\nu^3/\epsilon)^{1/4}$, will intensify it on a time-scale $t_\nu = (\nu/\epsilon)^{1/2}$ until it reaches a level of equipartition of energy with the smallest-scale (dissipation-range) ingredients of the turbulence, i.e. until

$$\langle \mathbf{B}^2 \rangle / \mu_0 \rho \sim (\epsilon \nu)^{1/2}.$$

I note that this estimate still attracts credence (see Kulsrud 1999, who writes "there is equipartition of the small-scale magnetic energy with the kinetic energy of the smallest eddy" in conformity with Batchelor's conclusion that "a steady state is reached when the magnetic field has as much energy as is contained in the small-scale components of the turbulence"). This energy level is smaller, it should be noted, by a factor $Re^{-1/2}$, than the overall mean kinetic energy $\langle \mathbf{u}^2 \rangle / 2$ of the turbulence.

There were however two reasons, themselves in mutual contradiction, to doubt Batchelor's conclusions. On the one hand, intensification of the magnetic field through the stretching mechanism may be expected to occur (albeit relatively slowly) on length-scales much larger than the Kolmogorov scale, and it is therefore arguable (as urged almost simultaneously by Schlüter & Biermann 1950) that the range of equipartition should extend ultimately to the full spectral range of the turbulence. On the other hand, intensification by stretching is naturally associated with decrease of scale in directions transverse to the field and hence with accelerated joule dissipation, and it was argued by Saffman (1964) that this effect would in all circumstances lead to ultimate decay of magnetic energy – as had been previously proved to be the case if all fields are assumed to be two-dimensional (Zeldovich 1957).

Thus, by 1965, all bets were open, and nothing was certain: in the presence of homogeneous isotropic turbulence, an initially random magnetic field with zero mean might grow to equipartition from an infinitesimal level, or might grow under some subsidiary conditions to some significantly lower level, or might grow for a while and then decay to zero. Into the gloom of this unresolved controversy, there penetrated a shaft of light from beyond the Iron Curtain, in the work of Steenbeck, Krause & Rädler published in 1966 in the

East German journal *Astronomie Nachrichten*. Instead of focusing on turbulent distortion of a magnetic field on small scales, these authors considered the possibility that the field might grow on scales *large* compared with the scale of the turbulence (the 'freedom' overlooked by Batchelor and others), and they used a two-scale analysis to investigate this possibility, thus giving birth to the subject of 'mean-field electrodynamics'. This seems so utterly natural now that one can only wonder why no one had thought of it before; perhaps it is always so with a breakthrough, which this undoubtedly was! What emerged was that provided the turbulence has the property of 'helicity' ('schraubensinn' or literally 'screw-sense' in the papers of Steenbeck, Krause & Rädler), then the magnetic field will always grow on a sufficiently large length-scale, this scale being determined by the 'magnetic Reynolds number' of the turbulence, which may be arbitrarily *small* (Moffatt 1970). So as it turned out, not only Batchelor, but also Schlüter & Biermann, and Saffman, were all wrong in different ways. The turbulent dynamo problem would never look quite the same again! Nevertheless, as Kulsrud (1999) has argued, arguments of the kind advanced 60 years ago by Batchelor are still relevant to consideration of the small-scale field and of saturation mechanisms when the magnetic Reynolds number is very large, as for example in the interstellar or intergalactic medium.

8.6 The decay of homogeneous turbulence

Much of Batchelor's early work was concerned with the idealized problem of homogeneous turbulence; that is, turbulence whose statistical properties are invariant under translations. The central problem addressed by Batchelor & Townsend (1947, 1948a,b) concerned the rate of decay of homogeneous turbulence, of the kind that could be produced by flow through a grid in a wind tunnel. It was perhaps natural to focus on this problem, which is so strongly influenced by the nonlinear transfer of energy from large to small scales, the process that had previously been investigated by the Soviet scientists Loitsyansky, Millionschikov and Kolmogorov (see Chapter 6).

The 1947 paper "Decay of vorticity in isotropic turbulence" by Batchelor & Townsend is of particular interest in this context. They note that the rate of change of mean-square vorticity in isotropic turbulence is proportional to the mean-cube of vorticity, which in turn is related to the skewness factor of the velocity derivative. Measurements of Townsend indicated that this skewness factor was approximately constant and independent of Reynolds number during decay of the turbulence, with the consequence that the contribution from the nonlinear inertial terms of the equation of motion to the rate of change of mean-square vorticity is apparently proportional to $\langle \omega^2 \rangle^{3/2}$. This term taken alone would lead to a blow-up of mean-square vorticity within a finite time. Batchelor & Townsend however stated that "the viscous contribution is always the greater [in comparison with the nonlinear contribution] but the contributions tend to equality as the grid Reynolds number increases". (The statement that the "viscous contribution is always the greater" clearly does not apply to the initial development of a field of turbulence concentrated at low wave-numbers in wave-number space; for such a field, the nonlinear contribution dominates and leads to initial rapid growth of mean-square vorticity, just as in the model problem treated by Taylor & Green 1937, now familiarly known as the 'Taylor–Green vortex'.)

There is still to this day great interest in the question as to whether, for an inviscid fluid, the mean-square vorticity (or indeed the pointwise distribution of vorticity) really can exhibit a singularity at finite time. Indeed, this remains a central unsolved problem in the mathematics of the Euler equations; it was debated at length at the meeting celebrating the 250th anniversary of the Euler equations, held in Aussois in the Haute Savoie, in 2008 (Eyink et al. 2008). The argument of Batchelor & Townsend (1947) rests on the semi-empirical observation of apparent approach to constancy of the skewness factor with increasing Reynolds number, but anything approaching a proof of this sort of result is still lacking.

In some ways, it seems now that the intense preoccupation in the post-war years with the problem of homogeneous isotropic turbulence with zero mean velocity was perhaps misguided. The intention was clearly to focus on the central problem of nonlinear inertial energy transfer, but in so doing the most intractable aspect of the problem was addressed, with no possibility (when $Re \gg 1$) of anything like a 'perturbative' approach. In shear flow turbulence (as opposed to homogeneous turbulence with zero mean) the (linear) interaction between the mean flow and the turbulent fluctuations provides a valid starting point for theoretical investigation, which is simply not available for the problem of homogeneous turbulence with zero mean. The intense and enduring difficulty of the latter problem is associated with the fact that all linearizable features have been stripped away, and the naked nonlinearity of the problem is all that remains.

Batchelor & Townsend nevertheless recognized (1948b) that nonlinear effects become negligible during the 'final period of decay' when eddies on all scales decay through direct viscous dissipation. They determined the asymptotic decay of turbulent energy (proportional to $t^{-5/2}$) during this final period. Batchelor's result depends on the assumed behaviour of the energy spectrum at small wavenumbers (Batchelor 1949a), and the later work of Batchelor &

Proudman (1956) revealed an awkward non-analytic behaviour of the spectrum tensor of homogeneous turbulence in the neighbourhood of the origin in wave-number space, induced by the long-range influence of the pressure field. The problem was taken up again by Saffman (1967) who showed that, under plausible assumptions concerning the means by which the turbulence is generated, the energy spectrum function could have a k^2 (rather than k^4) dependence near $k = 0$, with the consequence that the energy decays in the final period as $t^{-3/2}$ instead of $t^{-5/2}$. The form of the spectrum tensor near $\mathbf{k} = 0$ and its consequences for the final period of decay has been the subject of extended debate that continues to this day (Ishida, Davidson & Kaneda 2006). (See Chapter 12 for further discussion of this issue.)

Thus, even for this 'easiest' aspect of the problem of homogeneous turbulence, the situation turned out to be far more subtle than originally realized.

8.7 Batchelor's 1953 monograph,
The Theory of Homogeneous Turbulence

Batchelor's slim CUP monograph *The Theory of Homogeneous Turbulence* was published in 1953, and consisted largely of an account of his own work, much of it in collaboration with Townsend, prior to that date. It was the first book wholly and exclusively devoted to the subject of turbulence, and as such gave great impetus to research in this field. It was much later republished in the CUP "Cambridge Science Classics" series, a mark of its enduring value.

Batchelor sets out his general philosophy in the preface as follows:

Finally, it may be worthwhile to say a word about the attitude that I have adopted to the problem of turbulent motion, since workers in the field range over the whole spectrum from the purest of pure mathematicians to the most cautious of experimenters. It is my belief that applied mathematics, or theoretical physics, is a science in its own right, and is neither a watered-down version of pure mathematics nor a prim form of physics. The problem of turbulence falls within the province of this subject, since it is capable of being formulated precisely. The manner of presentation of the material in this book has been chosen, not with an eye to the needs of mathematicians or physicists or any other class of people, but according to what is best suited, in my opinion, to the task of *understanding the phenomenon*. Where mathematical analysis contributes to that end, I have used it as fully as I have been able, and equally I have not hesitated to talk in descriptive physical terms where mathematics seems to hinder the understanding. Such a plan will not suit everybody's taste, but it is consistent with my view of the nature of the subject matter.

Figure 8.3 The Fluid Dynamics group at the Cavendish laboratory, 1954; front row: Tom Ellison, Alan Townsend, G.I. Taylor, George Batchelor, Fritz Ursell, Milton Van Dyke. Philip Saffman is in the middle of the back row, Stewart Turner in the top right-hand corner, and Bruce Morton and Owen Phillips third and fifth from the left in the middle row.

Reading between the lines of this quotation, it is apparent that Batchelor already had serious reservations concerning the insight that 'pure mathematics', i.e. mathematics based on rigorous proof of precisely stated theorems, might be able to contribute to the problem of turbulence. Although he was by 1953 playing a central role in the Faculty of Mathematics at Cambridge, his wartime background in Aerodynamics still governed his real-world approach to scientific problems; in the spirit of his mentor G.I. Taylor, he would always give preference to an illuminating physical argument over a piece of abstract mathematical analysis; to such analysis, he was even on occasion quite hostile.

Batchelor himself provides a 'brief history of the subject' in his introductory chapter. In this, he places the origin of the study of homogeneous turbulence in the papers of G.I. Taylor (1935, 1938) in which "the fact that the velocity of the fluid in turbulent motion is a random continuous function of position and time" was first clearly recognized and developed. It was these papers that had particularly attracted Batchelor to the study of turbulence in the first place. Batchelor's immense admiration, if not awe, for the achievements of G.I. Taylor shines through his later writings on the subject, and particularly so in his biography *The Life and Legacy of G.I. Taylor*, published by CUP in 1996.

Figure 8.3 shows the Fluid Dynamics group of the Cambridge Cavendish Laboratory in 1954. Prominent in the front row are Townsend, Taylor and Batchelor; also Fritz Ursell, working on surface water waves, who later moved

Figure 8.4 George Batchelor, having been recently elected FRS, in his office at the old Cavendish Laboratory, October 1956.

to the Beyer Chair of Applied Mathematics at Manchester University; and Milton Van Dyke, working on the theory of hypersonic flow, who was that year visiting the Cavendish from NASA Ames. The Cavendish Laboratory was still at that time located in Free School Lane in the centre of Cambridge. G.I. Taylor was based there above the 'Balfour Room' which housed the wind tunnel used by Townsend for his experimental work. Batchelor was already planning the *Journal of Fluid Mechanics* (JFM) which was to be launched in 1956; Figure 8.4 shows him in his office at the Cavendish shortly after this launch, and shortly after his election as a Fellow of the Royal Society.

Figure 8.5 shows him in a group at the IXth Congress of Applied Mechanics in Brussels (1956), an important opportunity for disseminating information about the new journal. James Lighthill, whom Batchelor had engaged as one of the first Associate Editors of JFM, is on the right of this photo, one of few photos in which these two great figures of 20th-century fluid mechanics can be seen together. While each recognized the great talents of the other, they were poles apart in personality and style; their 'sparring' partnership through JFM was nevertheless amicable and endured for more than 20 years. Lighthill held the Lucasian Chair of Mathematics in DAMTP, Cambridge, from 1969

Figure 8.5　George Batchelor (second from left) at the IXth International Congress of Applied Mechanics in Brussels, 1956; James Lighthill on the right.

to 1978, but was content throughout that period to leave the running of the department to Batchelor, who was Head of the Department from its foundation in 1959 until his retirement in 1983.

8.8　Rapid distortion theory

G.I. Taylor had in 1935 considered the effect of an irrotational rapid distortion on a single Fourier component of a turbulent velocity field, the motivation being to understand the manner in which wind-tunnel turbulence might be suppressed by passage through a contracting section. Batchelor & Proudman (1954) took up this problem, and, by integrating over all the Fourier components of a turbulent field, determined the manner in which anisotropy is induced in an initially isotropic field of turbulence. If the distortion is sufficiently rapid relative to the timescale of the turbulent eddies, then a linear treatment is legitimate. This realization, and the fact that a linear treatment is distinctly better than no treatment at all, has led to many subsequent developments of rapid distortion theory and application in a wide range of contexts (see for example

Sagaut & Cambon 2008). The paper of Batchelor & Proudman led the way in this important branch of the theory of turbulence.

A closely related problem which may also be treated by linear techniques concerns the effect on wind tunnel turbulence of a wire gauze placed across the stream. This problem was treated by Taylor & Batchelor (1949), interestingly the only paper under their joint authorship. When the turbulence level is weak, the velocity field on the downstream side of the gauze is linearly related to that on the upstream side, and so the spectrum tensor of the turbulence immediately downstream of the gauze can be determined in terms of that on the upstream side. The transverse components of velocity are affected differently from the longitudinal components, so that turbulence that is isotropic upstream becomes non-isotropic, but axisymmetric, downstream, a behaviour that was at least qualitatively confirmed by Townsend (1951a). Batchelor (1946a) had previously developed a range of techniques appropriate to the description of axisymmetric turbulence, techniques that here found useful application.

8.9 Turbulent diffusion

Batchelor first addressed the problem of turbulent diffusion in a paper (1949b) published by the *Australian Journal of Scientific Research*, under the title "Diffusion in a field of homogeneous turbulence". This paper was clearly inspired by the seminal paper of Taylor (1921) in which the dispersion of a particle in a turbulent flow relative to a fixed point had been first considered. Batchelor extended this treatment to three dimensions, and, more importantly, showed that the mean concentration for a finite volume of marked fluid satisfied a diffusion equation with a time-dependent diffusion tensor, this being a generalization of Taylor's diffusion coefficient. In a subsequent series of papers (Batchelor 1950b, 1952a; Batchelor & Townsend 1956) Batchelor considered the relative diffusion of two particles, and established a theoretical link with Richardson's law of diffusion whereby the rate of increase of mean-square separation is proportional to the two-thirds power of the mean-square separation. With the hindsight of Kolmogorov's theory, this result can of course be obtained on dimensional grounds, when the particle separation is in the inertial range.

I have already referred to Batchelor's famous (1959) paper "Small-scale variation of convected quantities like temperature in turbulent fluid". It was in this paper that he recognized the critical importance of the Prandtl number v/κ, where κ is the molecular diffusivity of the convected scalar field. He argued that, when v/κ is large, scalar fluctuations persist on scales small compared with the Kolmogorov scale, in fact down to the 'Batchelor scale' $(\epsilon/v\kappa^2)^{1/4}$.

On scales between the Kolmogorov scale and the Batchelor scale, the velocity gradient is approximately uniform, and on this basis, Batchelor was able to determine the spectrum of the fluctuations of the scalar field in the corresponding range of wave-numbers k; he found this to be proportional to k^{-1}. It is worth remarking that the same technique applied to the corresponding passive *vector* problem leads to exponential growth of the energy of the vector field (Moffatt & Saffman 1964), reflecting in some degree the type of dynamo action that Batchelor had predicted in 1950.

In the companion paper (Batchelor, Howells & Townsend 1959) the small Prandtl number situation was considered; in this case, the 'conduction cut-off' occurs at wavenumber $(\epsilon/\kappa^3)^{1/4}$ (as previously found by Obukhov 1949), and the scalar spectrum was determined in the range of wavenumbers between the conduction cut-off and the viscous cut-off (the $k^{-17/3}$-law).

These two papers, coupled with those of Corrsin (1951) and Obukhov (1949), have provided the starting point for almost all subsequent treatments of the passive scalar problem, a problem which has attracted renewed attention, with respect to its intermittency characteristics, in recent years – see Chapter 10 concerning Kraichnan's contributions to this problem. It is worth noting that, although Kraichnan's model involved a 'delta-correlated' velocity field, i.e. one varying infinitely rapidly in time, as opposed to Batchelor's quasi-steady model, he still found a k^{-1} range when $\nu/\kappa \gg 1$, implying a certain robustness of this result.

I should not leave the topic of turbulent diffusion without mention of the interesting paper of Batchelor, Binnie & Phillips (1955) in which it was shown that the mean velocity of a fluid particle in turbulent pipe flow is equal to the conventional mean velocity (averaged over the cross-section). This result was tested experimentally using small neutrally-buoyant spheres injected into a pipe flow. This was perhaps the first Lagrangian measurement in turbulent flow. The paper is closely related in spirit to Taylor's famous (1954) paper treating the axial diffusion of a scalar in a pipe flow (laminar or turbulent).

8.10 Two-dimensional turbulence

Batchelor anticipated the potential interest of the two-dimensional problem in the final two pages of his 1953 monograph. This situation is very different from the three-dimensional case, because for two-dimensional turbulence, vortex stretching does not occur, and so the enstrophy (mean-square vorticity) as well as the energy are inviscid invariants of the flow. Batchelor recognized an important consequence of this additional constraint, namely that the net effect of inertial spreading of energy in wave-number space must be to transfer energy towards progressively *larger* scales – what later became known as an 'inverse cascade'. He further recognized that "from the original motion there will gradually emerge a few strong isolated vortices" and that "vortices of the same sign will continue to tend to group together", conclusions that have been amply confirmed by numerical simulations more than 30 years later. As Batchelor pointed out, this conclusion had been earlier anticipated by Onsager (1949) on the basis of statistical mechanics.

The issue of two-dimensional turbulence came up in the discussions at the 1961 Marseille Conference, and I believe that these discussions led both Batchelor and Kraichnan to investigate the situation further. Batchelor saw the possibility of numerical simulation, and put his research student Roger Bray to work on the problem; his PhD thesis "A study of turbulence and convection using Fourier and numerical analysis" was submitted and approved in 1966. Bray's computations were necessarily primitive, but the remarkable thing is that they were carried out at all in those early days when computing power was so low. On the basis of Bray's numerical results, and stimulated by Kraichnan's (1967) important paper on the subject (see Chapter 10), Batchelor finally returned to the problem with his 1969 paper, where he writes as follows:

> There appears to be an impression in the literature of fluid dynamics that the known differences between motion in two and three dimensions are so great that two-dimensional turbulence (which cannot normally be reproduced experimentally in our three-dimensional world owing to instability) bears no relation to three-dimensional turbulence. It is indeed true that particular properties of a turbulent motion depend strongly on the number of space dimensions; but the two basic properties of randomness and nonlinearity are present in two-dimensional turbulence, and these should ensure the applicability of concepts such as a cascade transfer process and statistical decoupling accompanying transfer between Fourier components. My own view is that there are enough common general properties of two- and three-dimensional turbulence to justify using a numerical investigation of two-dimensional turbulence as a means of testing the soundness of some of the plausible hypotheses about turbulent flow.

Such justification for a study of two-dimensional turbulence would hardly be needed now! An understanding of the phenomenon is essential in the context of atmospheric dynamics, where strong stable stratification tends to confine the velocity field to horizontal planes. Similarly coriolis forces in a rotating fluid, or Lorentz forces in a conducting fluid permeated by a strong uniform magnetic field can constrain turbulence to have a nearly two-dimensional structure.

Batchelor focused in this paper on the fact that in two-dimensional turbulence, it is enstrophy (rather than energy) that cascades towards the high wavenumbers at which dissipation by viscosity takes over. The rate of dissipation of enstropy η is now the important dimensional parameter in the inertial (cascade) regime, and dimensional analysis is then sufficient to determine an energy spectrum, $E(k) \sim k^{-3}$, a conclusion that may be subject, as pointed out by Kraichnan, to logarithmic correction.

It must be admitted that Batchelor's conclusion here cannot be universally valid (for all initial fields of two-dimensional turbulence). Saffman (1971) pointed out that, if the vorticity field $\omega(\mathbf{x}, \mathbf{t})$ consists of a random distribution of vortex patches, the vorticity being uniform inside each patch and zero elsewhere, then the vorticity spectrum is dominated for large wave-number k by contributions from the discontinuities in ω, and therefore falls off as k^{-2}; and correspondingly, $E(k) \sim k^{-4}$ in contrast to Batchelor's prediction. Saffman's conclusion is evidently correct when the vortex patches are far apart compared with their size, since then they interact only weakly, and the vorticity distribution retains its statistical character for a long time. If however the patches are close enough to interact significantly, the situation becomes much more complicated. Vorticity patches tend to merge when they are close enough; if they are of opposite sign this leads to partial or complete annihilation; if of the same sign then merging (involving viscous reconnection of isovorticity contours, or 'isovorts') leads to the growth of scale that is associated with the inverse energy cascade. The nature of the dynamics and resulting spectral form of two-dimensional turbulence is still to this day a matter of continuing controversy and debate.[1]

8.11 Later papers

During the next two decades, Batchelor's main research preoccupation was in suspension mechanics, or 'microhydrodynamics' as he himself named the subject. It would be inappropriate to describe his work in this area here except to say that he and the new research team that he gathered around him virtually created a new field of study during this period. Batchelor did return to turbulence twice during this period, on each occasion with a highly original contribution. The first of these was his 1980 paper "Mass transfer from small particles suspended in turbulent fluid". Here, 'small' means small in

---

[1] In the spirit of Richardson, may I propose a rhyme to capture the essence of two-dimensional turbulence:
An atmosphere stirred by convection
Responds to enstrophic injection:
  While filaments fade,
  An inverse cascade
Leads to rapid contour reconnection.

comparison with the inner Kolmogorov scale, so that again (shades of his 1959 paper!) he was able to represent the 'far field' as a uniform gradient velocity field. Batchelor assumed that the Péclet number was large and he solved the advection–diffusion equation in the 'concentration boundary layer' around the particle, in order to calculate the net rate of mass transfer from the particle. He identified two contributions to this mass transfer, the first related to the translational motion of the particle relative to the fluid, and the second related to the local velocity gradient. Batchelor argued that the first of these, for reasons associated with reflectional symmetry, is zero "in common turbulent flow fields", and he determined the second contribution in terms of a Nusselt number Nu, the result being $Nu = 0.55(a^2\epsilon/\kappa\nu^{1/2})^{1/3}$, where a is the radius of the particle (assumed spherical). The result is in very reasonable agreement with experiment. Batchelor's great skill in exploiting what is known of the small-scale features of turbulent flow is again evident in this paper. His neglect of the contribution to mass transfer due to particle slip is however debatable, particularly in turbulence that lacks reflectional symmetry, and still calls for further investigation.

Batchelor's final return to turbulence (in collaboration with V.M. Canuto & J.R. Chasnov) came with his 1992 paper "Homogeneous buoyancy-generated turbulence". Still, as ever, Batchelor felt most at home in a statistically homogeneous situation. But here a novel variation was conceived, namely a field of turbulence generated from rest by an initially prescribed, statistically homogeneous, random density distribution giving a random buoyancy force. The driven turbulence of course immediately modifies the distribution of buoyancy forces, and an interesting nonlinear interaction between velocity and buoyancy fields develops: the buoyancy field itself drives the flow, so we are faced here with an 'active' rather than 'passive' scalar field problem. As the authors state in their abstract, "the analytical and numerical results together give a comprehensive description of the birth, life and lingering death of buoyancy-generated turbulence". Alas, Batchelor was himself suffering a lingering decline from Parkinson's disease which afflicted him with growing intensity from about 1994 until his death at the age of 80 in the millennium year 2000.

8.12 George Batchelor as Editor and as Head of Department

Batchelor's involvement with the *Journal of Fluid Mechanics* was from the start a very personal one. Prior to 1956, papers in fluid mechanics had been scattered over the literature of applied mathematics and engineering mechanics, and in the journals of the various national academies, including the

Proceedings of the Royal Society. Batchelor saw a need for a journal that would bring together both experimental and theoretical aspects of the subject, and planned *JFM* accordingly, together with a small team of Associate Editors, George Carrier (Harvard), Wayland Griffith (Princeton), and James Lighthill (Manchester). A novel feature of the editorial process, maintained to this day, was that each editor and associate editor should have individual responsibility to accept (or reject) papers submitted to him: they each acted as individuals rather than as members of an editorial board, following agreed guidelines on the scope of the journal and the standards expected. The formula worked extremely well, as may be seen from the quality of the papers published in the early years of the journal.

Batchelor was aided by two Assistant Editors, Ian Proudman and Brooke Benjamin. Their task initially was to prepare accepted papers for the Press, i.e. to carry out the copy-editing process, a process in which George himself led the way, imposing strict rules of style: Fowler's *Dictionary of Modern English Usage* was his constant companion. I served myself as an Assistant Editor from 1961; one of my first tasks was the copy-editing of Kolmogorov's (1962) paper, by no means straightforward! One of Batchelor's guidelines was that the mathematical equations in a paper should be incorporated in such a way as to follow the normal rules of English grammar. Copy-editing was more than simply correcting grammar or punctuation: it sometimes involved complete reconstruction of sentences to clarify an author's intended meaning; of course, this required the author's approval at the proof stage, and extended correspondence could then ensue. But the results of this painstaking procedure are evident in the quality of the early volumes of the Journal, more difficult to maintain in later decades as the rate of publication remorselessly increased (with corresponding increase in the number of Associate Editors).

As I have already indicated, Batchelor was heavily committed, from 1959 on, not only to *JFM*, but also as Head of the new Department of Applied Mathematics and Theoretical Physics (DAMTP); this was his main administrative responsibility from 1959 till 1983, when, under the Thatcher regime, he was induced to take early retirement from his University Professorship. In the five-year period 1959–1964, there were just two Professors in DAMTP, the Lucasian Professor (and Nobel Laureate) Paul Dirac and the Plumian Professor of Astronomy Fred Hoyle. On his appointment as Head of DAMTP in 1959, Batchelor was promoted from his Lectureship to a Readership in Fluid Mechanics, and five years later to a newly established Professorship of Applied Mathematics. These first five years were crucial in the shaping of the

Department. A crisis arose in 1964 when Batchelor's appointment as Head of Department came up for renewal and Hoyle made a bid for the Headship. The matter was put to a vote within the Department, and Batchelor won by a convincing majority, this amounting to a vote of confidence in the quality of his leadership. Hoyle had sought the support of Dirac, but to no avail; in the event, he walked out of DAMTP in a fit of pique, and thereafter devoted his energies and his wayward genius to the foundation of the Institute of [Theoretical] Astronomy on Madingley Road. Dirac spent his time like a hermit in St John's College, above and beyond the fray! In his recent biography of Dirac, *The Strangest Man*, Graham Farmelo well describes Dirac's attitude to the new Department, of which, *faute de mieux*, he was a member: he wanted nothing to do with it, and wanted to be left alone in solitary contemplation in his College rooms! This state of affairs continued until Dirac's retirement in 1969; throughout this decade, his relationship with Batchelor was reserved, but never, so far as I was aware, other than courteous.

After the trauma of 1964, Batchelor settled more securely into his role as Head of Department. On the fluids side, he promoted the emerging field of geophysical fluid mechanics with the help of his brilliant student Adrian Gill, and in parallel with Brooke Benjamin, Owen Phillips and Francis Bretherton. The arrival of James Lighthill, who succeeded Dirac as Lucasian Professor in 1969, led to intense activity in biological fluid mechanics throughout the 1970s. As far as turbulence was concerned, it was MHD turbulence and turbulent dynamo theory, in which I was personally involved, with applications in geophysics and astrophysics, that became a principal focus of effort. Batchelor took a keen interest from the sidelines in these advances, but found it hard to relinquish the vorticity/magnetic field analogy that had been so central to his own thinking twenty years earlier.

8.13 International activity

Batchelor was an enthusiastic adherent of the International Union of Theoretical and Applied Mechanics (IUTAM), founded in 1948, and he soon became a familiar figure at its subsequent quadrennial congresses. For four of these – Stresa (1960), Munich (1964), Stanford (1968) and Moscow (1972) – he served on the Executive Committee of the Congress Committee. During the early 1960s he recognised a parallel need for greater scientific interaction specifically within Europe, and, together with Dietrich Küchemann (of the Royal Aircraft Establishment, Farnborough), he set up a European Mechanics

Committee and served as its founding Chairman from 1964 till 1987 (when he was succeeded in this role by the new Cambridge Professor of Applied Mathematics, David Crighton.)

This Committee established a series of Colloquia, known as 'EURO-MECHs', the first of which was held in Berlin in April 1965. By 1985 there were approximately 15 EUROMECHs each year; a total of 230 were held during Batchelor's chairmanship, many in the broad field of turbulence and its applications. His firm hand and meticulous attention to detail can be seen in the notes of guidance written for chairmen of Colloquia: three to five days' duration, participants from all countries of Europe "extending eastwards as far as Poland and Rumania", organizing committee of six members for each Colloquium, not more than 50 participants, topics chosen to be of theoretical or practical interest but not devoted to details of engineering applications, introductory lectures to be included, presentation of current and possibly incomplete work to be encouraged, no obligation to publish Proceedings, and so on. These Colloquia were run on a financial shoestring, most participants finding their own individual sources of funding, but this seems to have had little adverse effect; on the contrary, the freedom from the constraints often imposed by grant-giving bodies gave EUROMECHs a freshness and spontaneity that was widely valued. The foundation of EUROMECH following on that of *JFM*, raised the visibility and impact of research in fluid mechanics, and may be recognized as an important part of Batchelor's great legacy to the subject.

The inclusion of Poland was significant. Through IUTAM, Batchelor was a close friend of Wladek Fiszdon (of the Institute of Fundamental Technological Research, Warsaw), and had given him encouragement and help in the organization of the biennial Symposia on Fluid Mechanics in Poland that served so well to maintain communication between scientists from East and West during the long cold-war years. The link with Poland developed with the presence of Richard Herczynski, a younger colleague of Fiszdon, as a Senior Visitor in DAMTP in 1960/61.

In 1963 Batchelor attended the Fluid Dynamics Conference held in Zakopane, Poland – one of the first at which any Western scientists were present. He drove across Europe to this meeting with his wife and three young daughters, Adrienne, Clare and Bryony, and took me along also in the last space in the family car! Batchelor's contributed paper on dispersion of pollutant from a fixed source in a turbulent boundary layer (Batchelor 1964) is a beautiful piece of work, and deserves to be better known. At this conference, he had preliminary discussions with Fiszdon and Herczynski about the need for greater European cooperation in science. He attended several of the subsequent biennial meetings and was involved much later, through Herczynski, in the traumatic events for Polish scientists during the Solidarity years, always giving moral support when this was most needed.

8.14 Conclusion

Despite the intensive efforts of mathematicians, physicists and engineers over the last half century and more, turbulence remains, to paraphrase Einstein, "the most challenging unsolved problem of classical physics". The 1950s was a period when, under the inspiration of George Batchelor, there was a definite sense of progress towards a cracking of this problem. But he, like many who followed him, found that the problem of turbulence was challenging to the point of total intractibility. Nevertheless, Batchelor established new standards both in the rigour of the mathematical argument and the depth of physical reasoning that he brought to bear on the various aspects of the problems that he addressed. His 1953 monograph is still widely quoted as the definitive introduction to the theory of homogeneous turbulence, and his later contributions, particularly in the field of turbulent diffusion, have stood the test of time over the subsequent fifty years. There can be no doubt that any future treatise on the subject of turbulence will acknowledge Batchelor's major contributions to the field, coupled with those of his great mentor G.I. Taylor.

Acknowledgements This chapter is an expanded version of my paper entitled "G.K. Batchelor and the homogenisation of turbulence", published in *Ann. Rev. Fluid Mech.* **34**, 19–35 (2002). I have also drawn extensively from Moffatt (2002); see also Moffatt (2010). I thank Annual Reviews Inc. and the Royal Society for permission to reproduce material from these articles.

I am grateful to Adrienne Rosen and Bryony Allen, who provided the photographs of Figures 8.2, 8.3, 8.4 and 8.5.

References

Barenblatt, G.I. 2001. George Keith Batchelor (1920–2000) and David George Crighton (1942–2000): Applied Mathematicians. *Notices Amer. Math. Soc.* **48**, 800–806.

Batchelor, G.K. 1946a. The theory of axisymmetric turbulence. *Proc. Roy. Soc. A* **186**, 480–502.

Batchelor, G.K. 1946b. Double velocity correlation function in turbulent motion. *Nature* **158**, 883–884.

Batchelor, G.K. 1947. Kolmogoroff's theory of locally isotropic turbulence. *Proc. Cam. Phil. Soc.* **43**, 533–559.

Batchelor, G.K. 1949a. The role of big eddies in homogeneous turbulence. *Proc. Roy. Soc. A* **195**, 513–532.

Batchelor, G.K. 1949b. Diffusion in a field of homogeneous turbulence. I. Eulerian analysis. *Australian Journal of Scientific Research Series A – Physical Sciences* **2**, 437–450.

Batchelor, G.K. 1950a. On the spontaneous magnetic field in a conducting liquid in turbulent motion. *Proc. Roy. Soc. A* **201**, 405–416.

Batchelor, G.K. 1950b. The application of the similarity theory of turbulence to atmospheric diffusion. *Quarterly Journal of the Royal Meteorological Society* **76**, 133–146.

Batchelor, G.K. 1951. Magnetic fields and turbulence in a fluid of high conductivity. In: *Proceedings of the Symposium on the motion of gaseous masses of cosmical dimensions, Paris, August 16–19, 1949, "Problems of Cosmical Aerodynamics"*, Central Air Documents Office, 149–155.

Batchelor, G.K. 1952a. Diffusion in a field of homogeneous turbulence. II. The relative motion of particles. *Proc. Cam. Phil. Soc.* **48**, 345–362.

Batchelor, G.K. 1952b. The effect of homogeneous turbulence on material lines and surfaces. *Proc. Roy. Soc. A* **213**, 349–366.

Batchelor, G.K. 1953. *The Theory of Homogeneous Turbulence*. Cambridge University Press.

Batchelor, G.K. 1959. Small-scale variation of convected quantities like temperature in turbulent fluid. Part 1. General discussion and the case of small conductivity. *J. Fluid Mech.* **5**, 113–133.

Batchelor, G.K. 1964. Diffusion from sources in a turbulent boundary layer. *Arch. Mech. Stosowanech* **16**, 661–670.

Batchelor, G.K. 1967. *An Introduction to Fluid Dynamics*. Cambridge University Press.

Batchelor, G.K. 1969. Computation of the energy spectrum in homogeneous two-dimensional turbulence. *High-speed computing in fluid dynamics. The Physics of Fluids Supplement* **11**, 233–239.

Batchelor, G.K. 1980. Mass transfer from small particles suspended in turbulent fluid. *J. Fluid Mech.* **98**, 609–623.

Batchelor, G.K. 1992. Fifty years with fluid mechanics. *Proc. 11th Australian Fluid Mechanics Conf., Dec. 1992*, 1–8.

Batchelor, G.K. 1996. *The Life and Legacy of G.I. Taylor*, Cambridge University Press.

Batchelor, G.K. 1997. Research as a life style. *Appl. Mech. Rev.* **50**, R11–R20.

Batchelor, G.K. & Proudman, I. 1954. The effect of rapid distortion of a fluid in turbulent motion. *Quart. Journ. Mech. and Applied Math.* **7**, 83–103.

Batchelor, G.K. & Proudman, I. 1956. The large-scale structure of homogeneous turbulence. *Phil. Trans. Roy. Soc. London Series A. Mathematical and Physical Sciences, No. 949*, **248**, 369–405.

Batchelor, G.K. & Townsend, A.A. 1947. Decay of vorticity in isotropic turbulence. *Proc. Roy. Soc. A* **190**, 534–550.

Batchelor, G.K. & Townsend, A.A. 1948a. Decay of isotropic turbulence in the initial period. *Proc. Roy. Soc. A* **193**, 539–558.

Batchelor, G.K. & Townsend, A.A. 1948b. Decay of turbulence in the final period. *Proc. Roy. Soc. A* **194**, 527–543.

Batchelor, G.K. & Townsend, A.A. 1949. The nature of turbulent motion at large wave-numbers. *Proc. Roy. Soc. A* **199**, 238–255.

Batchelor, G.K. & Townsend, A.A. 1956. Turbulent diffusion. In: *Surveys in Mechanics: a collection of surveys of the present position of research in some branches of mechanics: written in commemoration of the 70th birthday of Geoffrey Ingram Taylor.* Cambridge University Press, 352–399.

Batchelor, G.K., Binnie, A.M. & Phillips, O.M. 1955. The mean velocity of discrete particles in turbulent flow in a pipe. *Proc. Phys. Soc. B* **68**, 1095–1104.

Batchelor, G.K., Canuto, V.M. & Chasnov, J.R. 1992. Homogeneous buoyancy-generated turbulence. *J. Fluid Mech.* **235**, 349–378.

Batchelor, G.K., Howells, I.D. & Townsend, A.A. 1959. Small-scale variation of convected quantities like temperature in turbulent fluid. Part 2. The case of large conductivity. *J. Fluid Mech.* **5**, 134–139.

Corrsin, S. 1951. On the spectrum of isotropic temperature fluctuations in an isotropic turbulence. *J. Appl. Phys.* **22**, 469–473.

Cowling, T.G. 1957. *Magnetohydrodynamics.* Interscience Publishers Inc.

Eyink, G., Frisch, U., Moreau, R. & Sobolevskii, A. (eds) 2008. Euler equations: 250 years on. *Physics D* **237**.

Favre, A. (ed) 1962. *Mécanique de la Turbulence.* No. 108, Editions du CNRS.

Frisch, U. 1995. *Turbulence, the Legacy of A.N. Kolmogorov.* Cambridge University Press.

Grant, H.L., Stewart, R.W. & Moilliet, A. 1962. Turbulence spectra from a tidal channel. *J. Fluid Mech.* **12**, 241–268.

Ishida, T., Davidson, P.A. & Kaneda, Y. 2006. On the decay of isotropic turbulence. *J. Fluid Mech.* **564**, 455–475.

Kolmogorov, A.N. 1941a. The local structure of turbulence in an incompressible fluid with very large Reynolds number. *C.R. Acad. Sci. URSS* **309**, 301–5.

Kolmogorov, A.N. 1941b. Dissipation of energy under locally isotropic turbulence. *C.R. Acad. Sci. URSS* **32**, 16–18.

Kolmogorov, A.N. 1962. A refinement of previous hypotheses concerning the local structure of turbulence in a viscous incompressible fluid at high Reynolds number. *J. Fluid Mech.* **13**, 82–85.

Kraichnan, R.H. 1959. The structure of isotropic turbulence at very high Reynolds numbers. *J. Fluid Mech.* **5**, 497–543.

Kraichnan, R.H. 1967. Inertial ranges in two-dimensional turbulence. *Phys. Fluids* **10**, 1417–1423.

Kulsrud, R.M. 1999. A critical review of galactic dynamos. *Ann. Rev. Astron. Astrophys.* **37**, 37–64.

Moffatt, H.K. 1970. Turbulent dynamo action at low magnetic Reynolds number. *J. Fluid Mech.* **41**, 435–452.

Moffatt, H.K. 2002. George Keith Batchelor, 8 March 1920–30 March 2000. *Biog. Mems. Fell. R.Soc. Lond.* **48**, 25–41 .

Moffatt, H.K. 2010. George Batchelor: a personal tribute, ten years on. *J. Fluid Mech.* **663**, 2–7.

Moffatt, H.K. & Saffman, P.G. 1964. Comment on "Growth of a weak magnetic field in a turbulent conducting fluid with large magnetic Prandtl number". *Phys. Fluids* **7**, 155.

Obukhov, A.M. 1949. Structure of the temperature field in a turbulent flow. *Izv. Akad. Nauk, SSSR, Geogr. i Geofiz.* **13**, 58–69.

Saffman, P.G. 1963. On the fine-scale structure of vector fields convected by a turbulent fluid. *J. Fluid Mech.* **16**, 542–572.

Saffman, P.G. 1967. The large-scale structure of homogeneous turbulence. *J. Fluid Mech.* **27**, 581–593.

Saffman, P.G. 1971. On the spectrum and decay of random two-dimensional vorticity distributions at large Reynolds number. *Studies in Applied Mathematics* **50**, 377–383.

Sagaut, P. & Cambon, C. 2008. *Homogeneous Turbulence Dynamics*. Cambridge University Press.

Schlüter, A. & Biermann, L. 1950. Interstellare Magnetfelder. *Z. Naturforsch.* **5a**, 237–251.

Steenbeck, M., Krause, F. & Rädler, K.-H. 1966. Berechnung der mittleren Lorentz-Feldstärke für ein elektrisch leitendes Medium in turbulenter, durch Coriolis-Kräfte beeinfluster Bewegung. *Z. Naturforsch.* **21a**, 369–376.

Taylor, G.I. 1921. Diffusion by continuous movements. *Proc. Lond. Math. Soc.* **20**, 196–212.

Taylor, G.I. 1935. Turbulence in a contracting stream. *Z. angew. Math. Mech.* **15**, 91–96.

Taylor, G.I. 1954. The dispersion of matter in turbulent flow through a pipe. *Proc. Roy. Soc. A* **223**, 446–68.

Taylor, G.I. & Batchelor, G.K. 1949. The effect of wire gauze on small disturbances in a uniform stream. *Quart. J. Mech. Appl. Math.* **2**, 1–29.

Taylor, G.I. & Green, A.E. 1937. Mechanism of the production of small eddies from large ones. *Proc. Roy. Soc. A* **158**, 499-521 .

Townsend, A.A. 1951a. The passage of turbulence through wire gauzes. *Quart. J. Mech. Appl. Math.* **4**, 308–329.

Townsend, A.A. 1951b. On the fine-scale structure of turbulence. *Proc. Roy. Soc. A* **208**, 534–542.

Townsend, A.A. 1990. Early days of turbulence research in Cambridge. *J. Fluid Mech.* **212**, 1–5.

Zeldovich, Ya. B. 1957. The magnetic field in the two-dimensional motion of a conducting turbulent fluid. *Sov. Phys. JETP* **4**, 460–462.

9

A.A. Townsend

Ivan Marusic and Timothy B. Nickels

9.1 Early years

Albert Alan Townsend was born on the 22nd of January 1917 in Melbourne Australia son of Albert Rinder Townsend and Daisy Townsend née Gay. At the time of his birth his father was a clerk in the accounts branch of the Department of Trade and Customs – he also served as secretary of the Commonwealth Film Censorship Board. His father went on to have a very successful career in the Commonwealth public service. As his career evolved he moved the family to Canberra in the ACT (Australian Capital Territory) which is the seat of the government in Australia. Albert and Daisy had three children: Alan, Elisabeth and Neil. In 1933 Albert Rinder Townsend was awarded the OBE.

Alan obtained his Leaving Certificate in 1933 from the Telopea Park High School with an outstanding pass, including first-class honours in mathematics, and the Canberra University College Council awarded him a scholarship of £120 a year to pursue a science course at Melbourne University. He completed his Bachelor of Science in 1936, graduating with first-class honours, and started his Master of Science. Just before his 20th birthday (1937) he graduated Master of Science, with honours in natural philosophy and pure mathematics. He was awarded the Dixson Research Scholarship and the Professor Kernot Research Scholarship.

He then carried out research on nuclear physics with L.H. Martin, E.H.S. Burhop and T.H. Laby (probably most famous for the scientific tables he wrote with Kaye) under the direct supervision of E.H.S. Burhop. In a letter to Professor R. Home concerning the early days of the Natural Philosophy Department of Melbourne he recalls:

> In my time, Laby did little research on his own account but he took a very close interest in the progress of all the work in the laboratory. In 1936, the action of Geiger–Müller counters was not well understood and their performance

Figure 9.1 Alan Townsend at the age of 21 years, having graduated MSc from the University of Melbourne.

unpredictable. Burhop and I found that only one-third of those we had made were reliable. Laby found us making up a large batch in the hope that some of them would be usable, and, finding out that we had no idea why some would work and others would not, he ordered us to study their operation in detail. Fortunately, that batch produced sufficient good counters for our needs and we kept out of his way until he had forgotten.

During this period he worked on experimental measurements of the absorption of neutrons which resulted in several papers including a paper in *Nature* (Burhop, Hill & Townsend, 1936) when he was still at the tender age of 19. His MSc thesis was entitled *Passage of High Energy Radiation through Matter*. At that time it was not possible to complete a PhD degree in Australia and so he was forced to look abroad (the PhD degree was not established in Australia until 1946). The obvious place was Cambridge since it was here that Laby and Burhop had studied and it led the world in nuclear physics.

9.2 Move to Cambridge

In 1938 at the age of 21 he was awarded an 1851 Exhibition scholarship (the youngest student of the University of Melbourne to have achieved the award at that time). He was also awarded a studentship by Emmanuel College and a scholarship entitling him to free return passage on the P&O Ship *The RMS Strathmore* from Australia to England. Emmanuel was Laby's old college. The scholarship value was only £250 per year which at that time was comfortable for a young man but not a princely sum (in 1938 the average salary in the UK was £209).

He went to work at the Cavendish laboratory, then under the directorship of the Australian Nobel laureate Lawrence Bragg. At this time he worked in the Cavendish High Voltage and Cyclotron Laboratories run by J.D. Cockcroft. Cockcroft and his student Walton had recently (April 1932) achieved fame by splitting the lithium atom with a proton beam (for which they were jointly awarded the Nobel Prize in physics in 1951). When Townsend arrived Cockcroft was working on the construction of a new high voltage Phillips set in the Old Library of the Cambridge Philosophical Society which was intended to produce two million volts based on the Cockcroft–Walton method of voltage amplification for use in cyclotron experiments. While he was waiting for his own apparatus to be constructed Townsend was occupied in assisting with the commissioning of the new set. In a letter to his mother in January 1939 Townsend recalls:

> There was a big thrill in the lab today as the two million volt set gave two million volts for the first time, a performance which was considered highly improbable before Christmas.

Once his own apparatus (a very accurate magnetic spectrometer) was constructed he set to work on his thesis research studying the β-ray spectra of light elements – work which resulted in a paper in the *Proceedings of the Royal Society* (Townsend, 1941).

9.3 War years

After the outbreak of war he spent six months working on radar (partly in the Shetland Islands). Cockcroft had played a leading role in the development of radar for coastal defence and had collected teams of physicists to build (by hand) three radar stations, one in the Shetlands and two in Fair Isle – Townsend was part of the team working in the Shetlands where his skill with instrumentation would have been invaluable. In a story related to his children Townsend remembers being part of a group of Cambridge research students taken to Bawdsey Manor in Suffolk to learn about radar so as to aid in its development. On arrival the students noted the large aerials and spent the evening working out the wavelengths and the relevant theory of radar (all of this supposedly secret information). The next morning when they were due to be lectured on the new breakthrough they flummoxed the security people by stating "we know the theory – just show us the wiring diagrams".

After that he requested permission to return to Australia to work for the CSIR (Council for Scientific and Industrial Research). He spent six months at the Mount Stromlo observatory working on metallic surfaces on glass and then another six months working on lubricants and bearings in the CSIR laboratory at the University of Melbourne (that was set up by F.P. Bowden). He then went to the Aeronautics lab at Fisherman's Bend in Melbourne to run the instruments section. It was here that he first met George Batchelor who was working in the Aerodynamics section; the two often collaborated on wind tunnel testing. One notable study was their investigation of the 'singing' of wind tunnel corner vanes, where they characterized this aero-elastic resonance effect and published a short letter in *Nature* (Batchelor & Townsend, 1945).

9.4 Return to Cambridge

In June 1945 (when the war was effectively over) he was granted permission to return to Cambridge to resume his studies at which time Batchelor convinced him to move into fluid mechanics under G.I. Taylor, rather than return to his studies in nuclear physics. Townsend's research experiences during this period are well summarized in his interesting and entertaining article entitled "Early days of turbulence research at Cambridge" in the *Journal of Fluid Mechanics*. Townsend (1990) recalls that his arrival to the Cavendish Laboratory in 1945

> can be entirely blamed on George Batchelor...I had interrupted my research in nuclear physics in 1939, intending to return and finish my scholarship, but I found myself signed on to do experimental work on turbulent flows. Why I accepted is still a mystery, as my ignorance of fluid dynamics, let alone turbulence, was almost total.

If they had hoped to work directly with Taylor on turbulence they were somewhat disappointed. Turner (1997), in his article on G.I. Taylor, notes that:

> George Batchelor (G.K.B.) and Alan Townsend (A.A.T.) arrived in 1945 specifically to become research students under G.I.'s supervision and to work respectively on theoretical and experimental problems in turbulence, a field in which they had been active at the Aeronautical Research Laboratory in Melbourne during the war. G.K.B. has admitted since that he was somewhat disconcerted to discover that G.I. did not immediately return to the subject of turbulent flow to which he had made such large contributions in the 1930s ...The research on turbulence was carried on almost independently by these two students, with G.I.

showing great interest in their results, but suggesting no specific program for their research. G.K.B. initially followed up Kolmogoroff's wartime papers on the statistical equilibrium of the small scale components of turbulent motions, and A.A.T. began work on turbulence behind grids in the wind tunnel in the Cavendish Laboratory.

The wind tunnel, designed by Taylor, was located in a large room on the ground floor of the old Cavendish Laboratory in Free School Lane. The next five years were an extremely productive time for Townsend. He and Batchelor set about examining and testing the theories of turbulence, in particular those of Kolmogorov and the Russian school that Batchelor was bringing to the attention of the West. Townsend published a total of eleven papers, five with Batchelor and six as sole author. Considering he was new to turbulence research when he arrived in 1945, this must be considered a remarkable output. These papers established a number of new ideas which he was further to develop throughout his career.

It was during this extremely productive time, in 1950, that he was appointed Assistant Director of Research of the Cavendish and married Valerie Dees. Valerie had come to Cambridge having served in the WAAF during the war and was working with G.C. Grindley in the Experimental Psychology lab in Cambridge. They got to know each other through mutual friends on a boating trip in the Norfolk Broads during which he proposed. They shared a great love of the outdoors, enjoying all manner of activities, especially camping and sports. Alan was a very keen sportsman enjoying, in particular, hockey, tennis and skiing. He once commented to his daughter that he became a British citizen only to save on visa costs for his many European ski trips with his Cavendish friends. He played hockey for the university and continued playing in the Cambridge third team until well into his sixties (Figure 9.2). He was also a keen tennis player serving as the university representative for the Lawn Tennis Association and was president of the Emmanuel College Tennis club from 1949 to 1968 (after which time it ceased to have a president). He was still playing regularly with other Fellows of the College in 1998 – at the age of 81.

9.5 Putting K41 to the test

As noted above, when Batchelor and Townsend arrived in Cambridge they began to collaborate on the study of the structure of small-scale turbulence. The timing for this work was right and was inspired by the statistical theory of isotropic turbulence that Taylor had developed one decade earlier, and the

Figure 9.2 Hockey. Townsend is on the end at the far right. In the background is Jesus College.

equations that T. von Kármán and L. Howarth derived in 1938. The concept of local isotopy of the small-scale eddies had also been advanced by A.N. Kolmogorov and A.M. Obukhov in 1941 and, while this was also independently formulated by C.F. von Weiszacker and L. Onsager, the attention to this was initially mostly confined to the Russian school led by Kolmogorov. Batchelor, however, was familiar with Kolmogorov's papers and saw that they potentially opened the way to advancing the isotropic theory to inhomogeneous and shear flows, and he was largely responsible for the dissemination of Kolmogorov's ideas to the West after the war. Heisenberg, having been prohibited from working on nuclear physics problems by the allied forces, returned to his research on turbulence, and in 1948 proposed a theory for the transfer of energy from low to high wavenumbers. In the same year L.S.G. Kovasznay proposed an alternative theory for the same problem and the topic saw considerable interest for many years. However, it was Townsend (together with the theoretical input from Batchelor) who conducted the initial pioneering experiments in this field.

In homogeneous isotropic turbulence the mean values of all functions of the components of the velocity fluctuations and their derivatives are independent of position and rotations and reflections of the axes defining the frame of reference. This leads to a considerable simplification of the problem and to a number of predictions concerning the behaviour of the turbulence. Townsend investigated this experimentally by using a turbulence-generating grid in the small (15 × 15 in.) but high flow quality wind tunnel. The key to making these measurements was to measure the small-scale fluctuating component of velocity and other quantities together with their time derivatives and correlations. For these tasks, Townsend's expertise was invaluable. His research on nuclear physics had given him a solid background in experimental techniques and instrumentation. Townsend's PhD students Bob Stewart and Harold Grant recall (Stewart & Grant, 1999):

> Townsend ... was a superb experimentalist with an extraordinary talent for experimental design and instrument development. He designed and built power supplies superior to anything on the market. In a day when it was normal to eliminate hum in apparatus by using high-pass filters, Townsend constructed low-noise amplifiers with flat response to less than 1 Hz. He was known to have been able to conceive of an experiment on one day, build the apparatus within a week, conduct the experiment the following week, and be writing the results the week after.

Townsend built the amplifier and compensator stages required for the hot-wire anemometry as well as the analogue devices required for the statistical analysis of the hot-wire signals. Hot-wire anemometry involves using fine platinum or tungsten wires (typically 5 microns in diameter) as thermal transducers with a small sensing length of less than 1 mm. Successfully using them for turbulence measurement requires circuitry that extends the frequency response of the sensor; at that early stage, this was as much an art as science. Townsend, however, mastered this technique. While others had previously used hot-wire anemometry (Dryden, Hall, Simmons and colleagues) their measurements were restricted to mean flow, mean-square intensities and double correlations. Townsend pioneered the use of electrical analysis to obtain time derivatives, triple correlations and other quantities that were required to test the theoretical prediction of isotropic turbulence.

Townsend's first turbulence paper on the decay of isotropic turbulence, co-authored with Batchelor, appeared in the *Proc. Royal Society* (Batchelor & Townsend, 1947), in the same year he submitted his PhD *Dissertation on β-ray spectra of light elements and turbulent flow*. It was at this time also that he was elected a Supernumerary Fellow and College Lecturer in Physics at Emmanuel College. He remained a Fellow of Emmanuel College for his whole life having been elected to a life Fellowship. Townsend always gave his affiliation in publications as Emmanuel College. This illustrates the important role the College played in his life and his career. He was Praelector of the College during 1968–1969 and Vice-Master from 1975 to 1979.

While working on the decay problem Townsend decided to investigate the validity of the Heisenberg inverse seventh power spectrum for the far viscous range. In Townsend (1990) he gives a characteristically modest description of the discovery of an important feature of turbulence as an unexpected result (Batchelor & Townsend, 1949):

> In a moment of inspiration, I decided that the power could be obtained by measuring the flatness factors of a number of high-order velocity derivatives, and I managed it for the first four. The kurtosis increased both with order of the derivative and with the Reynolds number of the turbulence. I then realised that the measurements said nothing about the spectrum, but they did show a considerable departure from the predictions of local similarity. The spatial intermittency of small-scale motion is now well known. Our trouble in writing up the work was to explain why we had done the measurements in the first place, pride preventing use of the words, "In a moment of exceptional stupidity, we measured . . .".

Spatial intermittency of the small-scale motion refers to the fact that the dissipation of energy (and other characteristics of the flow) is not uniformly distributed throughout the flow, as had been previously thought, but instead occurs in very intense events that are sparsely distributed. The spatial mean values then are averages of very large, very rare events. This is an important fundamental result concerning the nature of turbulence, and suggests immediately that the turbulent motion at these fine scales cannot be totally random and featureless.

Inspired by this result, Townsend then went on to try to understand what sort of underlying structure in turbulence could explain it. In Townsend (1951a) he considered models of the fine-scale motion consisting of

> a random distribution of vortex sheets and lines, in which the vorticity distribution is effectively stationary in time, due to balance between the opposing effects of vorticity diffusion by molecular viscosity, and vorticity production and convection by the turbulent shear.

He showed that the spectrum produced by a random array of sheets was a better fit to the available data than that produced by vortex lines (though they both gave reasonable results). In order to explain why sheets might be more prevalent he derived a remarkable analytical result. He related the principal rates of strain to the skewness of the velocity fluctuations (which was known, empirically, to be negative) and showed that this implied that the strain field must locally be of the form which produces sheets (that is, one compressive strain and two extensive). This particular result, though not unduly emphasized in the paper, has considerably influenced our understanding of turbulence since.

This work illustrates a common theme of Townsend's research. Unhappy with pure mathematical descriptions and models of turbulence, he was driven to try to understand turbulent flows in terms of their underlying physical structure. He was a pioneer in this approach to understanding turbulence, in particular in constructing mathematical models based on a physical picture of the underlying structure to explain the results of experiments.

9.6 Shear flows

Townsend's early work on isotropic turbulence made up only part of his research towards his PhD. On the same day that the Batchelor & Townsend (1947) paper was submitted (17 December 1946), Townsend also submitted a paper reporting the first extensive turbulence measurements in the turbulent wake of a circular cylinder (Townsend, 1947). The results were extraordinary at the time as they included turbulence statistics based on all three components of fluctuating velocity, and correlation derivatives for both the longitudinal and transverse correlations. This enabled further assessment of Kolmogorov's theories in a free-shear flow, while also determining the downstream behaviour of the wake. As Townsend (1990) recalls:

> My interest in the far wake of a cylinder began after a visit to R.A.E. Farnsborough with George [Batchelor], where we saw the work on achieving laminar flow over wing sections both in flight and in wind tunnels. A thin wire stretched normal to the wing about a chord length upstream was found to induce a wedge-shaped region of turbulent flow with a surprisingly sharp point. Curious to find the width of the disturbing wake, I stretched a 1.5 mm diameter wire across the entrance to the tunnel and started to measure turbulent intensities and velocity correlations.

The sharp turbulent wedge that Townsend noted was later known as the Emmons turbulent spot, based on H. Emmons' 1951 experiments in water table flows at Harvard, and continues to be studied together in the context of bypass transition in laminar boundary layers. This initial work on the turbulent wake sparked Townsend's interest in turbulent shear flows. While research on grid turbulence had been very useful for examining the theory, Townsend recognized early on that shear flows are much more common in practice and any theory of turbulence would need to be able to predict these flows as well.

Townsend made further measurements of the cylinder wake, in particular considering the distribution of the terms in the equation for the kinetic energy of the velocity fluctuations (Townsend, 1948). An analysis of the results showed that there was substantial energy transport *up* the intensity gradient across the jet. Townsend realized that this important observation invalidated local gradient diffusion models for the turbulence and implied the existence of large-scale motions. He discusses these in terms of *jets* of turbulent fluid emitted from the turbulent core but then states:

> While the conception of jets of turbulent fluid is more convenient for following the physical processes in the wake, the alternative but equivalent description that the turbulent motion consists of a motion of scale small compared with the mean flow superimposed on a slower turbulent motion whose scale is large compared with the mean flow may be used.

He further remarks:

> It is expected that this type of motion will occur in all systems of turbulent shear flow with a free boundary, such as wakes, jets and boundary layers.

This paper then marks the birth of his conception of *large eddies* which he used to great effect in modelling turbulent shear flows in the first and second editions of his book. These in effect are 'coherent structures' that became the topic of extensive study in the 1970s and continues to see considerable interest today. It is important to realize that many models of turbulent flow at that time were based on ideas such as eddy viscosity in which local stresses were related to local strains.

In his 1948 paper he also considered the nature of the entrainment process. Townsend describes:

> It may be put that the turbulent diffusion occurs in two stages, the mean jet move-ment doing the large scale diffusion and the jet turbulence performing small scale diffusion of the resulting pattern; that is, the jet motion acts to increase the sur-face area of the turbulent fluid, and so allow a comparatively slow linear diffusion at the bounding surfaces to achieve a large rate of volume increase.

Thus, using his concept of large eddies (here referred to as jets) he expounded a view of entrainment that, although controversial, is still very relevant today.

He developed and expounded many these ideas in the first edition of his monograph *The Structure of Turbulent Shear Flow* published in 1956. It was well received and partly on the basis of this and his many seminal papers he was elected a Fellow of the Royal Society in 1960.

9.7 The Townsend hypotheses

Alan Townsend's name is now synonymous with two prominent hypotheses in the theory of turbulent shear flows. These are the *Reynolds number simi-larity hypothesis*, which applies to all turbulent shear flows, and the *attached eddy hypothesis*, which applies to wall-bounded turbulence. Both stem from Townsend's general approach to the problem of inhomogeneous turbulence that started with his early observations of spatial and temporal coherence in wakes. As stated in the preface of his 1976 monograph, his desire was to

> develop a consistent view of the nature of turbulence from observations of simple flows and then use it to interpret and predict the behaviour of a variety of flows of more general interest.

9.7.1 The Reynolds number similarity hypothesis

The origins of the Reynolds number similarity hypothesis predate Townsend's interpretation, and rightly are most often attributed to Theodore von Kármán. Townsend built on the observations that the energy-containing structure of isotropic turbulence does not depend on the value of the fluid viscosity, and extended this observation to the fully turbulent region of general flows. His version of the hypothesis is articulated on page 89 of the first edition of *The Structure of Turbulent Shear Flow* published in 1956. Here Townsend refers to the region in fully turbulent flows, which includes almost all of the flow, where

> the mean [relative] motion and the motion of the energy-containing components of the turbulence are determined by the boundary conditions of the flow alone, and are independent of the fluid viscosity, except so far as a change in the fluid viscosity may change the boundary conditions.

For the authors of this chapter, this quote is extremely familiar, as we both were introduced to the topic of turbulence by A.E. Perry in his fourth-year fluid mechanics course at the University of Melbourne. Tony Perry would use an overhead projector to prominently display Townsend's quote on a large screen and have all the students in the class read it out aloud in unison (several times), and would emphasize the profound nature of the statement. While we students did not largely understand it at the time, Perry's passionate delivery made it clear that the hypothesis was important.

It is interesting that Townsend does not repeat the above statement explicitly in the second edition of *The Structure of Turbulent Shear Flow*, but rather the notion is discussed and extended throughout the book in a subtle and unassuming way. For this reason, future references to the hypothesis invariably include the addition of roughness as well as viscosity affecting the boundary conditions alone (Perry & Abell, 1977; Raupach et al., 1991; Jimenez, 2004; etc.), and carry the phrase "mean relative motion" (Perry & Abell, 1977) in place of "mean motion", as this is what Townsend meant. Stating "mean relative motions" (and hence mean velocity derivatives) also leads directly to an elegant derivation of the logarithmic law in the inertial (or equilibrium) layer of wall-bounded flows (Rotta, 1962). In the area of turbulent boundary layers developing on rough walls the hypothesis is particularly useful. It suggests that the direct effects of roughness are confined to a small region above the roughness elements whereas the outer part of the flow is influenced only by the change in the wall shear-stress. This is true in the limit of large Reynolds numbers and where the roughness elements are small relative to the boundary layer thickness. Although there is some dispute about its applicability in some particular situations, the bulk of the evidence indicates that it is true for flows which satisfy the necessary assumptions (e.g. Schultz & Flack, 2007; Nickels, 2010). Practically it means that in the outer part of the flow the mean velocity gradient and the stresses are unchanged when scaled with outer flow variables (the friction velocity the boundary layer thickness). This results in a considerable simplification when modelling these flows. The status of this hypothesis may be judged by the fact that it is now often simply referred to in the literature as 'Townsend's hypothesis' without requiring a citation of the original source.

9.7.2 The attached eddy hypothesis

Townsend's first paper on boundary layers was published in the *Proceedings of the Cambridge Philosophical Society* (Townsend, 1950). The paper does not appear to be well known, perhaps due to the local nature of the journal, but it encapsulates many of the ideas on boundary layer structure that he was later to develop. It is another seminal paper by Townsend which demonstrates his great physical insight. The paper reports careful hot-wire measurements in a turbulent boundary layer. In particular Townsend examined the terms in the equation for the kinetic energy of the velocity fluctuations. Noting that, near the wall, there was "a surprisingly strong flow of turbulent energy up the intensity gradient and towards the wall" he postulated that "the turbulence consists mostly of superimposed eddies" and

> the bulk of the energy-containing eddies are, in a sense, attached to the wall, and that the dependence of scale of distance from the wall is not a local effect but due to this attachment of most of the eddies.

He went even further in this paper postulating a possible form of these eddies that would be consistent with the measurements. As mentioned earlier, many models of turbulent flow at the time were based on the idea of gradient diffusion, local conditions determining local transport. One would then expect transport to be *down* the local gradient by analogy with molecular diffusion.

Townsend immediately saw that the transport *up* the gradient implied important non-local effects.

By 1976 Townsend had developed a unified view of incompressible flow turbulence that focused on understanding the main energy-containing eddies or motions. These eddies may be thought of as the velocity fields of some representative vortex structures. Townsend had successfully explained his earlier correlation measurements and those of Grant (1958) in plane wakes by modelling the flow in terms of large eddies, and an obvious extension now was to wall-bounded flows. Here, however, the key additional feature is the presence of a wall or surface, which acts as a continuous source (in the presence of a mean pressure gradient) of vorticity for the flow as it develops along the length of the surface. By studying the experimental data Townsend concluded that it was

> difficult to imagine how the presence of the wall could impose a dissipation length-scale proportional to distance from it unless the main eddies of the flow have diameters proportional to distance of their 'centres' from the wall because their motion is directly influenced by its presence. In other words, the velocity fields of the main eddies, regarded as persistent, organised flow patterns, extend to the wall and, in a sense, they are attached to the wall.

Therefore, in essence any eddy with a size that scales with its distance from the wall may be considered to be *attached* to the wall. Eddies farther from the wall are larger in size and hence their velocity fields still extend to the wall (of course the velocity fields of vortex structures extend to infinity but they decay rapidly at large distances). These eddies form the basis of the attached eddy hypothesis. The hypothesis itself is that the main energy-containing motion of a turbulent wall-bounded flow may be described by a random superposition of such eddies of different sizes, but with similar velocity distributions. These eddies are considered as statistically representative structures in that their geometry and strength are derived from an ensemble average of many different structures of similar scale.

Townsend then went on to form expressions for the contributions of a random superposition of attached eddies of different sizes to the correlation functions and, using the zero-penetration boundary condition at the wall, derived the distribution of eddy sizes with wall distance necessary to produce invariance of the Reynolds shear stress ($-\overline{uw}/U_\tau^2 \approx 1$) with distance from the wall, as observed in the equilibrium layer. (Here, u, v, w refer to the streamwise, spanwise and wall-normal components of fluctuating velocity respectively, and U_τ is the friction velocity.) This analysis effectively leads to a population density of eddies that varies inversely with size and hence with distance from the wall. That is, the number of eddies of size z per unit wall area is A/z, where A is a constant. These distributions of eddies also lead to predictions for the variation of the other components of the Reynolds stress tensor:

$$\overline{u^2}/U_\tau^2 = B_1 - A_1 \ln(z/R),$$
$$\overline{v^2}/U_\tau^2 = B_2 - A_2 \ln(z/R), \qquad\qquad (9.1)$$
$$\overline{w^2}/U_\tau^2 = B_3, \quad -\overline{uw}/U_\tau^2 = 1.$$

Here, z is the wall-normal direction and R is the outer length scale (pipe radius, channel half-height or boundary layer thickness depending on the wall-bounded flow under consideration). The terms A_1, A_2, and B_1, B_2, B_3 are constants, whose numerical values depends on the shape and details of the representative eddy. The above expressions hold for asymptotically high Reynolds number as the range of scales of eddies is effectively unbounded, but a natural Reynolds number dependence is obtained if one limits the size of the smallest attached eddy in viscous wall units (say, $100\nu/U_\tau$, following the findings of Kline et al., 1967). It is emphasized that the physical understanding of these relationships rests essentially on the fact that the normal to the wall component and the shear-stress component at a given height are mainly due to the influence of eddies with 'centres' at or close to that height, whereas the streamwise and spanwise fluctuations at a given distance from the wall are due to all eddies with heights greater than the wall-normal height. That is why these components increase as the distance from the wall is reduced: as we approach the wall, the number of eddies larger than our distance from the wall increases.

An important paper in the development of the attached eddy hypothesis is Townsend (1961), which focused mainly on self-similar boundary layer flows and the structure in the log (equilibrium) region. Here, Townsend described the differences with classical scaling in terms of active and inactive motions. This notion of active/inactive motions continues to be discussed extensively in the literature, with a strong influence due to Bradshaw's interpretation (Bradshaw, 1967) with the 'inactive' component viewed as an effectively irrotational part determined by the turbulence in the outer layer. However, it is not entirely clear that subsequent discussions in the literature are along converging lines due to the inference as to what produces the inactive component. It is clear from Townsend that the active/inactive concept is fully accounted for in the attached eddy description, where the motions due to the larger eddies (well above the point of interest) are mainly 'sweeping' or 'sloshing' motions, which carry very little shear stress since the scale of the flows is so large that they appear to be mostly parallel to the wall on the scale of the local distance from the wall. It is these flows that Townsend called inactive since they carry little shear stress. Townsend emphasized "the inactive flow at one level is an essential part of active flow at other higher levels". In other words, when measuring at a particular distance from the wall, z, the active motions come from eddies with

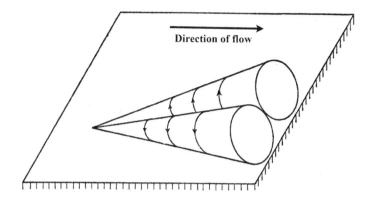

Figure 9.3 Townsend's sketch of a double-cone shaped attached eddy (Townsend, 1976).

centres located at or near z, whereas the inactive motions are due to eddies much larger than z. Which motions are considered active and inactive depends on the location of the measurement (see Nickels et al., 2007, for further explanation).

To proceed further with Townsend's attached eddy hypothesis requires specifying the detailed structure of the eddies, and this was later done by Perry and coworkers (Perry & Chong, 1982; Perry et al., 1986; Perry & Marusic, 1995). Townsend did offer an interpretation of the representative attached eddy in the form of conical eddies as shown in Figure 9.3. Townsend notes, however, that:

> Other possibilities exist, e.g. a distribution of shorter double-roller eddies of various sizes each with a lateral extent comparable with distance from the wall.

The further work by the Melbourne group considerably extended Townsend's hypothesis to be the basis of a mechanistic model to describe the kinematic state of wall-bounded flows, and has been further extended to include free-shear flows (Nickels & Perry, 1996) where a hierarchy of eddy scales is replaced with a single energy-containing eddy scale which is jittered randomly in space.

It should be noted that, while all of these ideas are contained in Townsend's book (Townsend, 1976), they may not be obvious to the casual reader (or indeed the careful expert reader). This highlights a general frustration for readers of Townsend's monographs. All the information is there (and further readings continue to reveal additional layers of depth) but it is not easily accessible with a quick first read, especially for readers new to the field. Embarrassment prevents the authors admitting how many times we had to read pages 150–156 of

Townsend (1976), where the attached eddy hypothesis is described. Only discussions with Townsend himself gave us the reassurance that we could stop. Following Townsend's thoughts has always been easier in direct discussion with the author. This is echoed by Stewart & Grant (1999), who note:

> He [Townsend] thinks not in words, but in three dimensions. Since he often cannot put his ideas into words, he frequently has difficulty promulgating them either in writing or in speech. One on one, however, he was a joy and inspiration to work with.

Related to this is an anecdote told by Tony Perry about the attached eddy hypothesis and k^{-1} law (where k is streamwise wavenumber). Perry & Abell (1975) presented the first serious assault, using dynamically calibrated hot-wires, to uncover the scaling laws of the streamwise fluctuating velocity in pipe flows, a topic where a compilation of the previous experimental data highlighted significant disparities and thus considerable uncertainty. The conclusions put forward by Perry & Abell (1975) supported Townsend's similarity hypothesis, and the classical law-of-the-wall description for the turbulence intensities in the logarithmic region where $\overline{u^2}/U_\tau^2$ should be invariant with Reynolds number. However, this latter result was in contradiction with Townsend's theories that were to appear in his book derived using the attached eddy hypothesis, where $\overline{u^2}/U_\tau^2$ followed a logarithmic function of wall-normal position (equation 9.1). Further, Perry & Abell were also puzzled as to why their u-spectra indicated a k^{-1} behaviour. Fortunately, Townsend was visiting Melbourne during this time and pointed out that Perry & Abell's data were completely consistent with his theories, and that it was all explained in the latest edition of his book. Perry was very familiar with Townsend's book but failed to see the connection and insisted that Alan show him where in the book this was written. Certainly, nowhere in Townsend (1976) is k^{-1} mentioned, while mention of it is made in the first edition (Townsend, 1956) in reference to the experimental data of Laufer (1955). With this personal 'translation' Perry was able to see that the k^{-1} behaviour for u-spectra admitted a log law for $\overline{u^2}$ and he saw the full implication of Townsend's two hypotheses. Perry & Abell reconciled their rough and smooth wall data by taking into account pipe development length issues, and published another paper (Perry & Abell, 1977) that revised their previous findings. Here they elegantly showed that the complete spectral behaviour across all wavenumbers in the logarithmic layer could be postulated using dimensional analysis and Townsend's hypotheses. Recent observations of the k^{-1} behaviour in a high Reynolds number boundary layer have also been presented in Nickels et al. (2005), and considerable attention to this issue continues.

Recent advances in the area of wall turbulence have taken advantage of the extensive new datasets obtained from multiplane particle-image velocimetry and volumetric data from numerical simulations, as reviewed by Ron Adrian, a PhD student of Townsend (Adrian, 2007). A key new feature of these studies has been the documentation of characteristic eddy shapes that take the form of hairpin-shaped vortices, which appear with a range of scales and a preferred streamwise organization where the individual vortices align to form hairpin packets. It is interesting to note that these observations were likely suspected by Townsend who relied solely on single-point time-series data. In describing the average eddy shape shown in Figure 9.3, he comments that the cone eddy could be described as "an organised assembly of simple attached eddies", which is in essence a vortex packet.

Townsend's attached eddy hypothesis has considerably shaped the field of wall turbulence thus far, and its full impact is probably yet to be realized. Concepts such as the blocking mechanism (Hunt & Graham, 1978), which have been used to explain wall turbulence statistics (Hunt & Carlotti, 2001; Mann, 1996), are inherent in the Townsend attached eddy approach. Moreover, it is worth noting that the Townsend formulations, equations (9.1), are contradictory to the classical law-of-the-wall description that have been, and continue to be, used, extensively in RANS (Reynolds averaged Navier–Stokes) calculation schemes. However, new high Reynolds number experiments and high-fidelity measurements strongly support Townsend's approach (Perry et al., 1986; Jimenez & Hoyas, 2008; Nickels et al., 2007; Marusic et al., 2010; Smits et al., 2011), and as the field enters the phase of dynamical descriptions required for a fuller understanding Townsend's ideas will undoubtably continue to shape the field (Marusic et al., 2010; Smits et al., 2011; Jimenez & Kawahara, 2011).

9.8 Turbulent shear flows and eddies

As mentioned, it was very early on that Townsend recognized the role of large eddies, or coherent structures, in turbulence. In his lectures he would refer to the large eddies as "cartwheels that could have long lifetime" (J.C.R. Hunt, private communication). Although Townsend found small anisotropy at very small scales, he showed emphatically how large-scale initial homogeneous anisotropy could persist for the lifetime of the decaying turbulence. This idea clashed with most numerical modelling approaches, as did Townsend's pioneering approach of establishing the relationship between eddies and the measured turbulence statistics. This was, and continues to be, a novel approach, as

Figure 9.4 Alan Townsend in his office at DAMTP, Cambridge, 1975.

it goes against the conventional thinking that underpins the majority of com-
mercial CFD (computational fluid dynamics) schemes. Townsend recognized
that the terms "organized eddy structure" and "turbulent flow" are themselves
descriptive of a "persistent duality in the development of theories of turbulent motion" (Townsend, 1987). This is because much of the modelling that
goes into conventional CFD schemes using the Reynolds averaged momentum
and turbulent kinetic energy, is based on concepts derived from randomness
of motion and on analogies with molecular motion (eddy viscosity concepts).
Townsend inferred that writing equations for mean values at a point in space,
implicitly assumes that the scale of the motion is small compared to the width
of the flow. In practice it was clear that the extent of flow coherence as measured by velocity correlations is usually as large as the flow is wide, and localized random flow concepts were inherently insufficient to capture the dominant
physics.

Townsend continued to work on the concept of coherent eddies throughout his career and chose a number of canonical flows to study. This involved a clear and systematic approach revealing the fundamental similarities and differences in the main eddying motions, and their associated statistical correlations. This was accompanied by beautiful examples of experimental design for answering specific scientific questions. These included using pressure gradients (Townsend, 1962, 1965b), and abrupt changes in surface roughness (Townsend, 1965a) to infer the behaviour of eddies in wall turbulence, and the use of distorting ducts to monitor the response of free-shear eddies in sheared turbulence (Townsend, 1980) and wakes (Elliot & Townsend, 1981).

The last journal paper Townsend published (Townsend, 1987) provides a succinct account of his viewpoints on organized eddy structures and turbulence, and how the concept of eddies can potentially be used to improve existing practical calculation schemes. Here, Townsend considers four flows that he has previously published papers on – the two-dimensional wake, wall turbulence, the plane mixing-layer and axisymmetric Couette flow. Emphasis is put on describing the differences in the nature and function of the organized eddies as herein lies the pertinent differences between these flows in different geometries. Townsend also shows how the different eddy structures can account for the entrainment processes in the wake, the observation of the burst-sweep cycle in wall turbulence, and jet engine noise originating from mixing layers. The common feature of the different eddies is their essential role in controlling rates of turbulent transport across the flow, and thus the quest for control of turbulence lies in the deliberate amplification or suppression of these eddies. Moreover, Townsend highlights that the performance of schemes for flow calculation can be made more rational from a knowledge of the organized eddies, and suggests this as a key area for future research.

9.9 Meteorological and other flows

In addition to study of turbulent shear flows, Townsend's research career in turbulence also included deriving equations to describe the behaviour of liquid helium-II (Townsend 1961, 1963), and a considerable body of work on turbulent convection and meteorological flows. Townsend's interest in convection and buoyancy seemed natural given the ongoing efforts in these fields at the Cavendish laboratory, including those of fellow Australians Bruce R. Morton and J. Stewart Turner, students of G.I. Taylor. Townsend, however, focused on the influence of these forces on turbulent motion. His first paper on the topic involved an analysis of the experimental data of D.B. Thomas, taken in the Cavendish laboratory, on the turbulent convection over a heated horizontal plate (Thomas & Townsend, 1957), and Townsend himself extended these experiments to more fully characterize the temperature fluctuation statistics (Townsend, 1959), data for which was very rare at the time. He continued to work on related topics throughout his career, later considering transient effects such as those associated with sea-breeze flows (Townsend, 1972, 1976). Townsend drew a distinction between convective turbulence and free and wall turbulence, where the flows "receive energy from the mean flow through the medium of eddies of specialized structure" (Townsend, 1976), to one where

> the source of energy is an actual or virtual store of potential energy and, in consequence, the mechanism of energy release is much simpler and can be carried by eddies of simple structure.

Townsend also considered turbulence under the conditions of stable stratification (with obvious applications to the atmosphere). In his first paper on the topic (Townsend, 1958a) he derived the relations for the levels of turbulence suppression that would take place with such buoyancy, and in a companion paper (Townsend, 1958b) quantified how radiative heat transfer would reduce the levels of these buoyancy forces. This helped explain observations of turbulence in the upper atmosphere, where a conventional analysis, which ignores radiative effects and is based only on levels of gradients of mean velocity and temperature, precludes its existence. His work on the stably-stratified atmosphere continued through the 1960s with analyses related to associated internal waves, together with studies of natural convection in the Earth's boundary layer. Consequently, during this period Townsend actively published in journals directly catering for the meteorological community (see, for example, Townsend, 1964, 1967; Panofsky & Townsend, 1964).

Townsend's studies on heat transfer and buoyancy appear as a new chapter in the second edition of his book (Townsend, 1976). As noted by Bradshaw (1976), there Townsend also presents new material including an elegant analysis using rapid distortion theory to show that the turbulent Prandtl number increases with total strain, and thus explaining the observed low values found in outer region of boundary layers or free-shear layers where total strain levels are low. The final chapter in Townsend (1976) follows from the discussions on buoyancy as they relate is a discussion on turbulent flows with mean streamline curvature, such as in Couette flow between concentric, rotating cylinders. Townsend makes the elegant observation that

> if the streamlines of the mean flow are appreciably curved, energy may be transferred between the mean flow and the turbulent motion in a way that resembles the transfer of energy by buoyancy forces in a stratified flows.

9.10 Concluding remarks

Alan Townsend's career was characterized by great physical insight into the problems he analysed. While he was adept with mathematics he was never obsessed with mathematical curiosities, or the pursuit of mathematical rigour at the expense of the usefulness of the results. He seemed to be always aware of the greater context of the problems and looked for solutions that were consistent with experimental data and of potential practical value in predicting turbulent flows. This does not detract from his many fundamental research contributions, but shows a wide appreciation of the field. This may well have been partly due to his beginnings as an outsider.

In his nature and personal dealings Alan was always a gentleman in the literal sense of the word (stripped of its class distinctions). He was always polite and pleasant in his dealings with other researchers, making no distinction between eminent Professors and first-year graduate students. When he did find the flaw in a presentation he seemed genuinely apologetic to the speaker (and was invariably correct).

Alan died peacefully in Cambridge on the 31st of August 2010, during the preparation of this chapter. His is a great loss to turbulence research. He will be sadly missed by his family, friends and anyone lucky enough to have known him.

Acknowledgements Many people were very helpful to the authors in compiling this chapter. First and foremost are Alan's children, Sally, Ann and Daniel. Ann generously provided Alan's letters and photographs and her time in discussing her recollections of his life. Sally provided several useful anecdotes. Professor Rod Home was also very helpful in providing material including transcripts of conversations with Alan concerning the history of Natural History at the University of Melbourne.

Author's note

Prior to the final editing of this chapter Tim Nickels passed away unexpectedly at the age of 44 on the 2nd of October 2010. The surviving author wishes to dedicate this chapter to his co-author, colleague and dear friend.

References

Adrian, R.J. 2007. Hairpin vortex organization in wall turbulence. *Phys. Fluids*, **19**, 041301.

Batchelor, G.K., and Townsend, A.A. 1947. Decay of vorticity in isotropic turbulence. *Proc. Roy. Soc. Lond. A*, **190**(1023), 534–550.

Batchelor, G.K., and Townsend, A.A. 1949. The nature of turbulent motion at large wave-numbers. *Proc. Roy. Soc. Lond. A*, **199**(1057), 238–255.

Batchelor, G.K., and Townsend, A.A. 1955. Singing corner vanes. *Nature*, **3930**, 236.

Bradshaw, P. 1967. The turbulence structure of equilibrium boundary layers. *J. Fluid Mech.*, **29**, 625–645.

Bradshaw, P. 1976. Review: The structure of tubulent shear flow, 2nd edition by A.A. Townsend. *J. Fluid Mech.*, **76**, 622–624.

Elliott, C.J., and Townsend, A.A. 1981. The development of a turbulent wake in a distorting duct. *J. Fluid Mech.*, **113**, 187–217.

Grant, H.L. 1958. The large eddies of turbulent motion. *J. Fluid Mech.*, **4**, 149–190.

Hunt, J.C.R., and Carlotti, P. 2001. Statistical structure at the wall of the high Reynolds number turbulent boundary layer. *Flow Turbul. Combust.*, **66**, 453–475.

Hunt, J.C.R., and Graham, J.M.R. 1978. Free-stream turbulence near plane boundaries. *J. Fluid Mech*, **84**, 209–235.

Jimenez, J. 2004. Turbulent flows over rough walls. *Ann. Rev. Fluid Mech.*, **36**, 173–196.

Jimenez, J., and Hoyas, S. 2008. Turbulent fluctuations above the buffer layer of wall-bounded flows. *J. Fluid Mech.*, **611**, 215–236.

Jimenez, J., and Kawahara, G. 2011. Dynamics of wall-bounded turbulence. In: Davidson, P.A., Sreenivasan, K.R., and Kaneda, Y. (eds), *The Nature of Turbulence*. Cambridge University Press.

Kline, S.J., Reynolds, W.C., Schaub, F.A., and Rundstadler, P.W. 1967. The structure of turbulent boundary layers. *J. Fluid Mech.*, **30**, 741–773.

Laufer, J. 1955. *The structure of turbulence in fully developed pipe flow*. NACA Tech. Rep. no. 1174.

Mann, J. 1994. The spatial structure of neutral atmospheric surface-layer turbulence. *J. Fluid Mech*, **273**, 141–148.

Marusic, I., McKeon, B.J., Monkewitz, P.A., Nagib, H.M., Smits, A.J., and Sreenivasan, K.R. 2010. Wall-bounded turbulent flows: Recent advances and key issues. *Phys. Fluids*, **22**, 065103.

Nickels, T.B. (ed). 2010. *IUTAM Symposium on Rough Wall Turbulence*. IUTAM Bookseries. Springer.

Nickels, T.B., and Perry, A.E. 1996. An experimental and theoretical study of the turbulent coflowing jet. *J. Fluid Mech.*, **309**, 157–182.

Nickels, T.B., Marusic, I., Hafez, S.M., and Chong, M.S. 2005. Evidence of the k^{-1} law in a high-Reynolds-number turbulent boundary layer. *Phys. Rev. Letters*, **95**, 074501.

Nickels, T.B., Marusic, I., Hafez, S.M., Hutchins, N., and Chong, M.S. 2007. Some predictions of the attached eddy model for a high Reynolds number boundary layer. *Phil. Trans. Roy. Soc. Lond. A*, **365**, 807–822.

Panofsky, H.A., and Townsend, A.A. 1964. Change of terrain roughness and the wind profile. *Quart. J. Roy. Met. Soc.*, **90**, 147–155.

Perry, A.E., and Abell, C.J. 1977. Asymptotic similarity of turbulence structures in smooth- and rough-walled pipes. *J. Fluid Mech.*, **79**, 785–799.

Perry, A.E., and Chong, M.S. 1982. On the mechanism of wall turbulence. *J. Fluid Mech.*, **119**, 173–217.

Perry, A.E., and Marusic, I. 1995. A wall-wake model for the turbulence structure of boundary layers. Part 1. Extension of the attached eddy hypothesis. *J. Fluid Mech.*, **298**, 361–388.

Perry, A.E., Henbest, S.M., and Chong, M.S. 1986. A theoretical and experimental study of wall turbulence. *J. Fluid Mech.*, **165**, 163–199.

Rotta, J.C. 1962. Turbulent boundary layers in incompressible flow. *Prog. Aero. Sci.*, **2**, 1–219.

Schultz, M.P., and Flack, K.A. 2007. The rough-wall turbulent boundary layer from the hydraulically smooth to the fully rough regime. *J. Fluid Mech.*, **580**, 381–405.

Smits, A.J., McKeon, B.J., and Marusic, I. 2011. High Reynolds number wall turbulence. *Ann. Rev. Fluid Mech.*, **43**, 353–375.

Stewart, R.W., and Grant, H.L. 1999. Early measurements of turbulence in the ocean: motives and techniques. *J. Atmos. Ocean. Tech.*, **16**, 1467–1473.

Townsend, A.A. 1941. Beta-ray spectra of light elements. *Proc. Roy. Soc. Lond. A*, **177**, 357–366.

Townsend, A.A. 1947a. Measurements in the turbulent wake of a cylinder. *Proc. Roy. Soc. Lond. A*, **190**(1023), 551–561.

Townsend, A.A. 1947b. The structure of the turbulent boundary layer. *Proc. Camb. Phil. Soc.*, **2**, 375–395.

Townsend, A.A. 1948. Local isotropy in the turbulent wake of a cylinder. *Aust. J. Sci. Res.*, **1**(2), 161–174.

Townsend, A.A. 1956. *The Structure of Turbulent Shear Flow*. Cambridge University Press.

Townsend, A.A. 1958a. The effects of radiative transfer on turbulent flow of a stratified fluid. *J. Fluid Mech.*, **4**, 361–375.

Townsend, A.A. 1958b. Turbulent flow in a stably stratified atmosphere. *J. Fluid Mech.*, **3**, 361–375.

Townsend, A.A. 1959. Temperature fluctuations over a heated horizontal surface. *J. Fluid Mech.*, 209–241.

Townsend, A.A. 1961. Equilibrium layers and wall turbulence. *J. Fluid Mech.*, **11**, 97–120.

Townsend, A.A. 1962. Natural convection in the Earth's boundary layer. *Quart. J. Roy. Met. Soc.*, **88**, 51–56.

Townsend, A.A. 1964. Natural convection in water over an ice surface. *Quart. J. Roy. Met. Soc.*, **90**, 147–155.

Townsend, A.A. 1965. Excitation of internal waves by a turbulent boundary layer. *J. Fluid Mech.*, **22**, 241–252.

Townsend, A.A. 1967. Wind and formation of inversions. *Atmos. Environ.*, **1**, 173–175.

Townsend, A.A. 1972. Mixed convection over a heated horizontal plane. *J. Fluid Mech.*, **55**, 209–227.

Townsend, A.A. 1976. *The Structure of Turbulent Shear Flow*. 2nd edn. Cambridge University Press.

Townsend, A.A. 1980. The response of sheared turbulence to additional distortion. *J. Fluid Mech.*, **98**, 171–191.

Townsend, A.A. 1990. Early days of turbulence research in Cambridge. *J. Fluid Mech.*, **212**, 1–5.

Turner, J.S. 1997. G.I. Taylor in his later years. *Ann. Rev. Fluid Mech.*, **29**, 1–25.

10

Robert H. Kraichnan

Gregory Eyink and Uriel Frisch

10.1 Introduction

Figure 10.1 Robert H. Kraichnan. Photo by Judy Moore-Kraichnan.

Robert Harry Kraichnan (1928–2008) was one of the leaders in the theory of turbulence for a span of about forty years (mid-1950s to mid-1990s). Among his many contributions, he is perhaps best known for his work on the inverse energy cascade (i.e. from small to large scales) for forced two-dimensional turbulence. This discovery was made in 1967 at a time when two-dimensional flow was becoming increasingly important for the study of large-scale phenomena in the Earth's atmosphere and oceans. The impact of the discovery

329

was amplified by the development of new experimental and numerical techniques that allowed full validation of the conjecture.

How did Kraichnan become interested in turbulence? His earliest scientific interest was in general relativity, which he began to study on his own at age 13. At age 18 he wrote at MIT a prescient undergraduate thesis, *Quantum Theory of the Linear Gravitational Field*; he received a PhD in physics from MIT in 1949 for his thesis, *Relativistic Scattering of Pseudoscalar Mesons by Nucleons*, supervised by Herman Feshbach. His interest in turbulence arose in 1950 while assisting Albert Einstein in search for highly nonlinear, particle-like solutions to unified field equations. He wrote to his long-time friend and collaborator J.R. Herring in the early 1990s how this happened:

> I realized I had no real idea of what I was doing and turned to Navier–Stokes as a nonlinear field problem where experiment could confront speculation. After initial surprise that turbulence did not succumb rapidly to field-theoretic attack, I have been trapped ever since. My overall research theme regarding turbulence has been to understand what aspects of turbulence can and cannot be described by statistical mechanics; that is, what characteristics follow from invariances, symmetries, and simple statistical measures, and what, in contrast, can be known only from experiment or detailed solution of the equations of motion. The principal tool I have used is the construction of a variety of stochastic dynamical systems that incorporate certain invariances and other dynamical properties of actual turbulence but whose statistics are exactly soluble. The similarities and differences between these model solutions and real turbulence help illuminate what is essential in turbulence dynamics. My interests have centered on isotropic Navier–Stokes and magnetohydrodynamic turbulence in two and three dimensions.

When Kraichnan started working in turbulence, the community was trying to turn A.N. Kolmogorov's 1941 (K41) ideas into a quantitative theory devoid of adjustable parameters. Kraichnan was the only one who truly succeeded in this endeavor, around the early 1960s, by among other things making extensive use of his training in theoretical physics and field theory. By that time it had however become clear that K41 is not the final word, because it misses intermittency effects that produce anomalous scaling: that is, scaling laws whose exponents cannot be obtained by dimensional considerations. Kraichnan's last major work proposed a fluid-dynamical framework for studying intermittency that quickly led the younger generation to identify the mathematical mechanism of intermittency and also to calculate by perturbation theory the anomalous scaling exponents for Kraichnan's passive scalar model.[1]

These were forty difficult years, especially the first twenty when K41 seemed to collapse, no *ab initio* theory was emerging and direct numerical simulation

---

[1] On K41, cf. Kolmogorov, 1941; Frisch, 1995.

had not yet reached Reynolds numbers high enough to supplement experimental data on intermittency. Thus a kind of crossing of the desert took place, with Kraichnan leading a small flock that found help mostly from geophysicists (at the Wood Hole Oceanographic Institute's GFD Program and at the National Center for Atmospheric Research, Boulder).

This crossing of the desert occurred at a time when Kraichnan took himself "far from the madding crowd". After having trod the traditional path of academia with positions at Columbia University and Courant Institute, he decided in the very early 1960s to become a self-employed turbulence consultant, solely funded by research grants. He moved from New York City to secluded mountains in New Hampshire, where he lived and worked for almost two decades. During those years he warmly welcomed visitors from all around the world and regularly participated in scientific meetings, workshops and schools. Eventually, he moved to New Mexico, close to the Los Alamos National Laboratory.

We shall take the reader on a tour of key contributions to turbulence by Kraichnan, organized in three main sections, roughly by chronological order: closure, realizability, the issue of Galilean invariance and MHD turbulence (Section 10.2); equilibrium statistical mechanics and two-dimensional turbulence (Section 10.3); intermittency and the Kraichnan passive-scalar model (Section 10.4). In the final section (Section 10.5), we first present two contributions of Kraichnan which do not fit naturally into the three main scientific sections: his very first published paper, on scattering of sound by turbulence (Section 10.5.1), and his prediction of the behavior of convection at extremely high Rayleigh numbers (Section 10.5.2). We then turn to Kraichnan's impact on computational turbulence and present concluding remarks.

The emphasis will be on understanding the flow of ideas and how they relate to the meandering history of the subject. We shall mostly avoid technical material. More details may be found in Kraichnan's numerous papers (over one hundred), published in journals such as *Phys. Rev., Phys. Fluids, J. Fluid Mech.* and *J. Atmosph. Sci.*, and in various reviews and textbooks by other authors,[2] but also in proceedings papers by Kraichnan, which are often more elementary and reader-friendly than those in journals.[3]

A word of warning: in what follows, standard notions used in turbulence theory, such as the Navier–Stokes and the magnetohydrodynamics (MHD) equations, statistical homogeneity and isotropy, energy spectra, etc. will be taken for granted.[4]

---

[2] Leslie, 1973; Monin & Yaglom, 1975: § 19.6; Orszag, 1977; Rose & Sulem, 1978.
[3] See Kraichnan, 1958c, 1972, 1975a.
[4] See, e.g., Monin & Yaglom, 1975; Frisch, 1995.

10.2 Closures: realizability, Galilean invariance and the random coupling models; MHD turbulence

After World War II the K41 theory became universally known and was seen to be mostly consistent with experimental data of that time. Attempts were then made to build a quantitative theory of homogeneous and isotropic turbulence, compatible with K41 but free of undetermined constants, such as the (Kolmogorov) constant appearing in the energy spectrum, and able to predict the functional form of the spectrum in the dissipation range. Particularly noteworthy was S. Chandrasekhar's attempt, based on the quasi-normal approximation (QNA). Introduced by M.D. Millionschikov, the QNA assumes that second- and fourth-order moments are related as in a normal (Gaussian) distribution. This is an instance of what is called *closure*: replacing the infinite hierarchy of moment equations derivable from the Navier–Stokes equations by a finite number of equations for suitable statistical quantities such as two-point correlations. Kraichnan was the first to point out that the simplest closures that come to mind, such as the QNA, can have *realizability* problems: basic probabilistic inequalities may be violated. A few years later, using the form of the QNA obtained by T. Tatsumi, Y. Ogura showed indeed that it can lead to an energy spectrum which is negative for certain wavenumbers.[5]

10.2.1 The direct interaction approximation and the Galilean problem

Kraichnan thus set out to obtain better closures, simultaneously free of adjustable parameters and realizable. He was the first to apply field-theoretic methods to non-equilibrium statistical fluid dynamics (Navier–Stokes and MHD): he observed that formal expansions in powers of the nonlinearity of the solutions of such equations, followed by averaging over random initial conditions and/or driving forces, generates terms that can be represented by Feynman diagrams. Eventually, P.C. Martin, E.D. Siggia and H.A. Rose would establish a detailed formal connection between statistical dynamics of classical fields, such as turbulence dynamics, and a certain quantum field theory (so-called MSR field theory).[6]

In tackling the closure problem from this point of view, Kraichnan was led to introduce a new object into turbulence theory, namely the (averaged infinitesimal) response function: that is, the linear operator giving the small change in

---

[5] Millionschikov, 1941; Chandrasekhar, 1955; see also Spiegel, 2011; Kraichnan, 1957b; Tatsumi, 1957; Ogura, 1963.
[6] Kraichnan, 1958c; Martin, Siggia & Rose, 1973.

the velocity field at time t, due to a small change in the driving force at times $t' < t$. In 1958, using a mixture of field-theoretic considerations and heuristic but plausible simplifying assumptions, Kraichnan proposed the *direct interaction approximation* (DIA).[7] When applied to homogeneous, isotropic and parity-invariant turbulence, the DIA takes the form of two coupled integro-differential equations for the two-point two-time velocity correlation function and the response function. These equations are written in Section 10.2.2 in abstract and concise notation, valid without assuming any particular symmetry. For homogeneous isotropic turbulence the DIA equations, which resemble in more complicated form the EDQNM equation (10.3) written below, can be found in Kraichnan's papers and in the references mentioned in the introduction. Such equations are easily extended to anisotropic and/or helical (non-parity-invariant) turbulence.

As Kraichnan observed himself, the DIA is not consistent with K41. In particular the inertial-range energy spectrum predicted by DIA is (to leading order)

$$E(k) = C'(\varepsilon v_0)^{1/2} k^{-3/2}, \tag{10.1}$$

instead of

$$E(k) = C \varepsilon^{2/3} k^{-5/3}, \tag{10.2}$$

where ε is the kinetic energy dissipation per mass, v_0 is the r.m.s. turbulent velocity, and C (the Kolmogorov constant) and C' are dimensionless constants. The discrepancy has to do with the way the larger eddies from the energy-containing range of wavenumbers interact with smaller eddies from the inertial range. Kraichnan expressed his doubts that the former merely

> convect local regions of the fluid bodily without significantly distorting their small-scale (high-k) internal dynamics.

This assumption would not permit the presence of v_0 in the inertial-range expression of the energy spectrum, except through an intermittency effect of a kind which was not seriously considered until the work by the Russian school in the early 1960s. Nevertheless, Kraichnan observed the following:

> As a counter-consideration to Kolmogorov's original argument in the *x*-space representation, it may be noted that the fine-scale structure of high Reynolds-number turbulence consists typically not of compact blobs but of a complicated tangle of extended vortex filaments and sheets.[8]

---

[7] Kraichnan, 1958a, 1958c, 1959a.
[8] Kraichnan, 1958a: §§ 9–10 and footnote 35.

For a while Kraichnan believed that the DIA may be asymptotically exact in the limit $L \to \infty$ for turbulence endowed with L-periodicity in all the spatial coordinates. However, after a few months, Kraichnan found that the DIA equations are actually the exact consequences of a certain random coupling model obtained from the hydrodynamical equations (Navier–Stokes or MHD) by a suitable modification of the nonlinear interactions (see Section 10.2.2). The presence of this model automatically guarantees the realizability of the DIA. The random coupling model has many properties in common with the original hydrodynamical equations, such as symmetries, energy conservation, etc. But Kraichnan noted that, when various correlation functions are expanded in terms of Feynman diagrams, the random coupling model retains only a subclass of all diagrams; in particular it misses the so-called vertex corrections which contribute (in field-theoretic language) to the renormalization of the nonlinear interaction.[9]

Six years later he discovered a more serious defect: the random coupling model and the DIA fail invariance under *random Galilean transformations*.[10] Ordinary Galilean invariance – for the Navier–Stokes equations – is the observation that if $(u(x, t), p(x, t))$ are the velocity and pressure fields which solve the Navier–Stokes equations in the absence of boundaries or with periodic boundary conditions, then $(u(x - Vt, t) + V, \, p(x - Vt, t))$ are also solutions for an arbitrary choice of the velocity V. Random Galilean invariance has the velocity V chosen randomly with an isotropic distribution and independent of the turbulent velocity-pressure fields. Isotropy ensures that the mean velocity remains zero. In the DIA, when a random Galilean transformation is performed, the mean square velocity v_0^2 increases and thus the spectrum given by (10.1) changes. On the one hand this is clearly inconsistent with the Galilean invariance of the Navier–Stokes equations. On the other hand, an influence of the energy-carrying eddies on the inertial range different from what K41 predicts (e.g. intermittency effects) cannot be ruled out.

But first, Kraichnan realized that the DIA had to be modified to restore random Galilean invariance and, if possible, without losing its other nice features. To explain how this can be done, we have to become slightly more technical. Kraichnan derives the following equation from the DIA for the evolution of the energy spectrum when the turbulence is homogeneous, isotropic and

---

[9] Kraichnan, 1958b, 1958c.

[10] Kraichnan, 1964b. Note that the random coupling model can be modified to preserve a Galilei symmetry group of dimension 3, for a model with N 3-dimensional velocity fields. See Kraichnan, 1964a: Appendix. However, this is insufficient to guarantee random Galilean invariance for the DIA equations.

parity-invariant:

$$\left(\partial_t + 2\nu k^2\right) E(k,t) = \int \int_{\Delta_k} dp dq\, \theta_{kpq} \times$$

$$b(k,p,q)\frac{k}{pq} E(q,t)\left[k^2 E(p,t) - p^2 E(k,t)\right] + F(k), \qquad (10.3)$$

$$\theta_{kpq} \equiv \frac{(\pi/2)^{1/2}}{v_0(k^2 + p^2 + q^2)^{1/2}}, \qquad b(k,p,q) = \frac{p}{k}(xy + z^3). \qquad (10.4)$$

Here $E(k,t)$ is the energy spectrum, $F(k)$ the energy input from random driving forces, Δ_k defines the set of $p \geq 0$ and $q \geq 0$ such that k, p, q can form a triangle, x, y, z are the cosines of the angles opposite to sides k, p and q of this triangle. The factor θ_{kpq} can be interpreted as a triad relaxation time. The particular form given in (10.4) is obtained from the DIA by an asymptotic expansion valid only when the wavenumber k is in the inertial range. This form implies that relaxation has a time scale comparable to the convection time $1/(kv_0)$ of an eddy of size $1/k$ at the r.m.s. velocity v_0. As Kraichnan notes, if v_0 is replaced "by, say, $[kE(k)]^{1/2}$, which may be considered the r.m.s. velocity associated with wavenumbers the order of k only", then K41 is recovered. Actually this choice, which also restores random Galilean invariance and is easily shown to be realizable, became later known as the *eddy-damped quasi-normal Markovian* (EDQNM) approximation. Note that the EDQNM requires the use of an adjustable dimensionless constant in front, say, of $[kE(k)]^{1/2}$, which can then be tuned to give a Kolmogorov constant matching experiments and/or simulations. Kraichnan clearly preferred having a systematic theory free of such constants and then seeing how well it can reproduce experimental data, not just constants, but also the complete functional forms of spectra. He eventually developed a version of the EDQNM, called the *test field model* in which the triad relaxation time is determined in a systematic way through the introduction of a passively advected (compressible) test velocity field and a procedure for eliminating unwanted sweeping effects.[11]

To overcome the difficulty with random Galilean invariance in a more systematic way, however, he first developed the *Lagrangian history direct interaction approximation* (LHDIA) and an abridged version thereof (ALHDIA) in which the correlation functions have fewer time arguments. LHDIA and ALHDIA make use of a generalized velocity $u(x, t|t')$ which has both Eulerian and Lagrangian characteristics. It is defined as the velocity measured at time

---

[11] Kraichnan, 1958c; for EDQNM, see Orszag, 1966, 1977; and also Rose & Sulem, 1978; Lesieur, 2008; for the test field model, see Kraichnan, 1971a, 1971b. S.A. Orszag died on 1st May, 2011.

t' in that fluid element which passes through x at time t, and it satisfies an extended form of the Navier–Stokes equation in the two time variables. When the DIA closure is written for this extended system, there appear integrals over the past history of the flow, which can be altered when working with the generalized field in such a way as to recover random Galilean invariance. Detailed numerical studies of the ALHDIA gave a full functional form of the spectra in very good agreement with experimental data. In particular the Kolmogorov constant was $C = 1.77$ while the best current value is 1.58, when ignoring intermittency corrections. Other detailed predictions were made for various Lagrangian quantities involving correlations of velocity increments, pressure-gradients, particle accelerations and also for two-particle dispersion. Many of these predictions are still to be subjected to experimental test.[12]

In the course of studying the DIA and all its 'children', Kraichnan had to develop quite sophisticated analytic and numerical tools. For example, he showed that the spectrum falls off exponentially in the far dissipation range and obtained the algebraic prefactor in front of the exponential. Such tools are transposable almost immediately to the study of other closures, for example of the kinetic theory of resonant nonlinear wave interactions.[13] They are also part of the Kraichnan legacy.

Many of Kraichnan's tools from statistical mechanics and field theory were unfamiliar to fluids scientists trained in the traditions of theoretical mechanics and applied mathematics, and some objects central to the DIA – particularly the mean response function – did not appear in previous exact moment hierarchies. The DIA was thus received with some scepticism by the scientific community, in particular in Cambridge (UK), but Kraichnan was given the opportunity to present his theory in the recently introduced *Journal of Fluid Mechanics*. Kraichnan's theory was discussed by Proudman at the famous Marseille 1961 conference. With hindsight, we see that he failed to notice the realizability of the DIA and lumped it together with the *quasi-normal approximation* (QNA); actually, it would take two more years until Ogura discovered the aforementioned problems with QNA. Much more interesting is that, at the end of his paper, Proudman (correctly) expressed doubts that DIA could cope with what was later termed dissipation-range intermittency, which would be first explained by Kraichnan six years later (see Section 10.4.1).[14] Eventually,

---

[12] Kraichnan, 1964b, 1965a, 1966a; for experimental data, see Grant, Stewart & Moilliet, 1962; for the Kolmogorov constant, see Donzis & Sreenivasan, 2010. Ott & Mann, 2000, corrected an error in Kraichnan's LHDIA theory of turbulent two-particle dispersion (Kraichnan, 1966b) and found from experiment that its prediction for the Richardson–Obukhov constant was too large by a factor of 10.

[13] Zakharov, L'vov & Falkovich, 1992; Nazarenko, 2011.

[14] Kraichnan, 1959a; Proudman, 1961; Ogura, 1963.

Kraichnan's efforts to develop an analytical theory, compatible with K41 and able to give detailed quantitative predictions, was remarkably successful and influential. Of course around the same time, at the Marseille conference, it became increasingly clear that K41 was not the final word for fully developed turbulence. Kraichnan's contributions to post-K41 theories will be described in Section 10.4.

10.2.2 The random coupling models and the $1/N$ expansion

Starting from the Navier–Stokes (NS) equations, Kraichnan derived the DIA using heuristic assumptions that, even if plausible, turned out to be partially false. Quite rapidly after this, Kraichnan showed that the DIA equations are actually the *exact* consequence in a certain limit of a stochastic model, called the *random coupling model* (RCM), which is obtained by a suitable modification of the Navier–Stokes equations. This is important not only because it ensures the realizability of the DIA but because it makes contact with – and largely anticipates – techniques still being developed in field theory and statistical mechanics. We shall try here to give a not-too-technical presentation of the RCM.[15]

Random coupling models can be written for a large class of quadratically nonlinear partial differential equations, which encompass the Burgers equation, the Navier–Stokes equations and the MHD equations. All these can be written in the following general form, where it is assumed that the solution is zero for $t < 0$:

$$\partial_t u(t) + B(u(t), u(t)) = Lu(t) + f(t) + u(0)\delta(t). \tag{10.5}$$

Here $u(t)$ (scalar or vector) collects all the dependent variables, $B(\cdot, \cdot)$ is a quadratic form (comprising the inertial and pressure terms for the case of Navier–Stokes), L is a time-independent linear operator (e.g. the viscous dissipation operator) and $f(t)$ is a prescribed zero-mean random force, taken as Gaussian for convenience; the prescribed zero-mean initial condition $u(0)$ in front of the Dirac distribution $\delta(t)$ is also taken as random Gaussian. Obviously, (10.5) can be rewritten as an equivalent integral equation

$$u(t) + \int_0^t dt' \, e^{L(t-t')} B(u(t'), u(t')) = \int_0^t dt' \, e^{L(t-t')} f(t') + u(0). \tag{10.6}$$

In somewhat more abstract form this may be written as

$$u + \mathcal{B}(u, u) = \mathcal{F}. \tag{10.7}$$

---

[15] Kraichnan, 1958c, 1961.

Let us now describe the random coupling model in the form used by Herring and Kraichnan. Imagine that N independent replicas of (10.7) are written with N independent fields, labeled u_α ($\alpha = 1, \dots, N$) and that these are then coupled as follows:

$$u_\alpha + \frac{1}{N} \sum_{\beta,\gamma} \phi_{\alpha\beta\gamma} \mathcal{B}(u_\beta, u_\gamma) = \mathcal{F}_\alpha, \qquad \alpha = 1, \dots, N. \qquad (10.8)$$

Here, the $\mathcal{F}_\alpha$ are N independent identically distributed (iid) replicas of the force $\mathcal{F}$ in (10.7); the 'random coupling coefficients' $\phi_{\alpha\beta\gamma}$ are a set of Gaussian random variables of zero mean and unit variance. These N^3 coefficients are taken as all independent, except for the requirement that $\phi_{\alpha\beta\gamma}$ be invariant under all permutations of (α, β, γ), a condition crucial for conservation of energy and other quadratic invariants. When the number of replicas tends to infinity, closed equations emerge for the two following objects: the correlation function $U \equiv \langle u_\alpha \otimes u_\alpha \rangle$ and the infinitesimal response function $G \equiv \langle \delta u_\alpha / \delta \mathcal{F}_\alpha \rangle$. Note that off-diagonal terms with $\alpha \neq \beta$ tend to zero as $N \to \infty$. The angular brackets denote ensemble averages and $\otimes$ indicates a tensor product: if we were dealing with a velocity field $u_i(\boldsymbol{x}, t)$, then $u \otimes u$ would stand for $u_i(\boldsymbol{x}, t) u_j(\boldsymbol{x}', t')$. It is also convenient to introduce auxiliary independent zero-mean Gaussian random fields $\hat{v}$ and v having the same correlation function U as any of the u_α. With such abstract notation the DIA equations take the compact form

$$U = \langle (G\mathcal{F}) \otimes (G\mathcal{F}) \rangle + 2G\langle \mathcal{B}(\hat{v}, v) \otimes G\mathcal{B}(\hat{v}, v) \rangle \qquad (10.9)$$

$$G - 4\langle \mathcal{B}(\hat{v}, G\mathcal{B}(\hat{v}, G)) \rangle = I, \qquad (10.10)$$

where I is the identity operator.[16]

The random coupling model can also be modified to give equations other than DIA. For example the phases $\phi_{\alpha\beta\gamma}$ can be made functions of the time with a white noise dependence: $\langle \phi_{\alpha\beta\gamma}(t) \phi_{\alpha\beta\gamma}(t') \rangle = \tau_0 \delta(t - t')$; this is the so-called *markovian random coupling model* (MRCM) which leads to an equation similar to (10.3) but with $\theta_{kpq} \equiv \tau_0/4$. The EDQNM equation can also be obtained in similar manner by taking the random phase dependent on triads of nonlinearly interacting wavevectors and delta-correlated in time with a coefficient proportional to the triad relaxation time θ_{kpq} chosen in agreement with K41, as explained in Section 10.2.1. Unfortunately, so far no random coupling model has been found for the Lagrangian variants of DIA.[17]

---

[16] For the random coupling model, see Kraichnan, 1958c, 1961; Herring & Kraichnan, 1972; for the compact form of the DIA, see Lesieur, Frisch & Brissaud, 1971.

[17] For variants of the random coupling model, see Kraichnan, 1958c, 1961; for the MRCM, see Frisch, Lesieur & Brissaud, 1974.

Kraichnan's work on the RCMs has many other interesting connections in mathematics and physics. For example, the simplest dynamics considered in Kraichnan's 1961 paper is the classical harmonic oscillator with a random frequency. The N-replica model in this case reduces to the study of a class of random $N \times N$ matrices (Toeplitz matrices with coefficients whose phases are randomly altered, consistent with Hermiticity). Following a suggestion of Kraichnan, it was shown by U. Frisch and R. Bourret that the DIA equation for the harmonic oscillator can also be obtained from an RCM employing random Wigner matrices selected out of the Gaussian unitary ensemble. This approach works for any linear dynamics with a random linear operator, such as the problems of turbulent advection of a passive scalar or the Schrödinger equation with random potential.

Kraichnan's discovery of the RCM in 1958 anticipated, in fact, decades-later developments in quantum field theory employing large-N models with quenched random parameters and random matrices. G. t' Hooft showed that $SU(N)$ gauge field theories for $N \to \infty$ re-sum a subset of all Feynman diagrams, the planar diagrams, and Yu.M. Makeenko and A.A. Migdal showed that the exact Schwinger–Dyson equations for Wilson loops reduce in that limit to a closed set of self-consistent equations like DIA. The most direct connection with Kraichnan's work is for random matrix models of large-N gauge theory in 0D and quantum gravity in 2D, where the self-consistent 'loop equation' is identical in form to the DIA equation for the harmonic oscillator with a non-Gaussian random frequency.[18] Kraichnan's RCM was not a rote application of standard field-theoretic techniques of the 1950s, but employed advanced ideas that were regarded as 'cutting edge' decades later.[19]

The random coupling models may have more than just historical interest, however. One of the successes of theoretical physics in the 1970s was the development of $1/N$ expansion methods to provide analytical tools to tackle nonperturbative problems with no other small parameter, such as anomalous scaling in critical phenomena and quark confinement in quantum chromodynamics. As we shall discuss in more detail in Section 10.4, turbulence anomalous scaling due to inertial-range intermittency is exactly this sort of problem. Moreover, recent numerical studies of large-N random coupling models for some simple dynamical systems model of turbulence, so-called 'shell models', have found that inertial-range anomalous dimensions vanish proportional to $1/N$. Kraichnan's pioneering work on the RCM together with modern $1/N$

---

[18] Kraichnan, 1961: § 9.

[19] On large-N gauge theory and quantum gravity, G. 't Hooft, 1974; Makeenko & Migdal, 1979; Migdal, 1983; Di Francesco, Ginsparg & Zinn-Justin, 1995.

expansion methods might provide a powerful tool for addressing the difficult problem of inertial-range turbulence scaling.[20]

10.2.3 MHD turbulence

Magnetohydrodynamics (MHD) was born with Alfvén's prediction of a new type of waves (now called *Alfvén waves*) in a conducting fluid in the presence of a uniform background magnetic field. As pointed out by von Neumann in his report on turbulence at the conference *Problems of Motion of Gaseous Masses of Cosmic Dimensions*, MHD was considered key in understanding the origin of various cosmic magnetic fields – beginning with that of the Earth – and the mechanism for accelerating cosmic rays. Immediately the question arose as to whether MHD turbulence at high kinetic and magnetic Reynolds numbers can be described using K41. The question was difficult because (i) there were no experimental data on MHD turbulence in the late 1940s, and (ii) dimensional analysis was of limited use, due to the presence of two fields, the velocity field *u* and the magnetic field *b*.[21] Important early papers on the statistical theory of turbulence were already dealing with MHD turbulence, even in their titles. This included Batchelor's paper on the analogy between vorticity and magnetic fields, Lee's 1952 paper (see Section 10.3) and Kraichnan's very first paper on the DIA, where he also handled the MHD case, but without explicitly stating that it leads to a $k^{-3/2}$ inertial-range spectrum, as in the purely hydrodynamic case.[22]

One year after discovering that the DIA fails random Galilean invariance, Kraichnan returned to MHD and found that the $k^{-3/2}$ law may actually apply to the three-dimensional MHD case when there is a significant random large-scale magnetic field of r.m.s. strength b_0. In a short research note he stated:

> In the present hydromagnetic case, it still may be argued plausibly that the action of the energy range on the inertial range is equivalent to that of spatially uniform fields. But, in contrast to a uniform velocity field, a uniform magnetic field has a profound effect on energy transfer.

Indeed, when the equations of MHD are rewritten in terms of the Elsässer variables $z^{\pm} \equiv v \pm b$, and a uniform background field of strength b_0 is present, it is found that:

---

[20] For DIA and Wigner matrices, see Frisch & Bourret, 1970. For $1/N$ expansion in critical phenomena, see Abe, 1973. For large-N shell models, see Pierotti, 1997; Pierotti, L'vov, Pomyalov & Procaccia, 2000.

[21] The magnetic field can be given the same dimension as the velocity field after division by $\sqrt{4\pi\mu\rho}$ where μ is the magnetic permeability and ρ the density: these are the units used here.

[22] Alfvén, 1942; von Neumann, 1949; Batchelor, 1950; Lee, 1952; Kraichnan, 1958a.

(i) in linear theory z^+ and z^- propagate as Alfvén waves in opposite directions with speed b_0 along the background field;

(ii) the only nonlinear interactions are between z^+ and z^- (no self-interactions).

It follows that a z^+-eddy and a z^--eddy of wavenumbers $\sim k$ will be able to significantly interact only during a coherence time $\sim 1/(kb_0)$. Thus the flux ε of total (kinetic plus magnetic) energy through the wavenumber k will be proportional to $1/b_0$, where b_0 is the r.m.s. magnetic field fluctuation (stemming from the energy range). Therefore ε and b_0 must appear in the combination εb_0. It then follows by dimensional analysis that both the kinetic and magnetic energy spectra are of the form

$$E(k) \sim (\varepsilon b_0)^{1/2} k^{-3/2}. \tag{10.11}$$

This is the same functional form as the DIA spectrum (10.1), and for a good reason: the large-scale magnetic field plays now the same role as the large-scale velocity field in the DIA. Of course, in the DIA this is a spurious role, whereas in MHD Alfvén waves make this effect real. It must be pointed out that Iroshnikov proposed the same MHD spectrum (10.11) slightly before Kraichnan, using a semi-phenomenological theory of interaction of three Alfvén waves and a diffusion approximation in k-space for the energy spectrum derived by arguments resembling those Leith used for two-dimensional turbulence. Iroshnikov's work did not penetrate much into the West until approximately 1990. Nowadays the energy spectrum (10.11) is called the *Iroshnikov–Kraichnan spectrum*.[23]

Finally we mention that the arguments of Iroshnikov and Kraichnan can be questioned because local anisotropies are ignored: in a small region in which the (large-scale) magnetic field is B, those small-scale eddies having wave-vectors perpendicular to B, or nearly so, are hardly affected by Alfvén waves. This, together with non-linear effects, leads to the emmergence of quasi-two-dimensional MHD turbulence. A more systematic theory of weakly interacting Alfvén waves predicts a k^{-2} spectrum for MHD turbulence. The problem of determining the spectrum of strong MHD turbulence – even at the level of a Kolmogorov mean-field theory, ignoring intermittency – is still quite open at this time.[24]

---

[23] Kraichnan, 1965b; Iroshnikov, 1963; Leith 1967.

[24] For anisotropies, see Kit & Tsinober, 1971; Rüdiger, 1974; Shebalin, Matthaeus, & Montgomery, 1983. For ideas of using resonant wave interactions, see Ng & Bhattacharjee, 1996; Goldreich & Sridhar, 1997. For the systematic theory of resonant wave interactions, see Galtier et al., 2000; Nazarenko, Newell & Galtier, 2001.

10.3 Statistical mechanics and two-dimensional turbulence

The essentially statistical character of turbulent flow had clearly been understood long before even the 1941 work of Kolmogorov, by such scientists as
Lord Kelvin, G.I. Taylor, N. Wiener, T. von Kármán, and J.M. Burgers. To this
day, probably the most successful statistical theory of any dynamical problem is the equilibrium statistical mechanics of J.W. Gibbs, L. Boltzmann and
A. Einstein for classical and quantum Hamiltonian systems. It was natural then
that the pioneers in turbulence theory would seek some guidance and inspiration from Gibbsian statistical mechanics. Notable was the attempt by Burgers
in a series of papers during 1929–1933 to apply the maximum entropy principle to turbulent flows. Unfortunately, it was pointed out by Taylor in 1935
that mean energy dissipation (and thus entropy production) does not vanish in
the limit of high Reynolds numbers for typical turbulent flows. Turbulence is
thus a fundamentally non-equilibrium problem to which the Boltzmann–Gibbs
formalism is not *directly* applicable. Nevertheless, a judicious application of
equilibrium statistical theory to hydrodynamics can still yield crucial hints and
ideas on the behavior of turbulent flows, and Kraichnan was one of the most
masterful practitioners of this art. He used Gibbsian statistical predictions in
his construction of closure theories and, most subtly, to help to divine the behavior of strongly disequilibrium turbulent cascades. This approach played an
especially important role in Kraichnan's development of his dual cascade picture of two-dimensional turbulence. For this reason, we shall treat these two
subjects together.

10.3.1 Equilibrium statistical hydrodynamics

Prior to Kraichnan's work, one of the most significant applications of Gibbsian
statistical mechanics to hydrodynamics was in fact to two-dimensional fluids,
by L. Onsager in a 1949 paper entitled "Statistical hydrodynamics". Onsager
noted that "two-dimensional convection, which merely redistributes vorticity"
leads to energy conservation at infinite Reynolds number. Thus, equilibrium
statistical mechanics may be legitimately applied to 2D Euler hydrodynamics,
if ergodicity is assumed (as usual). Onsager studied in detail the microcanoni-
cal distribution of a "gas" of N Kirchhoff point-vortices in an incompressible,
frictionless, 2D fluid. His most striking conclusion was that, for sufficiently
high kinetic energies of the point-vortices, thermal equilibrium states of *negative* absolute temperature occur. The point-vortices condense into a single
large-scale coherent vortex, which Onsager suggested could explain the "ubiquitous" appearance of such structures in quasi-2D flows.[25]

[25] Onsager, 1949.

A conservative system appears in any space dimension if the Euler equations or the MHD equations for ideal forced fluid flow are Galerkin-truncated, as first noted by Burgers and, later independently, by T.D. Lee and E. Hopf. For the case of space-periodic flow and in the abstract notation of Section 10.2.2 this amounts to replacing the nonlinear term in (10.5) by $P_G B(u(t), u(t))$, where P_G is the Galerkin projection. This operator is defined in terms of the spatial Fourier representation of $u(t)$ by setting to zero all Fourier components of wavenumbers $k > K_G$, the truncation wavenumber. The above authors showed then that, in suitable variables (based on the real and imaginary parts of the the Fourier amplitudes), the dynamics can be written as a set of ordinary real differential equations

$$\dot{q}_\alpha = F(q_1, \ldots, q_N), \qquad \alpha = 1, \ldots, N, \qquad (10.12)$$

with a Liouville theorem $\sum_\alpha \partial \dot{q}_\alpha / \partial q_\alpha = 0$, which expresses the conservation of volumes in the N-dimensional phase space. The hypothesis of ergodicity implies equipartition of energy among all degrees of freedom. In three dimensions, this led Lee and Hopf to a k^2 energy spectrum, with no divergence of the total energy because of truncation. Both recognized that, with viscosity added and absent truncation, things will be very different and that Kolmogorov's $-5/3$ spectrum is expected. In the MHD case, Lee obtained the same spectrum for the magnetic field and an equipartition between kinetic and magnetic energy for every Fourier harmonic. He conjectured this extends to non-equilibrium MHD turbulence, inferring that Kolmogorov's $-5/3$ spectrum should also hold in the MHD case.[26]

This was roughly the situation when Kraichnan entered the field of turbulence, and statistical mechanical ideas dominated much of his early thinking, too.[27] His first work on the application of equilibrium statistical mechanics to turbulence was in a hardly cited 1955 paper on compressible turbulence in which he rediscovered, in a slightly different context, the 1952 results of Lee. In particular Kraichnan found the Liouville theorem and equipartition solutions (here, the equipartition is between kinetic and potential acoustic energies). Kraichnan's preoccupation with statistical mechanics continued in his first DIA paper. He discussed there the Gibbs ensembles for the Galerkin-truncated equations but noted that "This artificial equilibrium case does *not* describe turbulence at infinite Reynolds number," when a forcing term and a viscous damping term are added.

There is, however, one very original and definitive result on equilibrium statistical mechanics obtained in Kraichnan's first paper on DIA. He

---

[26] Hopf, 1952; Lee, 1952.
[27] Kraichnan had discussions about statistical mechanics and the foundations of thermodynamics in the late 1950s with B. Mandelbrot (Mandelbrot, private communication, 2010).

established there a *fluctuation–dissipation theorem* (FDT; called by him, correctly, "fluctuation–relaxation") for conservative nonlinear dynamics in thermal equilibrium, generalizing previous results of H.B. Callen and co-workers. Kraichnan's derivation in an appendix to the 1958 paper was very ingenious. He showed that the FDT arises from the stability of the Gibbs measure for two replicas of the original dynamics under couplings that preserve the Liouville theorem and energy conservation. Recognizing the more general interest in this result, Kraichnan one year later published this argument separately for any dynamics that conserves both phase volume and an arbitrary energy function E. In the special case of the truncated Euler system considered in the 1958 DIA paper, Kraichnan's FDT reduces to the statement that the two-time correlation function U and the mean response function G are proportional, with the absolute temperature supplying the constant of proportionality. In this case, the DIA equations reduce to a single equation for G, which Kraichnan discussed in a separate section on "Nondissipative equilibrium" in his paper. This is the same as the "mode-coupling theory" applied to equilibrium critical dynamics twelve years later by K. Kawasaki (who was aware of and influenced by Kraichnan's earlier work on DIA).[28]

Kraichnan returned to the statistical mechanics of Galerkin-truncated ideal flow in a 1975 article "Remarks on turbulence theory". Although ostensibly a review article, this work – characteristically for Kraichnan – contained many original ideas and results not published elsewhere. Kraichnan pointed out that the "absolute equilibrium" for the inviscid truncated system in 3D had an interesting domain of physical applicability, describing the small thermal fluctuations of velocity in a fluid at rest. As an application of the equilibrium DIA equations, he showed that

> the truncated Euler system in thermal equilibrium exhibits a dynamical damping of low-wavenumber disturbances just like the viscous damping of the Navier–Stokes system at zero temperature. If k_{max} is taken as some kind of intermolecular spacing scale or mean free path, then the truncated Euler system constitutes a nontrivial model of a molecular liquid, with the equilibrium excitation corresponding to normal molecular thermal energy.

In the same paper he also suggested that the truncated Euler system could support an energy cascade just as in the Navier–Stokes system, for "a statistical ensemble whose initial distribution is multivariate-normal, with all energy concentrated in wavenumbers the order of k_0". In 1989 he and S. Chen went much further:

> the truncated Euler system can imitate NS fluid: the high-wavenumber degrees of freedom act like a thermal sink into which the energy of low-wave-number

---

[28] Kraichnan, 1958a, 1959b; Kawasaki, 1970.

modes excited above equilibrium is dissipated. In the limit where the sink wave-numbers are very large compared with the anomalously excited wavenumbers, this dynamical damping acts precisely like a molecular viscosity.

Actually, in 2005 very-high-resolution spectral simulations of the 3D Galerkin-truncated Euler equations showed that, when initial conditions are used that have mostly low-wavenumber modes, the truncated inviscid system tends at very long times to thermal equilibrium with energy equipartition, but it has long-lasting transients that are basically the same as for viscous high Rey-nolds-number flow, including a K41-type inertial range. In other words, the high-wavenumber thermalized modes behave as an artificial molecular micro-world.[29]

10.3.2 Two-dimensional turbulent cascades

All this is to set the stage for a major contribution made by Kraichnan in 1967, his theory of inverse cascade for two-dimensional (2D) turbulence. Equilib-rium statistical mechanical reasoning is one of the many strands of thought that Kraichnan wove together into a compelling argument for the existence of Kolmogorov-type cascades in 2D incompressible fluids. His remarkable paper in *Physics of Fluids* invoked also an analysis of triadic interactions under the assumption of scale-similarity, exact results on the instantaneous evolution of Gaussian-distributed initial conditions, various plausible statistical and physi-cal arguments, and even comparison with Kraichnan's simultaneous work on dynamics of quantum Bose condensation. As noted already in the introduction, Kraichnan's findings have considerable interest for understanding large-scale geophysical and planetary flows which, due to a combination of small vertical scale, rapid rotation and strong stratification, may be described by 2D Navier–Stokes or related equations.[30]

Before discussing Kraichnan's contributions, however, a brief digression on prior 2D turbulence work is in order. For many years, following the obser-vation by Taylor that there cannot be appreciable energy dissipation in high-Reynolds-number 2D flow, two-dimensional turbulence was not considered with much favor. Vortex stretching was regarded as the essence of turbulence and this effect is absent in 2D. Thus, it was often stated – and occasion-ally still is – that "there is no 2D turbulence". This situation changed in the late 1940s, when two-dimensional approximations were proposed for high-Reynolds-number geophysical flows, which had many of the attributes of

---

[29] The first quote is from Kraichnan, 1975a: p. 31, and the second from Kraichnan & Chen, 1989: p. 162; see also Forster et al., 1977. For the simulation of truncated Euler, see Cichowlas et al., 2005.

[30] The pioneering paper on 2D turbulent cascades is Kraichnan, 1967b.

turbulence (randomness, disorder, etc.) For example, J.G. Charney in 1948 formulated the quasi-geostrophic model, in which potential vorticity (like vorticity for 2D Euler) is conserved along every fluid element. A little later, J. von Neumann made a number of interesting remarks on 2D turbulence, in particular that it is expected to have far less disorder than in 3D, precisely because of vorticity conservation; but this material remained unpublished for a long time. At the very end of his 1953 turbulence monograph, G.K. Batchelor proposed to investigate the spottiness "of the energy of high wave-number components" using 2D turbulence. He observed that in a 2D ideal flow the conservation of the integral of (one half of) the squared vorticity – now called 'enstrophy', a term coined by C.E. Leith, one which we shall use liberally – prevents energy from solely flowing to high wavenumbers: some energy has to be transferred also to smaller wavenumbers (larger scales). Batchelor also concurred with Onsager about the tendency of small vortices of the same sign to merge into larger vortices. In the same year, R. Fjørtoft showed that, because of the simultaneous conservation of energy and enstrophy, it is impossible for 2D dynamics to change the amplitudes of only two (Fourier) modes. Within a triad of modes, he showed that the change in energy for the member of the triad with intermediate wavenumber is the opposite to that of the other two members and that the member with lowest wavenumber shows the largest energy change of the two extreme members.[31]

It is thus clear that there must be significant inverse transfer of energy in 2D. However, even in 3D there is some inverse transfer of energy for the case of freely decaying high Reynolds number flow, where the peak of the energy spectrum migrates to smaller wavenumbers. What about the K41 energy cascade and the presence of an inertial range over which the energy flux is uniform? Lee showed that a *direct* energy cascade is not possible in 2D, because it would violate enstrophy conservation.[32] No conjecture can be found in the literature before 1967 positing any type of 2D power-law cascade range.

Then came Kraichnan. The abstract of his first 1967 paper is worth quoting in full:

Two-dimensional turbulence has both kinetic energy and mean-square vorticity as inviscid constant of motion. Consequently it admits two formal inertial ranges, $E(k) \sim \epsilon^{2/3}k^{-5/3}$ and $E(k) \sim \eta^{2/3}k^{-3}$, where ϵ is the rate of cascade of kinetic energy per unit mass, η is the rate of cascade of mean-square vorticity, and the kinetic energy per unit mass is $\int_0^\infty E(k)\,dk$. The $-\frac{5}{3}$ range is found to entail backward energy cascade, from higher to lower wavenumbers k, together

---

[31] See Taylor, 1917: pp. 76–77; Charney, 1947; von Neumann, 1949: §§ 2.3–2.4; Batchelor, 1953: pp. 186–187; Fjørtoft, 1953.

[32] Lee, 1951.

with zero-vorticity flow. The -3 range gives an upward vorticity flow and zero-energy flow. The paradox in these results is resolved by the irreducibly triangular nature of the elementary wavenumber interactions. The formal -3 range gives a nonlocal cascade and consequently must be modified by logarithmic factors. If energy is fed in at a constant rate to a band of wavenumbers $\sim k_i$ and the Reynolds number is large, it is conjectured that a quasi-steady-state results with a $-\frac{5}{3}$ range for $k \ll k_i$ and a -3 range for $k \gg k_i$, up to the viscous cutoff. The total kinetic energy increases steadily with time as the $-\frac{5}{3}$ range pushes to ever-lower k, until scales the size of the entire fluid are strongly excited. The rate of energy dissipation by viscosity decreases to zero if kinematic viscosity is decreased to zero with other parameters unchanged.

This is followed by a detailed and very dense presentation of the arguments supporting such results. Other than the works already cited and known to Kraichnan, there were no experimental or numerical results which could guide him. Furthermore, he took up the formidable challenge of not using closure, although he stated:

> No use is made of closure approximations. However, the Lagrangian-history direct-interaction approximation, which yields Kolmogorov's similarity cascade in three dimensions, preserves the vorticity constraint in two dimensions and appears to yield the principal dynamical features inferred in the present paper.

Actually, some of the Fourier-space machinery developed by Kraichnan for closure, such as the use of the energy flux $\Pi(k)$ passing through wavenumber k and of the triad contributions to energy transfer $T(k, p, q)$, remains meaningful for homogeneous isotropic turbulence without recourse to closure. Of course, such quantities must then be expressed in terms of triple correlations of the velocity field and not solely in terms of the energy spectrum. Invoking a scaling ansatz for triple correlations, $T(ak, ap, aq) = a^{-(1+3n)/2} T(k, p, q)$, and also for the energy spectrum,[33] $E(k) \propto k^{-n}$, Kraichnan showed that for $n = 5/3$ one obtains an inertial range with zero enstrophy flux and non-vanishing energy flux and that for $n = 3$ one obtains an inertial range with zero energy flux and non-vanishing enstrophy flux.

These would be the famous dual cascades of 2D turbulence if Kraichnan could show that the former range has a negative energy flux, while the latter has a positive enstrophy flux. At this point, to predict the directions of the cascades, he made a very creative use of the equilibrium statistical mechanics of the Galerkin-truncated Euler equations, which he now calls "absolute (statistical) equilibria". In 2D, because of the simultaneous conservation of energy E and enstrophy Ω, it is easy to show that when the Euler equations are

---

[33] As Kraichnan pointed out, the latter is not used in his argument, thus allowing, in present-day language, for anomalous scaling.

Galerkin-truncated to a wavenumber band $[k_{min}, k_{max}]$ with $0 < k_{min} < k_{max}$, the absolute equilibrium $e^{-(\alpha E + \beta \Omega)}/Z$ is Gaussian with an energy spectrum

$$E(k) = \frac{k}{\alpha + \beta k^2}, \tag{10.13}$$

where α and $\beta > 0$ are constrained by the knowledge (say, from the initial conditions) of the total energy $E = \int_{k_{min}}^{k_{max}} E(k)\,dk$ and of the total enstrophy $\Omega = \int_{k_{min}}^{k_{max}} k^2 E(k)\,dk \geq k_{min}^2 E$. If $\Omega/(k_{min}^2 E)$ is very close to unity, it is seen that the 'inverse temperature' α must be negative and that the spectrum (10.13) displays a strong peak near k_{min}. The situation is in stark contrast to the case of 3D absolute equilibria: for 3D parity-invariant flow the absolute equilibrium energy spectrum is proportional to k^2, as we have seen, and always peaks at the highest wavenumber. These results suggested to Kraichnan that "a tendency toward equilibrium in an actual physical flow should involve an upward flow of vorticity and, therefore, by the conservation laws, a downward flow of energy". For good measure, Kraichnan gave two independent arguments to justify the same conclusions. Adapting earlier considerations of Fjørtoft, he noted that these cascade directions follow if the transfer in each triad is "a statistically plausible spreading of the excitation in wavenumber: out of the middle wavenumber into the extremes". Finally, he showed that such diffusive spreading in wavenumber indeed develops instantaneously for an initial Gaussian statistical distribution, applying expressions of W.H. Reid and Ogura for the quasi-normal closure in 2D.[34]

But are these formal similarity ranges physically realizable? Kraichnan next turned to this question. Since a $k^{-5/3}$ range gives a divergent energy at small wavenumbers and the k^{-3} range a logarithmically divergent enstrophy, cascade ranges of arbitrary extent require forcing at some intermediate wavenumber k_i. For finite ranges, Kraichnan noted, there must be some 'leakage' of energy input ε to high wavenumbers and of enstrophy input η to low wavenumbers. As either of the ranges increases in length it becomes a 'purer' cascade, due to the blocking effect of conservation of the dual invariant. The enstrophy cascade will proceed up to the cutoff wavenumber $k_d = (\eta/\nu^3)^{1/6}$ set by viscosity, with a vanishingly small energy dissipation $\varepsilon_d \sim \varepsilon(k_i/k_d)^2$ for $k_d \gg k_i$. If there is no minimum wavenumber k_0, Kraichnan concluded that an inverse cascade should proceed for ever to smaller and smaller wavenumbers which, on dimensional grounds, scale as a $\varepsilon^{-1/2} t^{-3/2}$, where t is the time elapsed. These cascade ranges are only plausible universal states, Kraichnan observed, if the cascade dynamics are scale-local, with the dominant nonlinear interactions among

[34] Reid, 1959; Ogura, 1963.

triads of wavenumbers, all of comparable magnitude. Kraichnan concluded that the 2D inverse cascade $-5/3$ range is local just as is the 3D direct cascade $-5/3$ range of Kolmogorov. The 2D direct enstrophy range is at the margin of locality, however, and must have logarithmic corrections to exact self-similarity. Echoing the earlier ideas of von Neumann, Kraichnan argued that the infinite number of local vorticity invariants in 2D suggests non-universality of spectral coefficients. He then noted:

> A further point is that the nonlocalness of the transfer in the -3 range suggests in itself that [the] cascade there is not accompanied by degradation of the higher statistics in the fashion usually assumed in a three-dimensional Kolmogorov cascade. This is consistent with a picture of the transfer process as a clumping-together and coalescence of similarly signed vortices with the high-wavenumber excitation confined principally to thin and infrequent shear layers attached to the ever-larger eddies thus formed.

This is the only point in the paper where Kraichnan speculates on physical-space mechanisms, clearly influenced by the statistical mechanics argument of Onsager.

Kraichnan considered finally in his 1967 paper the situation that the fluid is confined to a finite domain with a minimum wavenumber $k_0 \ll k_i$. He wrote:

> The conjecture is offered here that after the $-\frac{5}{3}$ range reaches down to wavenumbers $\sim k_0$ the downward cascade from k_i continues and the energy delivered to the bottom of the range piles up in the mode k_0. As the energy in k_0 rises sufficiently, modification of the $-\frac{5}{3}$ range toward absolute equilibrium is expected, starting at the bottom and working up to progressively larger wavenumbers.

This conclusion was supported by considering, once again, the 2D absolute equilibria; for $\Omega/(Ek_0^2)$ close to unity, Kraichnan showed that they have nearly all of the energy carried by the 'gravest' mode k_0. He pointed out an analogy with quantum Bose condensation which, apparently, played a key role in stimulating his whole analysis. Indeed, he wrote in the introduction:

> The present study grew out of an investigation of the approach of a weakly coupled boson gas to equilibrium below the Bose–Einstein condensation temperature. There is a fairly close dynamical analogy in which the number density and kinetic-energy density of the bosons play the respective roles of kinetic-energy density and squared vorticity.

This quantum phenomenon was discussed at length in a prior Kraichnan paper in 1967.[35]

---

[35] Kraichnan, 1967a.

Two very original concepts enter into turbulence theory with Kraichnan's landmark work. The first idea is that a single system may support two co-existing cascades with different spectral ranges: in 2D, the dual cascades of energy and enstrophy. The second idea is that there may be constant flux spectral ranges that correspond to an *inverse cascade*, from small to large scales. Kraichnan's concept of the 2D inverse energy cascade is very far from the Richardson–Kolmogorov vision of 3D turbulence, in which energy, introduced at large scales either through the initial conditions or by suitable forces or by instabilities, cascades to smaller scales and eventually dissipates by viscosity into heat. Two other groups were, however, pursuing ideas very closely related to those of Kraichnan, at this same time. These were G.K. Batchelor and R.W. Bray at Cambridge, UK, and V. Zakharov in the Soviet Union.

The 2D enstrophy cascade was proposed independently of Kraichnan, and even somewhat earlier, by Batchelor. This result is reported in the 1966 Cambridge PhD dissertation of Bray. At the beginning of §1.4 he gave his supervisor, Batchelor, credit for the idea that the enstrophy dissipation rate could remain finite in the limit of vanishing viscosity. This led Bray to suggest an enstrophy cascade with a k^{-1} spectrum and thus a k^{-3} energy spectrum. Bray attempted to check this theory by performing a 2D spectral numerical simulation, probably the first of its kind. These results were not made public, however, until a 1969 paper authored by Batchelor. The analysis of Batchelor and Bray is remarkably complementary to Kraichnan's. They considered only decaying 2D turbulence, not forced steady states. Most of their physical discussion was also in real space, not in spectral space, and focused on the analogy between stretching of vorticity-gradients in 2D and Taylor's vortex-stretching mechanism in 3D. Neither in Bray's thesis nor in Batchelor's paper was there any discussion of a separate $-5/3$ range in 2D with constant flux of energy to large scales (Batchelor was, by 1969, aware of Kraichnan's earlier paper and cited it in his work).[36]

The notion of two distinct power-law ranges already appears in the 1966 PhD thesis of Zakharov, on the weak turbulence of gravity waves. One of these ranges was identified as a direct cascade of energy to high-wavenumber, but the physical interpretation of the other was not clearly identified in the thesis. Shortly afterward, however, Zakharov hit upon the idea that the second power-law range corresponds to an inverse cascade of wave action or "quasi-particle number". It thus appears that Kraichnan and Zakharov arrived independently at the idea of an inverse cascade although Kraichnan, it seems, got the idea slightly earlier into print. Both Kraichnan and Zakharov also clearly realized

---

[36] Bray, 1966; Batchelor, 1969.

the applicability of these notions of dual cascades and inverse cascade to many other systems, including quantum dynamics of Bose condensation.[37]

The subject of 2D turbulence continued to interest Kraichnan until the early 1980s, at least, and he wrote several later papers which sharpened the predictions and clarified the physics of his 1967 theory. In a 1971 *J. Fluid Mech.* article he applied his TFM closure to both 2D inverse and 3D direct energy cascades and obtained quantitative results on spectral coefficients. Kraichnan also studied the 2D enstrophy cascade and, in particular, worked out the logarithmic correction mentioned in 1967. A 1976 paper in *J. of Atmosph. Sci.* disseminated these results to the community of meteorologists, with special attention to the phenomenological concept of 'eddy viscosity'. Kraichnan proposed there a new interpretation of the inverse cascade in terms of a negative eddy viscosity, an idea that goes back to V. Starr, and he gave a very simple heuristic explanation for this effect:

> If a small-scale motion has the form of a compact blob of vorticity, or an assembly of uncorrelated blobs, a steady straining will eventually draw a typical blob out into an elongated shape, with corresponding thinning and increase of typical wavenumber. The typical result will be a decrease of the kinetic energy of the small-scale motion and a corresponding reinforcement of the straining field.

This idea has been particularly influential in the geophysical literature, where it has often been invoked to explain inverse energy cascade.[38]

What is the empirical status of Kraichnan's dual cascade theory of 2D turbulence? A complete review would be out of place here, but we shall briefly discuss its verification with an emphasis on the most current work. Only very recently, in fact, has it become possible to observe both 2D cascades, inverse energy and direct enstrophy, in a single simulation. This has required herculean computations at spatial resolutions up to $32{,}768^2$ grid points. The simulations confirm Kraichnan's predictions for the $k^{-5/3}$ and k^{-3} ranges, with less accuracy for the latter due to finite-range effects. Earlier numerical simulations and laboratory experiments which have focused on a single range have, however, separately confirmed the predictions of the 1967 paper. A number of numerical studies of the enstrophy cascade with 'hyperviscosity' (powers of the Laplacian replacing the usual dissipative term) have reported observing the log-correction to the energy spectrum. The quasi-steady inverse cascade

---

[37] Zakharov, 1966. B. Kadomtsev informed Zakharov of Kraichnan's 2D paper sometime around 1969 (V. Zakharov, private communication, 2010).

[38] For Kraichnan's application of TFM closure to 2D turbulence cascades, see Kraichnan, 1971b, 1976c; for negative viscosity, see Starr, 1968; Kraichnan, 1976c; Kraichnan & Montgomery, 1980: §4.4; for vortex-thinning mechanism of inverse cascade, see Kraichnan, 1976c; Rhines, 1979; Salmon, 1980, 1998: p. 229. Another notable paper on 2D turbulence is Kraichnan, 1975b.

predicted by Kraichnan as a transient before energy from pumping reaches the largest scales has also been observed in both experiments and simulations, first by L. Smith and V. Yakhot. The constant flux $k^{-5/3}$ range is cleanly observed. However, contrary to Kraichnan's speculations in his 1967 paper, the cascade is not associated with 'coalescence' of vortices: indeed, the statistics of the velocity are quite close to Gaussian, and strong, coherent vortices do not appear until the energy begins to accumulate at the largest scales. Experiments and simulations on the statistical steady-state have instead found considerable evidence for Kraichnan's 1976 'vortex-thinning' mechanism of energy transfer even in the local cascade regime. In the situation without large-scale damping, there is 'energy condensation' at large scales as Kraichnan had supposed, but not confined to the gravest mode. Recent simulations show that condensation in a periodic domain appears as a pair of large, counterrotating vortices with a k^{-3} spectrum. These vortices are close to what is predicted by an equilibrium, maximum-entropy argument although the system is non-equilibrium, with continuously growing energy and constant negative energy flux.[39]

Perhaps the most interesting question, from the general scientific point of view, is the relevance of Kraichnan's ideas to planetary atmospheres and oceans. This question is complicated by the limited scale ranges that exist in those systems and the greater complexity of the dynamics. However, several recent observational studies have found evidence for both inverse energy and direct enstrophy cascades in the Earth's atmosphere and oceans.[40]

10.4 Intermittency

Intermittency is a rather general term referring to the spottiness of small-scale turbulent activity, be it at dissipation-range scales or at inertial-range scales. In the late 1940s Batchelor and A.A. Townsend observed intermittent behavior of low-order velocity derivatives; since such derivatives come predominantly from the transition region between the inertial and dissipation ranges, this inter-mittency cannot be directly taken as evidence that the self-similarity postulated for the K41 inertial range is breaking down.[41]

[39] For numerical simulation of simultaneous cascades, see Boffetta, 2007; Boffetta & Musacchio, 2010; for the log-correction in the enstrophy cascade, see Borue, 1993; Gotoh, 1998; Pasquero & Falkovich, 2002; for quasi-steady cascade, see Smith & Yakhot, 1993; for vortex thinning, see Chen et al., 2006; Xiao et al. 2008; for condensations, see Chertkov et al., 2007; Bouchet & Simonnet, 2009.

[40] For direct enstrophy cascade in the Earth's stratosphere, see Cho & Lindborg, 2001; for inverse energy cascade in the South Pacific, see Scott & Wang, 2005.

[41] Batchelor & Townsend, 1949.

10.4.1 Dissipation-range intermittency

Kraichnan was the first to explain intermittency in the far dissipation range or, equivalently, for high-order velocity derivatives. In slightly modernized form, his argument is as follows: suppose that the flow can be divided into macroscopic regions each having its energy dissipation rate ε and its energy spectrum $E(k)$. In the far dissipation range $E(k)$ falls off faster than algebraically. From DIA results or from von Neumann's analyticity conjecture regarding solutions of the Navier–Stokes equations, Kraichnan expected a fall-off proportional to $\exp(-k/k_d)$, where the dissipation wavenumber k_d is given, at least approximately, by the K41 expression $(\varepsilon/\nu^3)^{1/4}$. When $k \gg k_d$, even minute macroscopic fluctuations in ε, which are very likely as pointed out by L.D. Landau, will produce huge macroscopic fluctuations in $E(k)$ and thus strong intermittency in physical-space filtered velocity signals obtained by keeping only those Fourier coefficients which are in a high-k-octave of wavenumbers. This argument can be made more systematic by using singularities of the analytic continuation of the velocity field to complex space-time locations.[42]

10.4.2 Inertial-range intermittency

Much more difficult is the issue of intermittency at inertial-range scales and the problem of *anomalous scaling*: that is, scaling for which the exponents cannot be obtained by a dimensional argument, as in K41. In the early 1960s, A.M. Obukhov and his advisor Kolmogorov began to suspect that K41 must be somewhat modified because spatial averages ε_r of the local energy dissipation over balls with a radius r staying within the inertial range appeared to fluctuate more and more when r is decreased; they proposed a log-normal model of intermittency.[43] E.A. Novikov and R. Stewart and then A.M. Yaglom constructed *ad hoc* random multiplicative models to capture such intermittency and the corresponding scaling exponents. Mandelbrot showed that, in these models, the dissipation is taking place on a set with non-integer *fractal* dimension; in general such models are actually *multifractal*.[44]

For Kraichnan, who liked proceeding in a systematic way, keeping as much contact as possible with the true fluid-dynamical equations, inertial-range intermittency was a very difficult problem. Indeed, it had been known since 1966

---

[42] Kraichnan, 1967c, 1974a: p. 327; von Neumann, 1949; for complex singularities, see Frisch & Morf, 1981.

[43] See Chapter 6.

[44] Obukhov, 1962; Kolmogorov, 1962; Novikov & Stewart, 1964; Yaglom, 1966; for a review of the Russian work, see Chapter 6; Mandelbrot, 1968, 1974; for multifractality, see Parisi & Frisch, 1985. B. Mandelbrot, died on 14th October 2010.

that the full hierarchy of moment or cumulant equations derived for statistical solutions of the Navier–Stokes equation is compatible with the scale-invariant K41 theory in the limit of infinite Reynolds numbers. But Kraichnan was also aware that K41 is equally compatible with the Burgers equation, which definitely has no K41 scaling (because of the presence of shocks); he also noticed that the presence of the pressure in the incompressible Navier–Stokes was likely to reduce the intermittency one would otherwise expect from a simple vortex-stretching argument. Closure seemed incapable of saying anything about the breaking of the K41 scale-invariance (one major exception to this statement is discussed in Section 10.4.3).[45]

At first Kraichnan examined critically the toy models developed by the Russian school, observed that ε_r is not a pure inertial-range quantity and proposed to study intermittency in terms of more appropriate quantities, such as the local fluctuations of the energy flux associated with a wavenumber k in the inertial range. An estimate of this flux is u_r^3/r, where $r \sim 1/k$ and u_r is, say, the modulus of the velocity difference between two points separated by a distance r. With this in mind, he wrote:

> If we increase the intermittency by making the fluid into quiescent regions with negligible velocity and active regions, of equal extent, where u_r increases by $\sqrt{2}$, then the mean kinetic energy in scales order r is unchanged but the time constant decreases, and hence ε increases, by $\sqrt{2}$. This example suggests, first, that if Kolmogorov's theory holds in subregions of the fluid, then the constant $f(0)$ in the inertial-range law can be universal only if intermittency in the local dissipation ε_r, defined as average dissipation over a domain of size r, somewhat tends to a universal distribution. Second, if intermittency increases as scale size decreases, and Kolmogorov's basic ideas hold in local regions, then the cascade becomes more efficient as r decreases and $E(k)$ must fall off more rapidly than $k^{-5/3}$ if, according to conservation of energy, the overall cascade rate is r independent.[46]

A few years later this remark, together with ideas of Mandelbrot, became a key ingredient in the development of the β-model, a phenomenological model of intermittency that uses exclusively inertial-range quantities.[47]

Kraichnan pursued some of these ideas further himself in an influential 1974 paper in *J. Fluid Mech.* This paper is pure Kraichnan. A wealth of intriguing ideas are tossed out, very original model calculations sketched in brief, and clever counterexamples devised against conventional ideas. At least two contributions of this paper are now well known. First, Kraichnan proposed a

---

[45] For K41 compatibility of the Navier–Stokes equations, see Orszag & Kruskal, 1966; for the differences between Burgers and Navier–Stokes turbulence, see Kraichnan, 1974a, 1991.

[46] Kraichnan, 1972: p. 213.

[47] For using inertial-range quantities, see Kraichnan, 1972: p. 213, 1974a; Frisch, Sulem & Nelkin, 1978.

refined similarity hypothesis (RSH) alternative to that of Kolmogorov, which he based on inertial-range energy flux rather than volume-averaged energy dissipation. Later numerical and experimental studies have confirmed Kraichnan's RSH (and also that of Kolmogorov). In the same paper, Kraichnan gave what is now the standard formulation of the 'Landau argument' on intermittency and non-universality of coefficients in scaling laws. His argument is considerably clearer and more compelling than the brief remarks originally made by Landau in 1942. The crucial observation of Kraichnan is that only those K41 predictions which are linear in the ensemble-average energy dissipation $\langle \varepsilon \rangle$ – such as the Kolmogorov 4/5-th law – can be expected to be universally valid inertial-range laws. Other K41 predictions that depend upon fractional powers $\langle \varepsilon \rangle^p$ are not invariant under composition of sub-ensembles with distinct global values of mean-dissipation. There is at least as much Kraichnan in this argument as there is Landau.[48]

10.4.3 Passive scalar intermittency and the 'Kraichnan model'

The story of the Kraichnan model and of the birth of the first *ab initio* derivation of anomalous scaling is rather complex, spanning nearly three decades. Since it is understood that in this book the emphasis should be on what happened before 1980, we shall concentrate on the early developments, that begin in the late 1960s.

The transport of a passive scalar field (say a temperature field $T(x,t)$), advected by a prescribed incompressible turbulent velocity field $u(x,t)$ and subject to molecular diffusion with a diffusivity κ, is governed by the following *linear* stochastic equation:

$$\left[\partial_t + u(x,t) \cdot \nabla_x - \kappa \nabla_x^2 \right] T(x,t) = 0. \qquad (10.14)$$

The qualification 'passive' is used when there is no or negligible back-reaction on the turbulent flow of the field being transported. Examples of passive scalar transported fields are provided by the temperature of a fluid (when buoyancy is negligible), the humidity of the atmosphere, the concentration of chemical or biological species. Passive scalar transport has thus an important domain of applications and considerable efforts were made since the 1940s to gain an understanding at least as good as for turbulence dynamics. In particular Obukhov and, independently, S. Corrsin derived for passive scalars the counterpart of the $-5/3$ law, Yaglom derived an analog of the Kolmogorov's 4/5 law and Batchelor derived a k^{-1} passive scalar energy spectrum in a regime of fully developed

---

[48] For Landau's argument, see Landau & Lifshitz, 1987; Kraichnan, 1974a: § 2; Frisch, 1995: § 6.4.

turbulence with large Schmidt number (see below). It was thus quite natural for Kraichnan to see how well the closure tools he developed for turbulence in the 1950s and the 1960s were able to cope with passive scalar dynamics. He applied his LHDIA closure to the passive scalar problem, for example, reproducing Obukhov–Corrsin scaling with precise numerical coefficients.[49]

In 1968 Kraichnan realized that a closed equation can be obtained for a scalar field passively advected by a turbulent velocity with a very short correlation time, without any further approximation. The DIA closure is exact for this special system, reducing to a single equation for the scalar correlation function at two space points and simultaneous times. The mean Green function reduces to a Dirac delta because of the zero correlation-time assumption. This 1968 model is now usually called 'the Kraichnan model' [of passive scalar dynamics] and has assumed a paradigmatic status for turbulence theory, comparable to that of the Ising model in the statistical mechanics of critical phenomena. Its importance stems from a string of major discoveries by Kraichnan and others on the fundamental mechanism of intermittency, some of which will be described only briefly because they took place in the 1990s. Kraichnan showed that even when the velocity field is not at all intermittent, e.g. a Gaussian random field, the passive scalar (henceforth called 'temperature' for brevity) can become intermittent and this in several ways.[50]

A first mechanism, which applies in the far dissipation range, is basically the same as described in § 10.4.1 and will not concern us further.

A second mechanism identified by Kraichnan concerns the so-called Batchelor regime: when the Schmidt number v/κ is large, there is a range of scales for which the velocity field is strongly affected by viscous dissipation, but the temperature field does not undergo much diffusion; in this regime the velocity field can be locally replaced by a uniform random shear.[51] Tiny, well-separated temperature blobs are then stretched and squeezed in a way which is amenable to asymptotic analysis at large times. Actually doing this in a systematic way would have required all kinds of heavy-duty theoretical tools: path integrals, large deviation theory, fluctuations of Lyapunov exponents, etc.[52] It is then possible to show that the distributions of spatial derivatives of the temperature display a log-normal-type intermittency at zero diffusivity[53] and a weaker

---

[49] Obukhov, 1949; Corrsin, 1951; Yaglom, 1949; Batchelor, 1959; see also Chapters 8, 7 and 6. On LHDIA for passive scalars, see Kraichnan, 1965a: § 5–7.

[50] Kraichnan, 1968b, 1974b, 1994.

[51] Batchelor, 1959, and Chapter 8.

[52] See, e.g., the review by Falkovich, Gawędzki & Vergassola, 2001.

[53] This is a nontrivial variant of the obvious result that when $m(t)$ is a scalar Gaussian random function the solution of the differential equation $dq(t)/dt = m(t)q(t)$ with $q(0) = 1$ is log-normal.

form of intermittency in the regime with non-zero diffusivity. Actually, all this was done – and correctly so – by Kraichnan in a remarkable paper published in 1974, just after the paper on Kolmogorov's inertial-range theories.[54] This paper is a tour de force, combining very original analytical arguments and deep physical intuition to reach exact conclusions, without any assistance from the advanced mathematical methods that were later applied to this problem. Kraichnan's analysis was carried out for general space dimension d – following a suggestion of M. Nelkin – and one intriguing finding was that intermittency of the scalar vanished in the limit $d \to \infty$. Kraichnan's work, which was going to strongly influence subsequent, more formally rigorous, analyses, showed a thorough understanding of the mechanism of intermittency in the Batchelor regime.

The third mechanism identified by Kraichnan was rather close to one of the Holy Grails of turbulence theory, namely understanding inertial-range anomalous scaling and predicting the scaling exponents. In 1994 Kraichnan conjectured that when the velocity $u(x, t)$ is Gaussian with a power-law spectrum (K41 would be one instance) and with a very short correlation time (white-in-time), then for vanishingly small κ the structure functions of the temperature display anomalous scaling. This is a rather amazing proposal: how can a self-similar velocity field act on a transported temperature field to endow it with anomalous scaling and thus with lack of self-similarity? As we shall see, the qualitative aspects of Kraichnan's conjecture have been fully corroborated by later work.[55]

Now we shall have to become slightly more technical to explain how Kraichnan tackled this problem, starting with his 1968 work. Let us rewrite the temperature equation (10.14) in abstract form

$$\partial_t T(t) = MT(t) + \tilde{M}(t)T(t), \qquad (10.15)$$

where M is a linear deterministic operator (diffusion) and $\tilde{M}(t)$ a linear random operator (advection) with vanishing mean and 'very short correlation time'. More precisely, one performs the substitution

$$\tilde{M}(t) \longrightarrow \frac{1}{\epsilon}\tilde{M}\left(\frac{t}{\epsilon^2}\right), \qquad \epsilon \to 0, \qquad (10.16)$$

where $\tilde{M}(t)$ is statistically stationary. In the limit $\epsilon \to 0$ the temperature becomes a Markov process (in the time variable) and it may be shown that the

---

[54] Kraichnan, 1974b.
[55] Kraichnan, 1994.

mean temperature satisfies a *closed* equation, namely[56]

$$\partial_t \langle T \rangle = M \langle T(t) \rangle + \mathcal{D} \langle T(t) \rangle, \tag{10.17}$$

$$\mathcal{D} = \int_0^\infty \langle \tilde{M}(s)\tilde{M}(0) \rangle \, ds. \tag{10.18}$$

Similar closed equations can be derived for p-point moments of the temperature. In 1968 Kraichnan derived the equation for the two-point temperature correlation functions using this technique and found that the second-order temperature structure functions displayed scaling. The scaling exponent ζ_2 can actually be obtained by simple dimensional analysis. So far no evidence of anomalous scaling had emerged.

By a method similar to that used in 1968 for the two-point correlations of a passive scalar, Kraichnan derived in 1994 an equation for the structure function of order p. This equation is not closed (contrary to the equation for the p-point correlation function), but Kraichnan proposed a plausible approximate closure ansatz from which he derived the following scaling exponents ζ_p for the pth-order structure function:

$$\zeta_{2p} = \frac{1}{2}\sqrt{4pd\zeta_2 - 2 + (d - \zeta_2)^2} - \frac{1}{2}(d - \zeta_2). \tag{10.19}$$

Since ζ_{2p} is obviously not equal to $p\zeta_2$, as would be required by self-similarity, (10.19) implies anomalous scaling. One year later it was shown that there is indeed anomalous scaling, using a *zero modes* method, borrowed partially from field theory: the equation for the moments of order $2p$ has a linear operator L_{2p} acting on the $2p$-point correlation function and an inhomogeneous right-hand side involving correlation functions of lower order. The zero modes correspond to certain functions of $2p$ variables which are killed by L_{2p}. Actually determining the zero modes turned out to be quite difficult. In most instances it could be done only perturbatively, using as small parameter either the roughness exponent ξ of the prescribed velocity field or the inverse of the dimension of space d (as anticipated by Kraichnan's 1974 paper). The results agreed with numerical simulations, but did not agree with (10.19) except for a single value $\xi \doteq 1$. Kraichnan's prediction (10.19) must cross the numerical curve at one point, trivially, but it is possibly significant that Kraichnan's closure ansatz works best in the regime where the cascade dynamics is scale-local.[57]

---

[56] Hashminskii, 1966; Frisch & Wirth, 1997.

[57] Kraichnan, 1994. For zero-mode methods, see Gawędzki & Kupiainen, 1995; Chertkov et al., 1995; Shraiman & Siggia, 1995; and the review by Falkovich, Gawędzki & Vergassola, 2001. For simulations, see Frisch, Mazzino & Vergassola, 1998; Gat, Procaccia & Zeitak, 1998; Frisch et al., 1999. The whole story about anomalous scaling for passive scalars is recounted in www.oca.eu/etc7/work-on-passive-scalar.pdf.

10.5 Miscellany and conclusions

10.5.1 Scattering of sound by turbulence

In 1952 M.J. Lighthill published a landmark paper on the generation of sound by turbulence. The next year Kraichnan observed that the production of noise in this theory depends on a high power of the Mach number and that the "*scattering* [of sound by turbulence] is the most conspicuous acoustical phenomenon associated with very low Mach number turbulence". In his very first published paper, Kraichnan developed a systematic theory of the interaction of sound with nearly incompressible turbulence. This paper, together with further developments, was to be the basis of a non-intrusive ultrasonic technique for the remote probing of vorticity. The same year, 1953, and independently, Lighthill also published a theory of scattering.[58]

In his approach to the problem of interaction of sound and turbulence, Kraichnan assumed that the turbulence is incompressible and can be described by a divergence-free velocity, whereas the sound is given by a curl-free (potential) velocity. As done by Lighthill, Kraichnan assumed that density and pressure fluctuations are related by an adiabatic equation of state with a uniform speed of sound. Starting with the full compressible equations he performed a decomposition of the velocity

$$u = u^L + u^T \tag{10.20}$$

into a curl-free (longitudinal in the spatial Fourier space) and a divergence-free (transverse in Fourier space) part. (This is known as a Hodge decomposition in mathematics.) He then obtained a wave equation which has four terms. One term is linear in u^L, related to viscous stresses and is mostly negligible. The three remaining ones are quadratic and of type L–L, L–T and T–T. The T–T term is Lighthill's (quadrupolar) sound production term. The L–T term gives the scattering of a pre-existing sound wave by the turbulence. Kraichnan then worked out the angular distribution and frequency distribution of the scattered wave in terms of the four-dimensional Fourier transform of the shear velocity field. Explicit expressions for cross-sections were obtained for the case of a scattering from a region of isotropic turbulence.

Some remarks are in order. Kraichnan worked in relativity and quantum field theory for several years before engaging in hydrodynamics but this first published paper is about hydrodynamics;[59] the turbulent flow is here prescribed and defined as "characterized by the fact that although the detailed structure

---

[58] Lighthill, 1952, 1953; Kraichnan, 1953; for further developments, see, e.g., Lund & Rojas, 1989; Ting & Miksis, 1990; for vorticity probing, see, e.g. Baudet, Ciliberto & Pinton, 1991.

[59] His first relativity paper was to be published only two years later (Kraichnan, 1955b).

of the system is not known, suitable averages of certain quantities are known for a representative ensemble of similar systems". The article is unusually well written for a first paper and indicates considerable maturity of the young scientist who had already been active for six years, although he refrained from publishing.

10.5.2 High-Rayleigh number convection

Thermal convection is ubiquitous in technology and is amenable to controlled experiments where a fluid heated from below is placed between two horizontal plates. Within the so-called Boussinesq approximation, the dimensionless parameters are the Rayleigh number $Ra \equiv g\alpha\delta T h^3/(\nu\kappa)$ and the Prandtl number $Pr \equiv \nu/\kappa$. Here, g is the acceleration due to gravity, δT the vertical temperature difference across the fluid of height h, and α, ν and κ are the thermal expansion coefficient of the fluid, its kinematic viscosity and its thermal diffusivity. Turbulent thermal convection was and remains a central topic of the Woods Hole Oceanographic Institute Geophysical Fluid Dynamics summer program, with which Kraichnan had considerable interaction from the late 1950s. Around the same time he also had much interaction with E.A. Spiegel, who had been trained in astrophysical fluid dynamics: it is usually convective transport which allows the heat generated in the interiors of stars to escape. In the early days the easiest way to model astrophysical convection was through the mixing length theory, which follows ideas of Boussinesq and of Prandtl. In 1962 Kraichnan devoted a fairly substantial paper to thermal convection, which we cannot summarize in detail because of lack of space. We shall thus concentrate on his most orignal contribution, to what is now called 'ultimate convection', at extremely high Rayleigh numbers.[60]

One important question in high-Rayleigh number convection is the dependence upon Rayleigh and Prandtl numbers of the Nusselt number N, the heat flux non-dimensionalized by the conductive heat flux. In the 1950s C.H.B. Priestley found a dimensional argument which suggests that for high Rayleigh numbers $N \propto Ra^{1/3}$. As pointed out by Kraichnan

> [In] Priestley's theory . . . it is assumed that at sufficiently high Rayleigh numbers most of the change in mean temperature across the layer occurs in thin boundary regions, at the surfaces, where molecular heat conduction and and molecular viscosity are dominant. Elsewhere . . . convective heat transport and eddy viscosity are dominant.[61]

---

[60] Boussinesq, 1870; Prandtl, 1925; Kraichnan, 1962.
[61] Priestley, 1959, and references therein; Kraichnan, 1962: p. 1374.

However, at sufficiently high Rayleigh numbers the thermal boundary layer may be destroyed and another regime may emerge, which, as shown by Kraichnan, has an approximately square-root dependence on Ra. This can be partially derived by a simple dimensional argument due to Spiegel, which assumes that the heat flux depends neither on the viscosity nor on the thermal diffusivity and which gives[62] $N \sim (RaPr)^{1/2}$. Kraichnan's derivation makes use of the phenomenological theory of high-Reynolds number shear flow turbulence near a solid boundary, which gives a logarithmic correction proportional to $(\ln Ra)^{-3/2}$. Kraichnan also discussed the Prandtl number dependence of the various regimes. This allowed him to predict how high the Rayleigh number should be for the square-root regime to dominate: for a unit Prandtl number this threshold is around $Ra = 10^{21}$. It is generally believed that the threshold is significantly lower and depends on the Prandtl number and the boundary conditions. Successful attempts to observe this law may have been made with helium gas. Artefacts masquerading as a $Ra^{1/2}$ law cannot be ruled out. In Göttingen a two-meter high convection experiment using sulfur hexafluoride (SF6) at 20 times atmospheric pressure is under construction to try to capture Kraichnan's ultimate convection regime.[63]

10.5.3 Kraichnan and computers

Kraichnan, although basically a theoretician, was very far from being allergic to computers. Actually, not only was he a very talented programmer, but he got occasionally involved in writing system software and even in modifying hardware. Some of his closest collaborators, particularly S.A. Orszag, prodded by him, got deeply involved in three-dimensional simulations of Navier–Stokes turbulence. This was – and still is – called "direct numerical simulation" (DNS) because the original goal was to check on the validity of various closures by going *directly* to the fluid dynamical equations. Convinced that many features of high-Reynolds number turbulence should be universal, Kraichnan encouraged the use of the simplest type of boundary conditions (periodic) which allows the simple and efficient use of spectral methods. He also suggested using Gaussian initial conditions rather than more realistic ones. Curiously, although the thrust to do DNS started just after Kraichnan's discovery of the 2D inverse cascade, he strongly recommended focusing on 3D flow.

Considerable effort – this time often in collaboration with J.R. Herring – went into the numerical integration of various closure equations. Kraichnan

---

[62] It may be shown that this argument breaks down at large Prandtl numbers.

[63] Spiegel, 1971. For artefacts, see Sreenivasan (private communication, 2010). For helium gas experiments, see Chavanne et al., 2001; Niemela et al., 2000. For SF6, see http://www.sciencedaily.com/releases/2009/12/091203101418.htm

proposed using discrete wavenumbers in geometric rather than arithmetic progression. This allowed reaching very high Reynolds numbers. Kraichnan himself was actively involved in writing code for these investigations. His punched cards were shipped from New Hampshire to NASA Goddard Institute where the machine computations were performed during the 1960s and 1970s. Herring recalls that "Bob's programs very rarely contained any bugs".

10.5.4 Conclusions

Our survey has focused on three of Kraichnan's contributions to turbulence theory:

(1) spectral closures and realizability;
(2) inverse cascade of energy in 2D turbulence;
(3) intermittency of passive scalars advected by turbulence.

These are, arguably, his most significant achievements which have had the greatest impact on the field. Spectral closures of the DIA class still have numerous interesting applications when the questions under investigation do not depend crucially on deviations from K41. Even today an EDQNM calculation, for example, will often be the first line of assault on a difficult new turbulence problem. Furthermore, Kraichnan's criterion of realizability has become part of the standard toolbox of turbulence closure techniques. Realizability is necessary both for physical meaningfulness and, often, for successful numerical solution of the closure equations. Kraichnan's prediction of inverse cascade has been well verified by experiments and simulations and has relevance in explaining dynamical processes in the Earth's atmosphere and oceans. The concept of an inverse cascade has proved very fruitful in other systems, too, where similar fluxes of invariants to large scales may occur, such as magnetic helicity in 3D MHD turbulence, magnetic potential in 2D MHD turbulence, and particle number in quantum Bose systems. Finally, Kraichnan's model of a passive scalar advected by a white-in-time Gaussian random velocity has become a paradigm for turbulence intermittency and anomalous scaling – an 'Ising model' of turbulence. The theory of passive scalar intermittency has not yet led to a similar successful theory of intermittency in Navier–Stokes turbulence. However, the Kraichnan model has raised the scientific level of discourse in the field by providing a nontrivial example of a multifractal field generated by a turbulence dynamics. It is no longer debatable that anomalous scaling is *possible* for Navier–Stokes.[64]

---

[64] For the inverse magnetic helicity, see Frisch et al., 1975. For the inverse magnetic potential cascade, see Fyfe & Montgomery, 1976. For Bose condensates, see Semikoz & Tkachev, 1997.

A review of Kraichnan's scientific legacy within the length constraint of this book must be very selective. For example, we have not been able to discuss the numerous interactions Kraichnan had with many people in the USA and in other countries, particularly in France, Israel and Japan. Even focusing our discussion to his turbulence research prior to 1980, we have been forced to omit mention of a large number of problems to which Kraichnan made important contributions in that period. Furthermore, Kraichnan invested substantial time in other theoretical approaches that came after the 1980 cut-off for this chapter. We provide below just a few references to this additional work on turbulence by Kraichnan.[65] It may be that later generations will find that our survey has missed some of Kraichnan's most significant accomplishments. The richness of his œuvre can only be appreciated by poring, for oneself, over his densely written research articles, bristling with original ideas and novel methods. The reader who does so will be generously rewarded for his effort.

It is amusing to wonder what might be Einstein's assessment (from the welkins) of his former assistant. He would probably have to conclude that Kraichnan had a lot of *Sitzfleisch*.[66] Kraichnan's papers, numbering more than a hundred and spanning five decades, many of them formidably technical, bear ample witness to their author's iron determination and staying power. Turbulence is a dauntingly difficult subject where any significant advance is won by a hard-fought battle; and yet Kraichnan has left his record of victories throughout the field. Several international conferences held in his honor are testimony to the lasting impact of Robert H. Kraichnan.[67]

Regarding intermittency/anomalous scaling, note that there have been many incorrect 'proofs' of their absence in Navier–Stokes turbulence, to which the Kraichnan model is a counterexample; see, e.g., Belinicher & L'vov, 1987.

[65] Concerning the various topics that could not be covered in this review, see, for pressure fluctuations: Kraichnan, 1956a, 1956b, 1957a; for shear flow turbulence: Kraichnan, 1964a; for magnetic dynamo: Kraichnan & Nagarajan, 1967; Kraichnan, 1976a, 1976c, 1979b; for Vlasov plasma turbulence: Kraichnan & Orszag, 1967c; for predictability and error growth: Kraichnan, 1970b; Kraichnan & Leith, 1972; for Burgers: Kraichnan, 1968a, 1999; Kraichnan & Gotoh, 1993; for quantum turbulence: Kraichnan, 1967a; for path-integrals: Kraichnan, 1958a: § 4.3; Lewis & Kraichnan, 1962; for self-consistent Langevin models: Kraichnan, 1970c; for variational approaches: Kraichnan, 1958a: § 4.3; Kraichnan, 1979a; for Wiener chaos expansions: Kraichnan, 1979a; for Padé approximants: Kraichnan, 1968c, 1970a; for decimation: Kraichnan, 1985, 1988; for mapping closure: Kraichnan et al., 1989, Kraichnan, 1991; Kimura & Kraichnan, 1993; Kraichnan & Gotoh, 1993; for a critique of Tsallis statistics for turbulence: Gotoh & Kraichnan, 2004.

[66] The Germans have aptly called *Sitzfleisch* the ability to spend endless hours at a desk doing grueling work. *Sitzfleisch* is considered by mathematicians to be a better gauge of success than any of the attractive definitions of talent with which psychologists regale us from time to time (Gian-Carlo Rota, 1996: p. 64).

[67] Los Alamos (May 1998) for Kraichnan's 70th birthday and Santa Fe (May 2009) and Beijing (September 2009) after he left us in 2008.

Acknowledgements Many have helped us with their remarks and their own recollections. We are particularly indebted to B. Castaing, C. Connaughton, G. Falkovich, H. Frisch, T. Gotoh, J.R. Herring, C.E. Leith, H.K. Moffatt, S. Nazarenko, S.A. Orszag, I. Procaccia, H. Rose, E.A. Spiegel, K. Sreenivasan, B. Villone and V. Zakharov. GE's work was partially supported by NSF Grant Nos. AST–0428325 & CDI–0941530 and UF's by COST Action MP0806 and by ANR 'OTARIE' BLAN07-2_183172.

References

Abe, R. 1973. Expansion of a critical exponent in inverse powers of spin dimensionality, *Prog. Theor. Physics* **49**, 113–128.

Alfvén, H. 1942. Existence of electromagnetic-hydrodynamic waves, *Nature* **150**, 405–406.

Batchelor, G.K. 1950. On the spontaneous magnetic field in a conducting liquid in turbulent motion, *Proc. Roy. Soc. Lond. Ser. A* **201**, 405–416.

Batchelor, G.K. 1953. *The Theory of Homogeneous Turbulence*, Cambridge University Press, Cambridge, UK.

Batchelor, G.K. 1959. Small-scale variation of convected quantities like temperature in turbulent fluid. Part 1. General discussion and the case of small conductivity. *J. Fluid Mech.* **5**, 113–133.

Batchelor, G.K. 1969. Computation of the energy spectrum in homogeneous two-dimensional turbulence, *Phys. Fluids Suppl. II* **12**, 233–239.

Batchelor, G.K. & Townsend, A.A. 1949. The nature of turbulent motion at large wavenumbers. *Proc. R. Soc. Lond. A* **199**, 238–255.

Baudet, C., Ciliberto, S. & Pinton, J.F. 1991. Spectral analysis of the von Kármán flow using ultrasound scattering, *Phys. Rev. Lett.* **67**, 193–195.

Belinicher, V.I. & L'vov, V.S. 1987. The scale-invariant theory of developed hydrodynamic turbulence, *Zhurn. Eksp. Teor. Fiz.* **93**, 533–551.

Boffetta, G. 2007. Energy and enstrophy fluxes in the double cascade of two-dimensional turbulence, *J. Fluid Mech.* **589**, 253–260.

Boffetta, G. & Musacchio, S. 2010. An update of the double cascade in two-dimensional turbulence, *Phys. Rev. E* **82**, 016307.

Borue, V. 1993. Spectral exponents of enstrophy cascade in stationary two-dimensional homogeneous turbulence, *Phys. Rev. Lett.* **71**, 3967–3970.

Bouchet, F. & Simonnet, E. 2009. Random changes of flow topology in two-dimensional and geophysical turbulence, *Phys. Rev. Lett.* **102**, 094504.

Boussinesq, J. V. 1870. Essai théorique sur les lois trouvées expérimentalement par M. Bazin pour l'écoulement uniforme de l'eau dans les canaux découverts, *C. R. Acad. Sci. Paris* **71**, 389–393.

Bray, R.W. 1966. A study of turbulence and convection using Fourier and numerical analysis, PhD dissertation available in the library of the Department of Applied Mathematics and Theoretical Physics, Cambridge University, Cambridge [all chapters but the last on thermal convection are also available at `http://www.oca.eu/etc7/bray-phd1966.pdf`].

Chandrasekhar, S. 1955. A theory of turbulence, *Proc. R. Soc. Lond. A* **229**, 1–19.

Charney, J.G. 1947. The dynamics of long waves in a baroclinic westerly current, *J. Meteor.* **4**, 135–163.

Chavanne, X., Chillà, F., Chabaud, B., Castaing, B. & Hébral, B. 2001. Turbulent Rayleigh–Bénard convection in gaseous and liquid He, *Phys. Fluids* **13**, 1300–1320.

Chen, S., Ecke, R.E., Eyink, G.L., Rivera, M., Wan, M.-P., & Xiao, Z. 2006. Physical mechanism of the two-dimensional inverse energy cascade, *Phys. Rev. Lett.* **96**, 084502

Chertkov, M., Connaughton, C, Kolokolov, I. & Lebedev, V. 2007. Dynamics of energy condensation in two-dimensional turbulence, *Phys. Rev. Lett.* **99**, 084501.

Chertkov, M., Falkovich, G., Kolokolov, I. & Lebedev, V. 1995. Normal and anomalous scaling of the fourth-order correlation function of a randomly advected passive scalar, *Phys. Rev. E* **52**, 4924–4941.

Cho, J.Y.N. & Lindborg, E. 2001. Horizontal velocity structure functions in the upper troposphere and lower stratosphere 1. Observations, *J. Geophys. Res.* **106**, 10223–32.

Cichowlas, C., Bonaïti, C., Debbasch, F. & Brachet, M. 2005. Effective dissipation and turbulence in spectrally truncated Euler flows, *Phys. Rev. Lett.* **95**, 264502.

Corrsin, S. 1951. On the spectrum of isotropic temperature fluctuations in an isotropic turbulence, *J. Appl. Phys.* **22**, 469–473.

Donzis, D.A. & Sreenivasan, K.R. 2010. The bottleneck effect and the Kolmogorov constant in isotropic turbulence, *J. Fluid Mech.* **657**, 171–188.

Falkovich, G., Gawędzki, K. & Vergassola, M. 2001. Particles and fields in fluid turbulence, *Rev. Mod. Phys.* **73**, 913–975.

Fjørtoft, R. 1953. On the changes in the spectral distribution of kinetic energy for two-dimensional nondivergent flow, *Tellus* **5**, 225–230.

Forster D., Nelson, D.R. & Stephen, M.J. 1977. Large-distance and long-time properties of a randomly stirred fluid, *Phys. Rev. A* **16**, 732–749.

di Francesco, P., Ginsparg, P. & Zinn-Justin, J. 1995. 2D gravity and random matrices, *Phys. Rep.* **254**, 1–133.

Frisch, U. 1995. *Turbulence: The Legacy of A.N. Kolmogorov*, Cambridge University Press.

Frisch U. & Bourret, R. 1970. Parastochastics, *J. Math. Phys.* **11**, 364–390.

Frisch, U., Lesieur, M. & Brissaud, A. 1974. A Markovian random coupling model for turbulence, *J. Fluid Mech.* **65**, 145–152.

Frisch, U., Mazzino, A. & Vergassola, M. 1998. Intermittency in passive scalar advection, *Phys. Rev. Lett.* **80**, 5532–5537.

Frisch, U., Mazzino, A., Noullez, A. & Vergassola, M. 1999. Lagrangian method for multiple correlations in passive scalar advection, *Phys. Fluids* **11**, 2178–2186.

Frisch, U. & Morf, R. 1981. Intermittency in nonlinear dynamics and singularitites at complex times, *Phys. Rev. A* **23**, 2673–2705.

Frisch, U., Pouquet, A., Leorat, J. & Mazure, A. 1975. Possibility of an inverse cascade of magnetic helicity in magnetohydrodynamic turbulence, *J. Fluid Mech.* **68**, 769–778.

Frisch, U., Sulem, P.-L. & Nelkin, M. 1978. A simple dynamical model of intermittent fully developed turbulence, *J. Fluid Mech.* **87**, 719–736.

Frisch, U. & Wirth, A. 1996. Inertial-diffusive range for a passive scalar advected by a white-in-time velocity field, *Europhys. Lett.* **35**, 683–687.

Fyfe, D. & Montgomery, D. 1976. High-beta turbulence in two-dimensional magneto-hydrodynamics, *J. Plasma Phys.* **16**, 181–191.

Galtier, S., Nazarenko, S.V., Newell, A.C. & Pouquet, A. 2000. A weak turbulence theory for incompressible magnetohydrodynamics, *J. Plasma Phys.* **63**, 447–488.

Gat, O., Procaccia, I. & Zeitak, R. 1998. Anomalous scaling in passive scalar advection: Monte Carlo Lagrangian trajectories, *Phys. Rev. Lett.* **80**, 5536–5539.

Gawędzki, K. & Kupiainen, A. 1995. Anomalous scaling of the passive scalar, *Phys. Rev. Lett.* **75**, 3834–3837.

Goldreich, P. & Sridhar, S. 1997. Magnetohydrodynamic turbulence revisited, *Astrophys. J.* **485**, 680–688.

Gotoh, T. 1998. Energy spectrum in the inertial and dissipation ranges of two-dimensional steady turbulence, *Phys. Rev. E* **57**, 2984–2991.

Gotoh, T. & Kraichnan, R.H. 2004. Turbulence and Tsallis statistics, *Physica D* **193**, 231–244.

Grant, H.L., Stewart, R.W. & Moilliet, A. 1962. Turbulent spectra from a tidal channel, *J. Fluid Mech.* **12**, 241–268.

Hashminskii, R.Z. 1966. A limit theorem for the solutions of differential equations with random righthand sides, *Theor. Prob. Appl.* **11**, 390–406.

Herring, J.R. & Kraichnan, R.H. 1972. Comparison of some approximations for isotropic turbulence. In *Statistical Models and Turbulence*, Rosenblatt, M. & Van Atta, C., eds., Springer-Verlag, New York, 148–194.

Hopf, E. 1952. Statistical hydromechanics and functional calculus, *J. Rat. Mech. Anal.* **1**, 87–123.

't Hooft, G. 1974. A planar diagram theory for strong interactions, *Nucl. Phys. B* **72**, 461–473.

Iroshnikov, P.S. 1963. Turbulence of a conducting fluid in a strong magnetic field, *Astronomicheskii Zhurnal* **40**, No. 4, 742–750 (in Russian); translated in *Soviet Astronomy* **7**, 566–571 (1964). [The Russian version has the correct initial 'R' for his given name 'Ruslan'; this was incorrectly translated as a 'P'.]

Kawasaki, K. 1970. Kinetic equations and time correlation functions of critical fluctuations, *Ann. Phys. (N.Y.)* **61**, 1–56.

Kimura, Y. and Kraichnan, R.H. 1993. Statistics of an advected passive scalar, *Phys. Fluids A* **5**, 2264–2277.

Kit, L.G. & Tsinober, A.B. 1971. Possibility of creating and investigating two-dimensional turbulence in a strong magnetic field, *Magnitnaya Gidrodinamika,* **7**, 27–34.

Kolmogorov, A.N. 1941. The local structure of turbulence in incompressible viscous fluid for very large Reynolds number, *Dokl. Akad. Nauk SSSR* **30**, 299–303.

Kolmogorov, A.N. 1962. A refinement of previous hypotheses concerning the local structure of turbulence in a viscous incompressible fluid at high Reynolds number, *J. Fluid Mech.* **13**, 82–85.

Kraichnan, R.H. 1953. Scattering of sound in a turbulent medium, *J. Acoust. Soc. Am.* **25**, 1096–1104.

Kraichnan, R.H. 1955a. Statistical mechanics of an adiabatically compressible fluid, *J. Acoust. Soc. Am.* **27**, 438–441.

Kraichnan, R.H. 1955b. Special-relativistic derivation of generally covariant gravitation theory, *Phys. Rev.* **98**, 1118–1122.

Kraichnan, R.H. 1956a. Pressure field within homogeneous anisotropic turbulence, *J. Acoust. Soc. Am.* **28**, 64–72.

Kraichnan, R.H. 1956b. Pressure fluctuations in turbulent flow over a flat plate, *J. Acoust. Soc. Am.* **28**, 378–390.

Kraichnan, R.H. 1957a. Noise transmission from boundary-layer pressure fluctuations, *J. Acoust. Soc. Am.* **29**, 65–80.

Kraichnan, R.H. 1957b. Relation of fourth-order to second-order moments in stationary isotropic turbulence, *Phys. Rev.* **107**, 1385–1490.

Kraichnan, R.H. 1958a. Irreversible statistical mechanics of incompressible hydromagnetic turbulence, *Phys. Rev.* **109**, 1407–1422.

Kraichnan, R.H. 1958b. Higher order interactions in homogeneous turbulence theory, *Phys. Fluids* **1**, 358–359.

Kraichnan, R.H. 1958c. A theory of turbulence dynamics. In *Second Symposium on Naval Hydrodynamics*, Ref. ACR-38, Office of Naval Research, Washington, DC, pp. 29–44.

Kraichnan, R.H. 1959a. Structure of isotropic turbulence at very high Reynolds numbers, *J. Fluid Mech.* **5**, 497–543.

Kraichnan, R. H. 1959b. Classical fluctuation–relaxation theorem, *Phys. Rev.* **113**, 1181–1182.

Kraichnan, R.H. 1961. Dynamics of nonlinear stochastic systems, *J. Math. Phys.* **2**, 124–148.

Kraichnan, R.H. 1962. Mixing-length analysis of turbulent thermal convection at arbitrary Prandtl number, *Phys. Fluids* **5**, 1374–1389.

Kraichnan, R.H. 1964a. Direct-interaction approximation for shear and thermally driven turbulence, *Phys. Fluids* **7**, 1048–1062.

Kraichnan, R.H. 1964b. Kolmogorov's hypotheses and Eulerian turbulence theory, *Phys. Fluids* **7**, 1723–1734.

Kraichnan, R.H. 1965a. Lagrangian-history closure approximation for turbulence, *Phys. Fluids* **8**, 575–598.

Kraichnan, R.H. 1965b. Inertial-range spectrum of hydrodynamic turbulence, *Phys. Fluids* **8**, 1385–1387.

Kraichnan, R.H. 1966a. Isotropic turbulence and inertial-range structure, *Phys. Fluids* **9**, 1728–1752.

Kraichnan, R.H. 1966b. Dispersion of particle pairs in homogeneous turbulence, *Phys. Fluids* **9**, 1937–1943.

Kraichnan, R.H. 1967a. Condensate turbulence in a weakly coupled boson gas, *Phys. Rev. Lett.* **18**, 202–206.

Kraichnan, R.H. 1967b. Inertial ranges in two-dimensional turbulence, *Phys. Fluids* **10**, 1417–1423.

Kraichnan, R.H. 1967c. Intermittency in the very small scales of turbulence, *Phys. Fluids* **10**, 2081–2082.

Kraichnan, R.H. 1968a. Lagrangian-history statistical theory for Burgers equation, *Phys. Fluids* **11**, 266–277.

Kraichnan, R.H. 1968b. Small-scale structure of a scalar field convected by turbulence, *Phys. Fluids* **11**, 945–953.

Kraichnan, R.H. 1968c. Convergents to infinite series in turbulence theory. *Phys. Rev.* **174**, 240–246.

Kraichnan, R.H. 1970a. Turbulent diffusion: evaluation of primitive and renormalized perturbation series by Padé approximants and by expansion of Stieltjes transforms into contributions from continuous orthogonal functions. In *The Padé Approximant in Theoretical Physics*, Baker, G.A. & Gammel, J.L., eds., Academic Press, New York, pp. 129–170.

Kraichnan, R.H. 1970b. Instability in fully-developed turbulence, *Phys. Fluids* **13**, 569–575.

Kraichnan, R.H. 1970c. Convergents to turbulence functions, *J. Fluid Mech.* **41**, 189–218.

Kraichnan, R.H. 1971a. An almost-Markovian Galilean-invariant turbulence model, *J. Fluid Mech.* **47**, 513–524.

Kraichnan, R.H. 1971b. Inertial-range transfer in two- and three-dimensional turbulence, *J. Fluid Mech.* **47**, 525–535.

Kraichnan, R.H. 1972. Some modern developments in the statistical theory of turbulence. In *Statistical Mechanics: New Concepts, New Problems, New Applications*, Rice, S.A., Freed, K.F. and Light, J.C., eds., University of Chicago Press, Chicago, pp. 201–228.

Kraichnan, R.H. 1974a. On Kolmogorov's inertial-range theories, *J. Fluid Mech.* **62**, 305–330.

Kraichnan, R.H. 1974b. Passive-scalar convection by a quasi-uniform random straining field, *J. Fluid Mech.* **64**, 737–762.

Kraichnan, R.H. 1975a. Remarks on turbulence theory, *Adv. Math.* **16**, 305–331.

Kraichnan, R.H. 1975b. Statistical dynamics of two-dimensional flow, *J. Fluid Mech.* **67**, 155–175.

Kraichnan, R.H. 1976a. Diffusion of weak magnetic fields by isotropic turbulence. *J. Fluid Mech.* **75**, 657–676.

Kraichnan, R.H. 1976b. Eddy viscosity in two and three dimensions, *J. Atmos. Sci.* **33**, 1521–1536.

Kraichnan, R.H. 1976c. Diffusion of passive-scalar and magnetic fields by helical turbulence, *J. Fluid Mech.* **77**, 753–768.

Kraichnan, R.H. 1979a. Variational method in turbulence theory, *Phys. Rev. Lett.* **42**, 1263–1266.

Kraichnan, R.H. 1979b. Consistency of the alpha-effect turbulent dynamo, *Phys. Rev. Lett.* **42** 1677–1680.

Kraichnan, R.H. 1985. Decimated amplitude equations in turbulence dynamics. In *Theoretical Approaches to Turbulence*, Dwoyer, D.L., Hussaini, M.Y. and Voigt, R.G., eds., Springer-Verlag, New York, pp. 91–135.

Kraichnan, R.H. 1988. Reduced description of hydrodynamic turbulence, *J. Stat. Phys.* **51**, 949–963.

Kraichnan, R.H. 1991. Turbulent cascade and intermittency growth, *Proc. Roy. Soc. Lond. A* **434**, 65–78.

Kraichnan, R.H. 1994. Anomalous scaling of a randomly advected passive scalar, *Phys. Rev. Lett.* **72**, 1016–1019.

Kraichnan, R.H. 1999. Note on forced Burgers turbulence, *Phys. Fluids* **11**, 3738–3742.

Kraichnan, R.H. & Chen, S. 1989. Is there a statistical mechanics of turbulence? *Physica D* **37**, 160–172.

Kraichnan, R.H., Chen, H. & Chen, S. 1989. Probability distribution of a stochastically advected scalar field, *Phys. Rev. Lett.* **63**, 2657–2657.

Kraichnan, R.H. & Gotoh, T. 1993. Statistics of decaying Burgers turbulence, *Phys. Fluids A* **5**, 445–457.

Kraichnan, R.H. & Leith, C.E. 1972. Predictability of turbulent flows, *J. Atmos. Sci.* **29**, 1041–1058.

Kraichnan, R.H. & Montgomery, D. 1980. Two-dimensional turbulence, *Rep. Prog. Phys.* **43**, 547–618.

Kraichnan, R.H. & Nagarajan, S. 1967. Growth of turbulent magnetic fields, *Phys. Fluids* **10**, 859–870.

Kraichnan, R.H. & Orszag, S.A. 1967. Model equations for strong turbulence in a Vlasov plasma, *Phys. Fluids* **10**, 1720–1736.

Landau, L.D. & Lifshitz, E.M. 1987. *Fluid Mechanics*, 2nd ed., Pergamon Press, Oxford.

Lee, T.D. 1951. Difference between turbulence in a two-dimensional fluid and in a three-dimensional fluid, *J. Appl. Phys.* **22**, 524–524.

Lee, T.D. 1952. On some statistical properties of hydrodynamic and hydromagnetic fields, *Quarterly Appl. Math.* **10**, 69–72.

Leith, C.E. 1967. Diffusion approximation to inertial energy transfer in isotropic turbulence, *Phys. Fluids* **10**, 1409–1416.

Lesieur, M. 2008. *Turbulence in Fluids*, 4th ed., Springer, Heidelberg.

Leslie, D.C. 1973. *Developments in the Theory of Turbulence*, Oxford University Press, Clarendon.

Lesieur, M., Frisch, U. & Brissaud, A. 1971. Théorie de Kraichnan de la turbulence. Application à l'étude d'une turbulence possédant de l'hélicité, *Ann. Géophys. (Paris)* **27**, 151–165.

Lewis, R.M. & Kraichnan, R.H. 1962. A space-time functional formalism for turbulence, *Commun. Pure Appl. Math.* **15**, 397–411.

Lighthill, M.J. 1952. On sound generated aerodynamically. I. General theory, *Proc. R. Soc. Lond. A* **211**, 564–587.

Lighthill, M.J. 1953. On the energy scattered from the interaction of turbulence with sound or shock waves, *Proc. Camb. Phil. Soc.* **49**, 531–551.

Lund, F. & Rojas, C. 1989. Ultrasound as a probe of turbulence, *Physica D* **37**, 508–514.

Makeenko, Y.M. & Migdal, A.A. 1979. Exact equation for the loop average in multicolor QCD, *Phys. Lett. B* **88**, 135–137.

Mandelbrot, B.B. 1968. On intermittent free turbulence. In *Turbulence of Fluids and Plasmas. New York, April 16–18*, Brooklyn Polytechnic Inst., New York (abstract).

Mandelbrot, B.B. 1974. Intermittent turbulence in self-similar cascades: divergence of high moments and dimension of the carrier, *J. Fluid Mech.* **62**, 331–358.

Martin, P.C., Siggia, E.D. & Rose, H.A. 1973. Statistical dynamics of classical systems, *Phys. Rev. A* **8**, 423–437.

Migdal, A.A. 1983. Loop equations and $1/N$ expansion, *Phys. Rep.* **102**, 199–290.

Millionschikov, M. 1941. On the theory of homogeneous isotropic turbulence, *Dokl. Akad. Nauk SSSR* **32**, 615–618.

Monin, A.S. & Yaglom, A.M. 1975. *Statistical Fluid Mechanics*, Vol. 2, MIT Press, Cambridge, MA.

Nazarenko, S.V. 2011. *Wave Turbulence*, Lect. Notes in Phys., Vol. 825 Springer, Heidelberg.

Nazarenko, S.V., Newell, A.C. & Galtier, S. 2001. Non-local MHD turbulence, *Physica D* **152**, 646–652.

Neumann, J. von. 1949. Recent theories of turbulence. In *Collected Works (1949–1963)* **6**, 437–472, ed. A.H. Taub. Pergamon Press, New York (1963).

Ng, C.S. & Bhattacharjee, A. 1996. Interaction of shear-Alfven wave packets: implication for weak magnetohydrodynamic turbulence in astrophysical plasmas, *Astrophys. J.* **465**, 845–854.

Niemela, J.J., Skrbek, L., Sreenivasan, K.R. & Donnelly, R.J. 2000. Turbulent convection at high Rayleigh numbers, *Nature* **404**, 837–841.

Novikov E.A. & Stewart R. 1964. Intermittency of turbulence and spectrum of fluctuations in energy-dissipation, *Izv. Akad. Nauk. SSSR. Ser. Geofiz* **3**, 408–413.

Obukhov, A.M. 1949. Structure of the temperature field in a turbulent flow, *Izv. Akad. Nauk SSSR, Ser. Geogr. i Geofiz.* **13**, 58–69.

Obukhov, A.M. 1962. Some specific features of atmospheric turbulence, *J. Fluid Mech.* **13**, 77–81.

Ogura, Y. 1963. A consequence of the zero-fourth-cumulant approximation in the decay of isotropic turbulence, *J. Fluid Mech.* **16**, 33–40.

Onsager, L. 1949. Statistical hydrodynamics, *Nuovo Cim. Suppl.* **6**, 279–287.

Orszag, S.A. 1966. Dynamics of fluid turbulence, Princeton Plasma Physics Laboratory, report PPL-AF-13.

Orszag, S.A. 1977. Statistical theory of turbulence. In *Fluid Dynamics, Les Houches 1973*, eds. R. Balian & J.L. Peube. Gordon and Breach, New York, pp. 237–374.

Orszag, S.A. & Kruskal, M.D. 1966. Theory of turbulence, *Phys. Rev. Lett.* **16**, 441–444.

Ott, S. & Mann, J. 2000. An experimental investigation of the relative diffusion of particle pairs in three-dimensional turbulent flow, *J. Fluid Mech.* **422**, 207–223.

Parisi, G. & Frisch, U. 1985. On the singularity structure of fully developed turbulence. In *Turbulence and Predictability in Geophysical Fluid Dynamics, Proceedings of the International School of Physics 'E. Fermi', 1983, Varenna, Italy*. Eds. M. Ghil, R. Benzi & G. Parisi, North Holland, Amsterdam, pp. 84–87.

Pasquero, C. & Falkovich, G. 2002. Stationary spectrum of vorticity cascade in two-dimensional turbulence, *Phys. Rev. E* **65**, 056305.

Pierotti, D. 1997. Intermittency in the large-N limit of a spherical shell model for turbulence, *Europhys. Lett.* **37**, 323–328

Pierotti, D., L'vov, V.S., Pomyalov, A. & Procaccia, I. 2000. Anomalous scaling in a model of hydrodynamic turbulence with a small parameter, *Europhys. Lett.* **50**, 473–479.

Prandtl, L. 1925. Bericht über Untersuchungen zur ausgebildeten Turbulenz, *Zs. angew. Math. Mech.* **5**, 136–139.

Priestley, C.H.B. 1959. *Turbulent Transfer in the Lower Atmosphere*, University of Chicago Press, Chicago.

Proudman, I. 1961. On Kraichnan's theory of turbulence. In *Mécanique de la Turbulence. Colloques Internat. CNRS* **108**, 107–112, ed. A. Favre, CNRS, Paris.

Reid, W.H. 1959. Tech. Rept. No.23, Division of Applied Mathematics, Brown University.

Rhines, P.B. 1979. Geostrophic turbulence, *Ann. Rev. Fluid Mech.* **11**, 401–441.

Rose, H.A. & Sulem, P.-L. 1978. Fully developed turbulence and statistical mechanics, *J. Phys. France* **39**, 441–484.

Rota, G.-C. 1996. *Indiscrete Thoughts*, Birkhäuser, Boston.

Rüdiger, G. 1974. The influence of a uniform magnetic field of arbitrary strength on turbulence, *Astron. Nachr.* **295**, 275–284.

Salmon, R. 1980. Baroclinic instability and geostrophic turbulence, *Geophys. Astrophys. Fluid Dyn.* **15**, 167–211.

Salmon, R. 1998. *Lectures on Geophysical Fluid Dynamics*, Oxford University Press.

Scott, R.B. & Wang, F. 2005. Direct evidence of an oceanic inverse kinetic energy cascade from satellite altimetry, *J. Phys. Ocean.* **35**, 1650–1666.

Semikoz, D.V. & Tkachev, I.I. 1997. Condensation of bosons in the kinetic regime, *Phys. Rev. D* **55**, 489–502.

Shebalin, J.V., Matthaeus, W.H. & Montgomery, D. 1983. Anisotropy in MHD turbulence due to a mean magnetic field, *J. Plasma Phys.* **29**, 525–547.

Shraiman, B.I. & Siggia, E.D. 1995. Anomalous scaling and small scale anisotropy of a passive scalar in turbulent flow, *C.R. Acad. Sci.* **321**, 279–284.

Smith, L.M. & Yakhot, V. 1993. Bose condensation and small-scale structure generation in a random force driven 2D turbulence, *Phys. Rev. Lett.* **71**, 352–355.

Spiegel, E.A. 1971. Convection in stars, *Ann. Rev. Astron. Astrophys.* **9**, 323–352.

Spiegel, E.A. 2011. Chandrasekhar's lecture notes on the theory of turbulence (1954). *Lect. Notes in Phys.* Vol. 810 Springer, Heidelberg.

Starr, V.P. 1968. *Physics of Negative Viscosity Phenomena*, McGraw-Hill, New York.

Tatsumi, T. 1957. The theory of decay processes of incompressible isotropic turbulence. *Proc. R. Soc. Lond. A* **239**, 16–45.

Taylor, G.I. 1917. Observations and speculations on the nature of turbulence motion, Reports and Memoranda of the Advisory Committee for Aeronautics, no. 345; reproduced in *G.I. Taylor's Scientific Papers*, Batchelor, G.K., ed., Cambridge University Press, 1960, Vol. II, paper no. 7, pp. 69–78.

Ting, L. & Miksis, M.J. 1990. On vortical flow and sound generation, *Siam. J. Appl. Math.* **50**, 521–536.

Xiao, Z., Wan, M., Chen, S. & Eyink, G.L. 2008. Physical mechanism of the inverse energy cascade of two-dimensional turbulence: a numerical investigation, *J. Fluid Mech.* **619**, 1–44.

Yaglom, A.M. 1949. Local structure of the temperature field in a turbulent flow, *Dokl. Akad. Nauk SSSR* **69**, 743–746.

Yaglom, A.M. 1966. The influence of fluctuations in energy dissipation on the shape of turbulence characteristics in the inertial interval, *Dokl. Akad. Nauk. SSSR* **166**, 49–52.

Zakharov, V.E. 1966. *Some aspects of nonlinear theory of surface waves* (in Russian), PhD Thesis, Budker Institute for Nuclear Physics for Nuclear Physics, Novosibirsk, USSR.

Zakharov, V., L'vov, V. & Falkovich, G. 1992. *Kolmogorov Spectra of Turbulence*, Springer-Verlag, Berlin.

11

Satish Dhawan

Roddam Narasimha

11.1 Introduction

Satish Dhawan was born on 25 September 1920 in Srinagar, Kashmir, the home town of his mother Lakshmi. His father, Devidayal, was from the North Western Frontier Province; both parents came from professional families, full of doctors, lawyers and academics – Devidayal retired as a respected judge of the High Court in Lahore, now in Pakistan. Satish's education began under private tutors at home, as his father kept getting transferred in his early career from one town to another in the North West (Kipling country to Indo-British readers). He completed his Indian education at the University of Punjab in Lahore with an unusual combination of degrees: BA in physics and mathematics (1938), MA in English literature (1941) and BE (Hons.) in mechanical engineering (1945). In 1946 he sailed to the USA on a government scholarship, and obtained an MS in aeronautical engineering from the University of Minnesota the following year. (The summer of 1947 saw much turmoil in the subcontinent preceding its imminent partition, and Satish's parents reluctantly left Lahore for India – never to return – a week before the formal end of colonial rule.) In the USA Satish moved to the California Institute of Technology where, with Hans W. Liepmann as his adviser, he obtained the degree of Aeronautical Engineer in 1949 and a PhD in aeronautics and mathematics in 1951. Dhawan made a strong impression, scientifically and otherwise, on everybody he came in contact with at Caltech. In an obituary he wrote in 2002, Liepmann noted Dhawan's "unusual maturity in judging both scientific and human problems", his very good sense of humour, and the way he was "immediately accepted and respected by [the] highly competent and proud group of young scientists" who worked with Liepmann at the time.

At the end of tenure of the scholarship Dhawan returned to India and joined the Indian Institute of Science (IISc) as a Senior Scientific Officer. He rose

373

Figure 11.1 (Left) Satish Dhawan at Nandidurg, a small hill resort, approximately 45 km from Bangalore; *c*. 1955. (Right) *c*. 1985.

rapidly at the Institute, becoming Head of the Department of Aeronautical Engineering in 1955 and Director of the Institute in 1962, the latter at the unusually young age (certainly for India) of 42 years.

Dhawan's scientific career divides into four distinct phases. First there were his years at Caltech, where his main contribution was the direct measurement of skin friction in boundary layers, but there was also other work that dealt with aspects of turbulent boundary layers in high-speed flows. The second phase was marked by intense activity as a faculty member in aeronautical engineering at IISc (Figure 11.1, left). Here he set up the High Speed Aerodynamics and low-speed Boundary Layer Laboratories, and initiated experimental fluid dynamics research especially in transitional boundary layers and other wall-bounded turbulent flows. The third phase started when he became Director of IISc in 1962. He continued advising research students, but his energies were consumed by his ambitious project for the academic and administrative transformation of the Institute. His impact here has been long-lasting. [A postage stamp released in 2009 on the occasion of the centenary of the Institute contains pictures of its several icons: the founder (Jamsetji Tata, Parsi businessman and industrialist from Bombay), his spiritual ally (Swami Vivekananda), his princely supporter (the Maharaja of Mysore), the Institute's first director (British chemist Morris Travers), its greatest scientist (C.V. Raman), its greatest alumnus (the biophysicist G.N. Ramachandran) and (presumably) its greatest director (Satish Dhawan).] The fourth phase began in 1972 when he was appointed head of the Indian space programme, which he ran for nine years even as he continued as Director of IISc. (His last doctoral student had graduated in 1971.) He thereafter proceeded to build the Indian Space Research Organization into one of India's most successful technological enterprises – an achievement that was so admired by his countrymen that in the public eye it somewhat overshadows his career as an academic scientist and leader.

A more detailed account of Dhawan's life, work and personality will be found in Narasimha (2002).

11.2 The Caltech years

In retrospect, it would appear that Hans Liepmann's young and lively fluid dynamics group at GALCIT (then Guggenheim, now Graduate Aeronautical Laboratories at Caltech) was the perfect match for Satish Dhawan's abilities, inclinations and temperament: its emphasis on serious experimental research on basic problems, coupled with active contacts with the aircraft industry and an easy Californian informality, turned out to be just his cup of tea.

As Liepmann wrote of those years in that same obituary: "It was a marvelous time! Almost everything we touched was new and exciting." The very first research project Dhawan was involved in, namely shock–boundary-layer interaction (Liepmann, Roshko & Dhawan, 1951, LRD henceforth; work completed in 1949), was one such.

Shock–boundary-layer interaction The primary object of interest in the project was not the turbulent boundary layer *per se*, but the phenomenon of shock reflection from a solid surface, and in particular the difference in its characteristics between the cases when the boundary layer on the surface is laminar and when it is turbulent. This study was conducted at a time when knowledge of even laminar boundary layers in compressible flows was still in its infancy. In fact the LRD report points out that one may ask

> why a complicated phenomenon such as an interaction between a shock wave and boundary layer is investigated before the boundary layer in uniform supersonic flows has been studied carefully.

The authors defend this admittedly "illogical approach" by saying that the problem arose naturally from earlier investigations of their own in transonic flow (Liepmann, 1946) and those of Ackeret, Feldmann & Rott (1946) in supersonic flow. These two reports, together with LRD, constitute in fact the first systematic studies of the phenomenon. The effects were seen as dramatic and clearly of importance even in the general state of inadequate knowledge in supersonic boundary layers at the time. The reason was that the presence of a viscous layer at the wall can so strikingly alter the classical inviscid flow picture of reflection because it introduces an additional boundary condition of no slip at the surface. Even the boundary layer approximation becomes questionable, because of large pressure gradients across the flow.

Clearly, viscosity can diffuse the effect of the shock wave upstream in subsonic flow near the surface, and the associated adverse pressure gradient can even cause separation of the flow. Between the laminar–turbulent and attached–separated states in the boundary layer, and weak–strong, normal–oblique in the shock wave, many different combinations become possible. The phenomenon may in particular involve more than one shock system. The pioneering work at Caltech and Zurich shed much light on the many physical processes operating in the different flow combinations.

The difference in effect between turbulent and laminar boundary layers had been observed in transonic flows, but there the incoming shock at some distance from the boundary layer actually depends on the boundary layer itself as well as on the flow field, and therefore cannot be controlled independently. On the other hand in supersonic flow, in particular in the interaction on a flat plate, the flow phenomena are clearer, and the LRD report provides both pressure distributions and schlieren pictures of the flow that were so revealing that they quickly found their way into textbooks (e.g. Schlicting, 1955, pp. 300, 301).

The major result of the investigation is in Figure 11.2, which shows the dramatic difference between the pressure distributions on the surface depending on whether the boundary layer thereon is laminar or turbulent. In the laminar case the effect is felt even 50 boundary-layer thicknesses upstream of the shock, whereas in the turbulent case it is felt only over 5 boundary-layer thicknesses, making it closer to the inviscid limit. As a secondary investigation there is a brief digression on transition, which is visualized using an oil film technique. This enabled sketches of turbulent wedges in laminar flow and also of the patterns resulting from tripping the boundary layer using a wire.

Shock–boundary-layer interactions were extensively studied in the subsequent two decades, both experimentally and theoretically. In laminar flows the physical mechanisms responsible became largely understood, following the work of Lighthill (1953) and others. The shorter region of upstream influence in turbulent boundary layers could be interpreted as due to the much thinner subsonic viscous layer they possess in comparison to that in laminar layers. Correspondingly, interaction in much of the turbulent layer away from the wall is inviscid in character. As in any turbulent flow the problem of closure presents itself here too, and a variety of models have been used with variable results. A review of the status three decades after the first systematic experimental studies of the phenomenon was provided by Adamson & Messiter (1980).

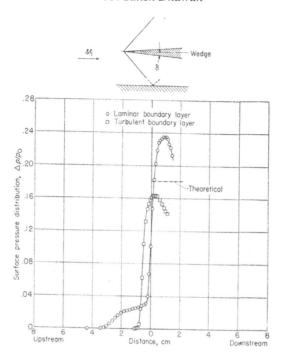

Figure 11.2 Pressure distribution in shock reflection from a flat surface with laminar and turbulent boundary layers on surface. From Liepmann et al. (1951).

Direct measurement of skin friction This was the subject of Dhawan's doctoral thesis. Dhawan once told me that von Kármán would ask how it was that such a technologically and scientifically important parameter as the skin friction, for which there was a prediction by the generally successful boundary-layer theory in laminar flow, had not yet been directly measured. Liepmann and Dhawan set about tackling that question. Dhawan's answer is described in the NACA Report 1121 (completed in 1951, published in 1953). This report – very well and clearly written, by the way – starts by noting how both wave drag and induced drag are better understood, theoretically as well as experimentally, than skin friction drag and boundary layer separation. It then reviews both direct and indirect methods of measuring friction drag, and mentions the early efforts of Froude and Kempf, and of Schultz-Grunow (1940), to measure directly the tangential force on a free-moving or floating element of the surface. The great advantage of the method, as seen in the report, was that it did not assume the validity of boundary-layer (or any other) theory – and to that extent would provide an independent test of theory. This might now seem a surprising statement to make, but it must be remembered that, for a remarkably long time, boundary-layer theory was seen as no more than an intriguingly successful engineering approximation; it was only in the 1950s that it began to be recognized as the leading term in an asymptotic solution of the Navier–Stokes equations as Reynolds number tends to infinity.

Dhawan then proceeds to describe his design for the skin friction balance, using a variable reluctance transformer for sensing the motion of the free element. He then makes a variety of checks to ensure that the velocity profile on the moving element is the same as on the flat plate in the immediate neighbourhood. The effects of the gaps, the presence of any circulating flow, corrections for the (small) residual pressure gradients, the effect of vibrations etc. are all either shown to be negligible or taken into account. The boundary layer was tripped at the leading edge of the plate by a wire roughness element. It was reported that causing transition by raising free-stream turbulence did not affect the measured skin friction.

The final results of the measurements are shown in Figure 11.3. The experiments with turbulent boundary layers were found to be in 'excellent agreement' with the theory of von Kármán (1936), by which was meant the log-law expression

$$c_f^{-1/2} = A + B \log_{10}(c_f Re)^{1/2}, \tag{11.1}$$

where c_f is the local skin friction coefficient (wall stress in free-stream dynamic pressure units), $Re = Ux/v$ is the Reynolds number based on distance x from the leading edge and free-stream velocity U, and A and B are constants. However, Dhawan notes that the values that fit his data, $A = -0.9$, $B = 5.06$, were different from von Kármán's values, $A = 1.7$ and $B = 4.15$, and so "this agreement is believed to be fortuitous to some extent". [Incidentally the quoted values of B correspond to von Kármán 'constants' (or coefficients) of 0.455 and 0.554 respectively – both higher than the current range of accepted values, but Dhawan's is closer.] He further adds that these differences in the constants

are perhaps to be attributed to the fact that A and B are not absolute constants but depend somewhat on the conditions of the experiment, for example, on Reynolds number and so forth.

This issue still continues to be debated (see e.g. Marusic et al., 2010). Dhawan finally concludes that "the logarithmic formula of von Kármán is a fair approximation for incompressible flow".

Interestingly, the same experimental data of Dhawan are reproduced by Schlichting (1955, Figure 7.11), except that the theoretical turbulent skin

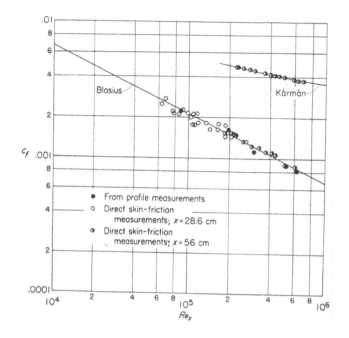

Figure 11.3 Local skin friction coefficient in laminar and turbulent boundary layers. In the former estimates from velocity profile measurements are also included. From Dhawan (1951).

friction curve plotted is due to Prandtl,

$$c_f = 0.0592 \, Re^{-1/5}, \tag{11.2}$$

based on a power-law approximation rather than the log-law. (The agreement is again excellent.) Interestingly the power-law versus log-law issue has also continued to be debated right down to present times. The issue can only be settled with data over a large Reynolds number range, and recent work appears to favour the log-law approach.

Dhawan then goes on to present some data at compressible subsonic speeds, $M = 0.2$ to 0.8. His general conclusion is that at these speeds the skin friction coefficient is quite close to but definitely lower than the predictions of incompressible theory.

Towards the end of the report Dhawan describes two sets of preliminary measurements – in supersonic and transitional flows, respectively. The former could only be done at two Mach numbers, 1.24 and 1.44. There were difficulties with pressure gradients on the plate and detached shock waves at the leading edge, among others. The measurements are again close to von Kármán's theory, but Dhawan is (also again) very cautious: despite the closeness, "it cannot be concluded that incompressibility lowers the skin friction coefficient by an amount given by von Kármán's empirical curve".

Interestingly, there are preliminary measurements of c_f in the region of transition from laminar to turbulent flow as the Mach number increases from 0.24 to 0.6 (simultaneously with an accompanying increase in the Reynolds number from about 4.3×10^5 to a little more than 9×10^5). These appear to have been the first ever direct measurements of c_f in the transition region of a boundary layer. The data covered the range from laminar nearly-Blasius values to close to the von Kármán fully turbulent ones. Once again Dhawan is careful to call the measurements "quite qualitative", but notes that they cannot be explained by a steady transition at a critical Reynolds number. Instead, he points out, Dryden (1936) had already observed intermittent laminar/turbulent flow in the transition region, and Liepmann (1943) had substantiated this observation. He then mentions the work of Emmons on turbulent spots, published around the same time (in fact a month later in July 1951), as providing a reasonable explanation of the observations.

These two exploratory studies led to the more detailed investigations of Donald Coles (1953) on skin friction in supersonic flow at the Jet Propulsion Laboratory, and the work done in Bangalore on the transition region in low-speed boundary layers, as I shall shortly describe.

But before doing so I cannot resist the temptation to remark on how Dhawan's NACA report brings out the character of his research. Incidentally the work described there was never published in a journal. There was perhaps no need, for in those times NACA reports probably had a wider readership than any aeronautical journal did. One of their advantages was that they were actually reports – not the highly condensed journal papers of today. So we can see Dhawan's method at work: ingenious in design, meticulous in execution and cautious in interpretation. I have often thought that the same qualities were responsible for his success as a scientific leader.

11.3 At Bangalore

Dhawan's first task on his return to India in 1951 was to establish a High Speed Aerodynamics Laboratory, for which he built two blow-down supersonic tunnels, respectively 1 in. by 3 in. and 5 in. by 7 in. test-section size. He also built a 20 in. by 20 in. low-speed boundary layer tunnel.

I joined the Institute in 1953 for an IISc Diploma (the equivalent of a Master's) in aeronautical engineering, and got to know Dhawan personally as I helped him calibrate the 1 in. by 3 in. tunnel and designed its nozzles. At the end of the two-year course Dhawan asked me if I would like to do research with him, and I jumped at the chance and registered for the Associateship (a Master's by research). I had done a short research project earlier with him measuring boundary layer velocity profiles in a small low-speed tunnel, and his suggestion that I should work on boundary layer transition coincided with my own thoughts. I helped him in setting up the boundary layer lab, built a hot-wire amplifier and started making turbulence measurements.

Why did Dhawan suggest I work on transition? It is clear from what has been said before that the subject was very much on his mind when he made his first preliminary direct skin friction measurements in the transition zone at Caltech. But, characteristically for Dhawan, there was a strong 'Indian' reason as well. He described this in a lecture he gave at the first Asian Congress of Fluid Mechanics, held in Bangalore in 1980 (Dhawan, 1981). The Department had a closed-circuit low-speed wind tunnel with a 5 ft. by 7 ft. elliptic test-section, with tunnel Reynolds number in an awkward range (1.5×10^6 per ft.) because of transition. Tests were just then being made on a model of the HF24, a fighter being designed at Hindustan Aircraft (more of this later), under the leadership of the well-known German designer Kurt Tank (ex Messerschmitt). Dhawan wanted to explore if we could understand a little better how IISc wind tunnel results could be scaled up to flight Reynolds numbers.

These two motivations were just the kind that appealed to Dhawan, and drove the transition programme at Bangalore, although it quickly became a research project entirely in its own right.

Now, intermittent velocity signals, i.e. those showing the presence of turbulent patches (sometimes called bursts), of varying duration and between-patch time intervals, had frequently been observed in the transition region by various investigators (Dryden, 1936; Liepmann, 1943). One physical picture inspired by such observations was that there might be a sharp, fluctuating, jagged front separating laminar from turbulent flow. Emmons (1951) proposed a radically different picture in which laminar flow breaks down at isolated points, at each of which a turbulent 'spot' is born. This replaced the laminar-turbulent 'front' by a set of 'islands' of turbulence in a laminar sea. (The idea was radical because it permitted a laminar state *downstream* of the turbulence in the island. Reynolds (1883) had already observed 'flashes' in his pipe where the same phenomenon occurs; but it was perhaps easier to accept it in a confined duct than on a semi-open surface.) Emmons further proposed that these spots grow as they move downstream, eventually leading to fully turbulent flow as they cover the entire surface. The fraction of time that the flow was turbulent at any point on the surface (say station x from the leading edge of the surface in two-dimensional flow) was called the intermittency γ at x. Emmons formulated a statistical theory which related γ to the rate at which spots were born on the surface, denoted by $g(x)$ (number per unit area, unit time at x), and their propagation characteristics, assumed linear in time and in both space directions on the surface. In retrospect we can see that the Emmons model can be derived by taking spot formation as a Poisson process.

The major problem in making further progress was the choice of the function g, which may also be called the mean spot formation rate, equivalently the probability density that a spot will be born at x, t. Several alternative proposals for g were in fact discussed in Emmons (1951) and a companion paper by Emmons & Bryson (1952) (each referring to the other as to be shortly published; the second considered swept plates and cones as well). Will g increase with x (because the flow is increasingly unstable), say like x^m (with $m > 0$) – or decrease (because flow becomes turbulent in any case)? Should g remain zero below a critical 'transition Reynolds number or location' (i.e. for $x < x_c$, say)? After some discussion of these issues Emmons concluded: "As a first and simplest assumption, g will be taken as independent of position on the flat plate". In preferring not to assume an $x_c > 0$ he appears to have been influenced by the observations of Charters (1948), who reported wind tunnel experiments showing 'transverse contamination'. This is a turbulent wedge spreading inwards into the plate from its leading edge where it is in contact with the tunnel side wall.

Emmons & Bryson (1952) went on to work out results for the case $g \sim x^m$. This leads to an expression for γ of the form

$$\gamma = 1 - \exp\left\{-\text{const.}x^{m+3}\right\} = 1 - \exp\left\{-\ln 2(x/\bar{x})^{m+3}\right\}, \qquad (11.3)$$

where $\bar{x}$ locates the halfway point at which $\gamma = 0.5$. (The second equality, a simple consequence of the first, appears only in my IISc thesis, Narasimha, 1958.) However, even in Emmons & Bryson the final choice made is $m = 0$, i.e. g is independent of x.

In this view laminar and turbulent flow coexist everywhere, but in different proportions determined by γ. At the leading edge $\gamma = 0$, and far downstream $\gamma \to 1$. Along with the front idea, that of a privileged transition onset location also became tenuous in this view (but see below). However, there were no experimental data on γ, so the major target in both the Emmons papers was prediction of local and total skin friction drag coefficients, on which some results had been published.

Just as I was absorbing all this Dhawan passed on to me a copy of NACA TN 3489 by Schubauer & Klebanoff (1955; SK henceforth), which he had just received by mail. Their brilliant confirmation of Emmons' ideas about turbulent spots and his basic assumptions about their linear propagation characteristics was a powerful trigger for our later work. SK presented the first intermittency measurements, and was read with great interest by both of us. It was immediately decided to make similar measurements in the IISc boundary layer tunnel. We also quickly saw that SK contained an unarticulated puzzle: why was it that, in spite of confirming Emmons' physical ideas so convincingly, the authors did not compare their intermittency measurements with Emmons' theory, but instead fitted them to a 'Gaussian integral', i.e. an error function curve? When I made the comparison, the disagreement between the two for the intermittency was found to be huge. For example, SK reported that in natural transition in their famous quiet tunnel at the National Bureau of Standards (NBS) $\bar{x}$ was at 6.25 ft; γ was down to 0 already at $x = 5.25$ ft $= 0.84\bar{x}$. However, at this value of $x/\bar{x}$ expression (11.3) gives (with $m = 0$) a value of γ as high as 0.34 – way beyond any uncertainty in the excellent SK measurements. Basically, therefore, the transition region was much sharper than Emmons would predict relative to the distance to the halfway point. The SK data, and the more extensive datasets we quickly acquired at Bangalore, seemed to be suggesting that Emmons' proposal on spot generation across the whole surface of the plate was questionable. Furthermore, the discovery of the 'calming effect', as SK called it, showed that each spot left behind a trail of highly stable flow, in which further breakdown into spots was unlikely. The function $g(x)$ may therefore be expected to decrease with increasing x.

After some false starts, it turned out that the simplest explanation of the data was actually to go to the other extreme and say that all spots were born at one streamwise location x_t, but randomly in time and in spanwise location; i.e. g is proportional to a Dirac delta function at x_t. Upstream of x_t laminar flow is not sufficiently unstable or disturbed, downstream the spots stabilize the flow.[1] This makes x_t a strong candidate for being identified with the onset of transition. The rest of Emmons' picture stays intact, and (11.3) is replaced by

$$\gamma = \begin{cases} 0 & \text{for } x \leq x_t, \\ 1 - \exp\left\{-0.41(x - x_t)^2/\lambda^2\right\} & \text{for } x \geq x_t, \end{cases} \qquad (11.4)$$

(Narasimha, 1957), where λ is the distance between points with $\gamma = 0.25$ and 0.75. The results agree with observation very well (Figure 11.4) – investing a precise new meaning to an old length in the problem (namely x_t). In this new picture transition does *not* occur everywhere on the plate, and the concept of an onset location is not only resurrected but indeed found to play a key role.

---

[1] This is not completely true: Wygnanski et al. (1979) found that Tollmien–Schlichting type waves are excited in the wing-tip regions of isolated spots.

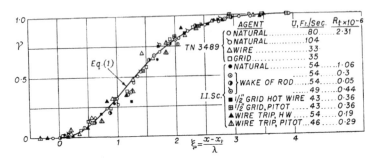

Figure 11.4 Data on intermittency distributions. Full curve (equation (11.4) of text) is spot-based theory assuming concentrated breakdown at x_t; $\xi = (x - x_t)/\lambda$ where λ is distance between the points where $\gamma = 0.25, 0.75$. From Dhawan & Narasimha (1958), reproduced with permission of Cambridge University Press.

Dhawan & Narasimha (1958; DN henceforth) tested the consequences of this idea in detail.

Let us now return to the chief target of the Emmons papers, namely drag. For the local skin friction he had only the velocity profile estimates of Burgers and Van der Hegge Zijnen published in 1934; agreement was poor, and Emmons concluded: "More local coefficient data are badly needed". That, of course, was precisely what Dhawan was doing right then on the other coast. Emmons also compared his theory with data for total drag, finding reasonable agreement with Gebers' work but not that of Schoenherr. Emmons & Bryson present results for different assumptions on g (with $1 < m + 3 < 4$). The striking feature of all these predictions is that the fully turbulent values are always approached from below, even for the local skin friction coefficient.

As x_t was an additional length scale that Emmons' proposals did not have, DN first collected data on the extent of the transition zone λ, and found the rough correlation $Re_\lambda = 5\,Re_{x_t}^{0.8}$ between the Reynolds numbers based on λ and x_t respectively and the free-stream velocity U. The data therefore suggest that $\lambda/x_t \to 0$ like $Re_{x_t}^{-0.2}$, i.e. the transition zone gets proportionately sharper as transition occurs later. The above relation was later improved to $Re_\lambda = 9\,Re_{x_t}^{3/4}$ (Narasimha, 1985), with the attractive and reasonable implication that the spot formation rate n (per unit span, time) at x_t scales with the local boundary-layer thickness δ_t and viscosity ν, as it might be expected to do in an instability-dominated scenario: $n\delta_t^3/\nu$ at x_t is $O(1)$. We may contrast this with the non-dimensional spot formation rates in Emmons & Bryson, based on U and ν as scales, which were in the range 10^{14} to 10^{20} depending on free-stream turbulence levels.

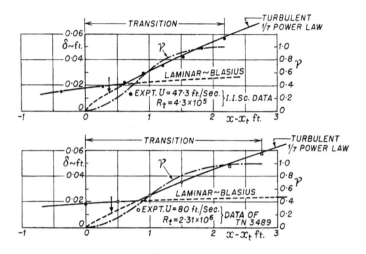

Figure 11.5 Boundary-layer thickness in the transition zone, given (except just downstream of onset) by the fully turbulent value corresponding to a virtual origin near x_t. From Dhawan & Narasimha (1958), reproduced with permission of Cambridge University Press.

The hypothesis of concentrated breakdown offered simple physical explanations for a variety of striking observations in the transition region. For example, the total boundary-layer thickness will in this picture be that of the laminar layer starting from the leading edge or the turbulent layer starting from x_t, whichever is greater. This was found to be the case in DN (Figure 11.5). More interestingly, the hypothesis predicts an overshoot in the skin friction coefficient above the curve for a turbulent boundary layer starting from the leading edge (Figure 11.6). The reason is that towards the end of the transition zone $\gamma \approx 1$, but the local turbulent boundary layer corresponds to a Reynolds number proportional to $x - x_t$ and not to x; the lower Reynolds number has the higher c_f, which (because $\gamma \approx 1$) makes the dominant contribution to the total skin friction: hence the overshoot. The work of Coles (1953, 1954a,b),[2] which followed Dhawan's Caltech thesis, provided direct measurements of

---

[2] Incidentally Coles found that between the three of them Dhawan, Schultz–Grunow and Kempf had made direct skin friction measurements over more than three decades (roughly one each, by the way!) in Reynolds number, and displayed this in a figure with a two-page spread in *ZAMP* (Coles, 1954b). He "admitted" that his own work was "inspired by the remarkable consistency of these local friction data … with a scatter so small as to be almost unprecedented in boundary layer research". He introduced one major improvement over Dhawan's method by using a nulling technique for force measurement.

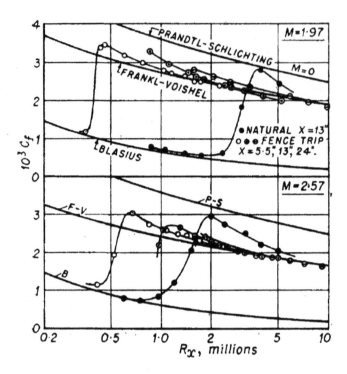

Figure 11.6 Direct measurements of local skin friction coefficient on a flat plate
during transition from laminar to turbulent flow. Experimental data from Coles
(1954), replotted on a log-linear scale, with the Frankl–Voishel theory for turbulent skin friction in supersonic flow added. Note the overshoot above turbulent
theory (for boundary layer originating at the leading edge), as predicted by spot
theory based on concentrated breakdown at x_t (DN). From Dhawan & Narasimha
(1958), reproduced with permission of Cambridge University Press.

c_f in supersonic flows, often covering the transition zone. They left no doubt
about the presence of overshoots in c_f (Figure 11.6).

Finally, the DN picture predicts that early in the transition zone the displacement thickness δ^* does not grow monotonically; in fact it has a minimum. This
dip is at first surprising, but the reason from the DN view is clear. For in this
view there is a place where the total boundary-layer thickness δ is the same
for both the laminar layer originating at $x \approx 0$ *and* the turbulent originating
at $x = x_t$ (see Figure 11.5). And the γ-weighted mean velocity profile at this
station, reflecting its turbulent part, *has* to be fuller than the pure laminar flow.

It follows that δ^* has to be lower. Estimates of the δ^* from SK data confirm the prediction (Figure 11, Narasimha, 1985).

On the whole the broad characteristics of then-recent observations fitted the DN picture very well. The intermittency data could, I am sure, be made to fit other curves, including some in the family considered by Emmons and Bryson, with an appropriate choice of constants (including x_c). Such attempts were indeed made later, for example by Abu-Ghannam & Shaw (1980). However, the conceptual framework of DN not only explains the main features in simple terms, but has other advantages. For example, it implicitly retains the run-up to transition onset through instability, which the assumption of constant g basically renders irrelevant. It also enables predictions of local skin friction or total drag without ambiguities about what the c_f values within the spots are — as all the spots are born at x_t and have (so to speak) about the same age at any downstream station, their average would to a first approximation be the same as in the plane turbulent boundary layer with origin at x_t. Thus the DN picture fitted all the observations in the transition zone available at the time into a coherent scheme which was based on Emmons spots but differed crucially on their postulated birthplace, i.e. on the assignment of the *a priori* probability for $g(x)$.

Assuming concentrated breakdown at one location does however seem extreme, so DN calculated the intermittency when spot formation occurred with a Gaussian distribution of various widths centred at x_t. It became clear that, while there was a band in x (estimated as being no wider than $\lambda/3$ by Narasimha, 1985) in which spots might form, introducing one more length scale describing the *width* of the band was generally not worthwhile. This view thus resolved an old question going back to Prandtl and the rest: is transition abrupt or gradual? Answer: it is *abrupt* at onset, occurs at isolated points within a narrow band, then evolves *gradually* to an asymptotic state of full-time turbulence.

This work continued for many years at Bangalore with other students, some working under Dhawan's supervision, applying the above ideas to pipes, channels, boundary layers on axisymmetric bodies and under pressure gradients etc. A review of this work is provided by Narasimha (1985).

11.4 Dhawan's approach to building engineering science

Dhawan's thinking on doing as well as promoting research consisted of several strands. At one level was the desire to establish a tradition of research in India in the kind of engineering science that had grown so rapidly in the West. A second strand had to do with the needs of the nascent aircraft industry in India at the time. The privately owned Hindustan Aircraft Limited (HAL) had been established in Bangalore in December 1940 to service the needs chiefly of the US Air Force in terms of maintenance and other support for aircraft operating in the South East Asia Command during the War. In 1942 IISc established a Department of Aeronautical Engineering with Dr Vishnu Madhav Ghatage (PhD 1936 with Prandtl in Göttingen) leading it. But Ghatage quickly moved to head the design office in the new industry in 1944. By the time Dhawan returned to India in 1951 HAL had already designed and built a primary trainer which was getting into service; it later embarked on more ambitious projects with a jet trainer and the HF-24 fighter. Dhawan had spent a year in this industry before he went to the USA; he established contacts with it on his return and was familiar with the design and development issues that the new projects posed. He went about extracting from among these a set of basic problems that could be pursued at the Institute.

The third strand emerged a little later when the National Aeronautical Laboratory (NAL), a national research and development centre, embarked on the construction of high-speed wind tunnels; Dhawan carried out pilot project studies at the Institute, and many of those who worked with him in this project went on to build the bigger facilities at NAL and lead research and development there.

This philosophy in research was briefly explained by him in his 1980 lecture (Dhawan, 1981). One concrete example of it has already been mentioned: the transition studies were inspired by the needs of wind tunnel testing of aircraft models for HAL. As thin high speed airfoils were being developed for the fighter project, one of his students (S.K. Ojha) studied separation bubbles. Testing the intake for the same aircraft prompted suggestions about the possibility of using blowing to improve its performance, and this led to the work of S.P. Parthasarathy on turbulent wall jets. And so on.

After he became Director of the Institute Dhawan had less time for research, but continued to advise research students who worked in the same spirit, as briefly described in Dhawan (1981). However, he rarely published anything.

Curiously, in the 1980 lecture Dhawan does not dwell much on the scientific motivation for the projects he chose, but its presence and role in his group's work cannot be doubted. From the way he formulated problems for his research students, he clearly thought of combining basic research and applications as the most appropriate route to establish a robust tradition in engineering science in India. As already remarked, the transition work of 1958 went ahead to study the *phenomena* (rather than aeronautical applications) by research in pipes, channels and axially aligned cylinders, and on reverse transition from turbulent to laminar flow at both high and low speeds. He clearly derived much satisfaction from all this work, for he concluded his 1980 lecture by saying: "It

Figure 11.7 Dhawan with young students and colleagues at a party.

called for considerable ingenuity and dedication to work in adverse conditions, and the results reported in many papers in journals, not mentioned here, bear testimony to the spirit of scientific investigation". But he and his students found that it was actually a lot of fun as well (Figure 11.7).

Among the many distinguished fluid dynamicists at the 1980 Congress[3] was Zhou Pei-Yuan from China (Figure 11.8). Here we had, on the same platform, two doyens of fluid mechanics in Asia: both with PhDs from Caltech, both working on turbulent flows but one an experimenter and the other a theoretician, and both holding positions of national responsibility in their respective countries. The third, unfortunately a late arrival at the Congress and so absent at the time the picture was taken, was of course Itiro Tani from Japan.

---

[3] I notice that, without my having planned it, the first Asian Congress of Fluid Mechanics held at Bangalore in 1980 figures in this account several times. So (at the editors' suggestion) I should say a few words as background. It was the outcome of several encounters between Hiroshi Sato and me at various conferences held in the USA and Europe in the 1970s; Sato also had spent some time in JPL. On one of those occasions I remarked that it was crazy that we always met on the other side of the planet. Sato agreed, and said we should do something about it. We wrote a joint letter to every Asian fluid dynamicist we knew, and were overwhelmed by the enthusiastic response to the idea of a fluid mechanics meeting in Asia. I offered to host it in Bangalore, so the Congress took place during 8–13 December 1980. The idea was warmly welcomed by all the three Asian doyens I have mentioned, and their support played a key role in its success. Among others who attended were Donald Coles, Gerry Whitham, Fazle Hussain, Ruby Krishnamurti, and many leading Asian fluid dynamicists. Reviewing for *JFM* in 1982 the volume of invited lectures given at the Congress (Narasimha & Deshpande, 1982), George Batchelor declared that the Congresses were "off to a good start". Since 1980 they have regularly met once every two or three years in some Asian city.

Figure 11.8 Two doyens: Satish Dhawan and Zhou Pei-Yuan, at Bangalore in 1980. Both got their PhDs from Caltech, both worked on turbulence (one experimentally, the other theoretically), and both were given high national responsibilities in their respective countries.

Dhawan passed away quietly during the night of 3 January 2002 at home. His education integrated science, technology and the humanities; so did his life. A combination of his undoubted technical gifts, unquestioned integrity and great personal charm enabled him to combine doing and promoting science, work for state and society, and manage megatechnology while championing little science. He left a precious legacy for his country at a special period in its history.

Acknowledgements I am grateful to four charming and accomplished Dhawans for having kindly and generously spoken to me about Satish and his family: his wife Nalini, son Vivek and daughters Amrita and, in particular, Jyotsna. I thank the editors for their helpful comments on a draft of this essay.

References

Abu-Ghannam, B.J. and Shaw, R. 1980. Natural transition of boundary layers – the effects of turbulence, pressure gradient and flow history. *J. Mech. Engg. Sci.* **22**, 213–228.

Ackeret, J., Feldmann, F. and Rott, N. 1946. Inst. Aerodyn. ETH, Report no. 10.

Adamson, Jr., T.C. and Messiter, A.F. 1980. Analysis of two-dimensional interactions between shock waves and boundary layers. *Ann. Rev. Fluid Mech.* **12**, 103–138.

Coles, D.E. 1953. *Measurements in the boundary layer on a smooth flat plate in super-sonic flow.* PhD thesis, Caltech.

Coles, D.E. 1954a. Measurements of turbulent friction on a smooth flat plate in supersonic flow. *J. Aero Sci.* **21**, 433–448.

Coles, D. 1954b. The problem of the turbulent boundary layer. *ZAMP* **5**, 182–203.

Dhawan, S. 1953. Direct measurements of skin friction. NACA Report 1121.

Dhawan, S. 1981. A glimpse of fluid mechanics research in Bangalore 25 years ago. *Proc. Ind. Acad. Sci. (Engg. Sci.)* **4**, 95–109.

Dhawan, S. and Narasimha, R. 1958. Some properties of boundary layer flow during transition from laminar to turbulent motion. *J. Fluid Mech.* **3**, 418–436.

Dryden, H.L. 1936. Air flow in the boundary layer near a plate. NACA Report 562.

Dryden, H.L. 1953. Review of published data on the effect of roughness. *J. Aero. Sci.* **20**, 477–482.

Emmons, H.W. 1951. The laminar-turbulent transition in a boundary layer – Part I. *J. Aero. Sci.* **18**, 490–498.

Emmons, H.W. and Bryson, A.E. 1952. The laminar-turbulent transition in a boundary layer (Part II). *Proc. 1st US Natl. Cong. Appl. Mech.*, 859–868.

Kármán, von, T. 1936. The problem of resistance in compressible fluids. *Att. dei Convegni 5*, R. Accad. d'Italia, pp. 222–277.

Liepmann, H.W. 1943. Investigations in laminar boundary-layer stability and transition on curved boundaries. NACA ACR 3H30.

Liepmann, H.W. 1945. Investigation of boundary layer transition on concave walls. NACA ACR 4J28.

Liepmann, H.W. 1946. The interaction between boundary layer and shock waves in transonic flow. *J. Aero. Sci.* **13**, 623–637.

Liepmann, H.W. 2002. Remembering Satish Dhawan. *Engineering & Science (Caltech)* **65**(4), 41–43.

Liepmann, H.W., Roshko, A. and Dhawan, S. 1951. On reflection of shock waves from boundary layers. NACA Report 1100.

Marusic, I., McKeon, B.J., Monkewitz, P.A., Nagib, H.M., Smits, A.J. and Sreenivasan, K.R. 2010. Wall-bounded turbulent flows at high Reynolds numbers: recent advances and key issues. *Phys. Fluids* **22**, 065103.

Narasimha, R. 1957. On the distribution of intermittency in the transition region of a boundary layer. *J. Aero. Sci.* **24**, 711–712.

Narasimha, R. 1958. *A study of transition from laminar to turbulent flow in the boundary layer of a flat plate.* AIISc thesis, Dept. Aero. Eng., Ind. Inst. Sci., Bangalore.

Narasimha, R. 1985. The laminar–turbulent transition zone in the boundary layer. *Prog. Aerospace. Sci.* **22**, 29–80.

Narasimha, R. 2002. Satish Dhawan. *Current Science* **82**, 222–225.

Narasimha, R. and Deshpande, S.M. 1982. *Surveys in Fluid Mechanics.* Indian Academy of Sciences, Bangalore.

Reynolds, O. 1883. An experimental investigation of the circumstances which determine whether the motion of water shall be direct or sinuous and of the law of resistance in parallel channels. *Phil. Trans. Roy. Soc. A* **174**, 935–982.

Schlichting, H. 1955. *Boundary Layer Theory*. Pergamon Press.

Schubauer, G.B and Klebanoff, P.S. 1955. Contributions on the mechanics of boundary-layer transition. NACA Tech. Note 3489.

Shultz-Grunow, F. 1940. Neues Reibungswiderstandsgesetz für glatte Platten. *Luftfahrtforschung*, **17**, 239–246. Available in English as NACA Tech. Mem. 986 (1941).

Wygnanski, I.J., Haritonidis, J.H. and Kaplan, R.E. 1979. On a Tollmien–Schlichting wave packet produced by a turbulent spot. *J. Fluid Mech.* **92**, 505–528.

12

Philip G. Saffman

D. I. Pullin and Daniel I. Meiron

12.1 Introduction

Philip G. Saffman was a leading theoretical fluid dynamicist of the second half of the twentieth century. He worked in many different sub-fields of fluid dynamics and, while his impact in other areas perhaps exceeded that in turbulence research, which is the topic of this article, his contributions to the theory of turbulence were significant and remain relevant today. He was also an incisive and, some might conclude, a somewhat harsh critic of progress or what he perceived as the lack thereof, in solving 'the turbulence problem'. This extended to his own work; he stated in a preface to lectures on homogeneous turbulence (Saffman, 1968) that

> the ideas ... are new and hopefully important, but are speculative and quite possibly in serious error.

In this article, we will try to survey Saffman's thinking and contribution to turbulence research from the mid 1950s, when he began to mature as a scholar, until the late 1970s when he moved away from the study of turbulence to concentrate on the related but separate area, of the dynamics of isolated and interacting vortices. Although, for the most part, the evolution of his ideas and their application to turbulence in this period developed both thematically and chronologically together, where there are departures we will tend to focus on the former.

12.1.1 Early life

Philip Geoffrey Saffman (19 March 1931 – 17 August 2008) was born in England to parents of Russian and Lithuanian heritage. His father was a solicitor; his mother had a strong proclivity for 'fresh air', and at one point this determined her choice of school for her son. Saffman was one of three brothers,

Figure 12.1 Philip G. Saffman (19 March 1931 – 17 August 2008). Photo circa 1975. Source: Caltech Archives.

and the only one of the Saffman children that did not choose to pursue the legal profession. He and his siblings were raised in the Leeds area of England, but were evacuated to Blackpool during WWII, a move which appeared to have little effect on his early years. Saffman was something of a prodigy, completing his high school certificate and taking the Entrance Scholarship Exam to Cambridge University by the age of fifteen. Cambridge, however, would not admit him, not because of his age, but because he had yet to complete his compulsory military service for which of course he was too young. So he informally attended lectures at Leeds University and learned to play golf. He then entered the Royal Air Force where he became a teleprinter operator and learned to type, a skill that he found useful in the later age of computers, email and computational fluid dynamics. Saffman entered Cambridge in 1950, where his aversion to chemistry propelled him to pursue a major in mathematics.

12.1.2 Graduate study at Cambridge

Completing the Mathematical Tripos examination in 1953, he continued on to graduate study under the supervision of G.K. Batchelor. While Batchelor was on sabbatical for a term, G.I. Taylor acted as a formal substitute supervisor, although their interaction was minimal, and it was not until later that they worked together on viscous fingering in Hele-Shaw flow and what was to become known as the Saffman–Taylor instability. In 1955, Saffman was elected to a Trinity Prize Fellowship which allowed him to continue research after completion of his PhD in 1956. He then accepted a tenure track position as assistant lecturer in mathematics at Cambridge which he held until 1960 when he moved to Hermann Bondi's department at King's College London. At around the time of a 1961 meeting in Marseilles (see below), he met Janos Laufer who invited him to visit Laufer's fluid mechanics group at the Jet Propulsion Laboratory in Pasadena, California. Saffman spent six months at JPL in 1963. There he met G.B. Whitham and H. Liepmann, who encouraged him to consider moving permanently to Caltech, which he did in 1964 as Professor of Fluid Mechanics. Saffman remained at Caltech for the remainder of his academic career. He was was elected a Fellow of the Royal Society in 1990 and appointed von Kármán Professor at Caltech in 1995.

The work described in his 1956 thesis, titled *Some Problems on the Motion of Drops, Bubbles and Particles in Fluids*, was apparently motivated by Batchelor's casual and perhaps flippant observation that rising bubbles in a glass of champagne or carbonated water sometimes did not seem to move strictly vertically but rather upwards in a swirling motion. Saffman's PhD work examines three separate phenomena, two of which were the rise of small bubbles of air in water and the slow motion of small particles in viscous sheared flows. The last and perhaps most interesting, done in collaboration with J.S. Turner and subsequently published in the first issue of the *Journal of Fluid Mechanics* (Saffman and Turner, 1956), considered the coalescence of nearly equal-sized drops in clouds.

It was long known that condensation from the vapor phase was not sufficiently rapid for small water droplets, first nucleated on fine dust particles, to grow to raindrop-sized droplets. The basic idea was that larger drops formed via collisions of more-or-less equally sized small droplets, these collisions being driven by the relative motion of random droplets caused by turbulent motion of the air. Since the droplet sizes are at least an order of magnitude smaller than the Kolmogorov scale, it is to be expected that the relative motion of two drops, and hence the binary collision frequency of drops will be governed by the small-scale turbulent motion. Essentially, it was the spatial variation of the small-scale velocity field which drives the collision dynamics. Utilizing ideas from the kinetic theory of gases, Saffman obtained explicit expressions for the collision rate that accounted for both collisions when drops were moving with the local air velocity and also when drop motion relative to the air was included. The key result depended on various parameters including the mean longitudinal velocity gradient and the mean square acceleration within the driving turbulent motion. An estimate of the former was provided by Taylor (1935) while acceleration statistics were derived from Batchelor's then recent work on pressure fluctuations in turbulence (Batchelor, 1951).

Saffman's PhD work did not then contribute to turbulence research, but it used both well-known and recent advances in understanding of turbulence kinematics and dynamics. It was clear that Saffman was well versed in these ideas. While Taylor had long left turbulence research to work in other areas, Batchelor was highly active in investigations of turbulence dynamics and diffusion, having published his monograph on the theory of homogeneous turbulence in 1953, the same year that Saffman commenced graduate study. Batchelor had recently worked on pressure/acceleration statistics (Batchelor, 1951), Lagrangian motion (Batchelor, 1952), the large-scale structure of turbulence (Batchelor and Proudman, 1956) along with other topics, and had collaborated with A.A. Townsend in writing an influential survey on contemporary turbulence research (Batchelor and Townsend, 1956) (see accompanying articles on A.A. Townsend and G.K. Batchelor, Chapters 9 and 8). As was customary in supervisor/student relations of the period, Batchelor did not place his name on his student Saffman's papers. Indeed, they did not publish jointly, and there is no evidence that they collaborated on research projects, even though, as will be seen, they later worked on closely related topics. Interviews of Saffman by Shirley Cohen for the Caltech Archives in 1989 and 1999 (Cohen, 1999), which provides much of the anecdotal material above on Saffman's early life and Cambridge experience, contains detailed discussion of his collaboration with Taylor and others. It seems clear that Saffman's relations with Batchelor remained cordial but somewhat formal. His work with Taylor is described as a collaboration where Taylor wanted someone to do the mathematics, for which Saffman was eminently well-qualified.

12.1.3 Lectures in homogeneous turbulence

In June 1966, Saffman presented a set of ten 45-minute lectures at the International School of Nonlinear Physics and Mathematics held in Munich at the Max Planck Institute (Saffman, 1968). Although much has transpired in turbulence theory since these lectures were delivered, they remain a remarkable resource. Many of the problems that are discussed in these notes regrettably remain open. It will not be our purpose here to provide a detailed overview of the content of the lectures. Rather we will discuss a few of the problems as a way of highlighting the issues that were critical at the time the lectures were delivered.

12.2 The problem of turbulent diffusion

12.2.1 Fundamental aspects of turbulent diffusion

Saffman's interest in turbulent diffusion was longstanding. For example, he began his 1966 lectures by pointing out that the main reason one is interested in turbulence is the enhancement of diffusion of mass and momentum by a turbulent flow. Although this is intuitively understood, he pointed out that, at that time, there is no rigorous proof that for a turbulent three-dimensional velocity field the distance between two fluid particles must always increase. If two particles (call them P and Q) are close together and thought of as at opposite ends of a straight, infinitesimal line element, then one can write the equation for the (infinitesimal) distance

$$l = x_P - x_Q \tag{12.1}$$

as

$$\frac{dl_j}{dt} = A_{ij}(t)l_i, \tag{12.2}$$

where

$$A_{ij} = \frac{\partial u_i}{\partial x_j} \tag{12.3}$$

is the velocity gradient tensor evaluated at the point P. For turbulence, A_{ij} is a random function of time whose diagonal components are constrained by the incompressibility condition. It is of interest to know under what conditions, if any, the ensemble average (in some sense) of element lengths will be stretched; that is,

$$\frac{d}{dt}\langle |l| \rangle \geq 0, \tag{12.4}$$

where $\langle \cdot \rangle$ is some ensemble average. Since 1966 attempted proofs have been proposed (Cocke, 1969; Orszag, 1970; Etemadi, 1990) for infinitesimal separation, that the average line element length will increase under certain conditions.

12.2.2 Behavior of turbulent diffusivity

Saffman's first direct contribution to turbulence, which predated the 1966 lectures, was his study of the effect of the molecular diffusivity coefficient κ on the mean-square displacement, or dispersion of fluid particles in a given direction from their original position, in the presence of a turbulent velocity field. This is essentially (12.4) in one coordinate direction but with finite separation. In his seminal paper on turbulent diffusion, Taylor (1922) had shown that this could be expressed as

$$\overline{Y^2(t)} = 2 \int_{t_0}^{t} dt' \int_{t_0}^{t'} \overline{u(t')\,u(t'')}\,dt'', \tag{12.5}$$

where $u(t)$ is the fluid element velocity in the specified direction, and the integrand is understood as a Lagrangian autocorrelation of a fluid element at separated times t'' and $t' > t''$. For statistically stationary turbulence, this is

$$\overline{Y^2(t-t_0)} = 2\overline{u^2} \int_{0}^{t-t_0} (t-t_0-\tau) S_p(\tau)\,d\tau, \qquad S_p(\tau) \equiv \frac{\overline{u(t)\,u(t+\tau)}}{\overline{u^2}}, \tag{12.6}$$

and where $S_p(\tau)$ is the normalized Lagrangian autocorrelation function. For the dispersion of a conserved substance or quantity (such as temperature in a mildly heated fluid) that is subject to molecular diffusion but which has no effect on the turbulence dynamics, when the effects of molecular and turbulent diffusion are considered to be independent, (12.6) must be modified by the addition of the strictly diffusive term to give a total dispersion D for the substance (or heat) (Saffman, 1960):

$$D^2 = \overline{Y^2} + 2\kappa(t-t_0). \tag{12.7}$$

Earlier, Townsend (1954) had considered the growth in the width of a thermal wake behind a line source of heat in turbulence. Arguments based on these results suggested that, for times such that $t-t_0$ is small compared to the inverse root-mean-square vorticity of the turbulence,

$$D^2 = \overline{Y^2} + 2\kappa(t-t_0) + \frac{5}{9}\kappa(t-t_0)^3\,\overline{\omega^2} + \text{higher-order terms}, \tag{12.8}$$

where $\overline{\omega^2}$ is the mean-square vorticity. This indicates that the rate at which a volume of heated fluid increases is itself increased by the effect of turbulence. That is, the effect of molecular diffusivity is to accelerate the dispersion over and above the additive independent effects of turbulence and molecular diffusion.

Saffman (1960) considered this interaction problem, starting from a substance Lagrangian autocorrelation function that included both the continuum fluid motion and the underlying random molecular motion,

$$\mathcal{R}(\tau) = \left\langle \overline{[U(t) + q(t)] \, [U(t + \tau) + q(t + \tau)]} \right\rangle, \tag{12.9}$$

where $U(t)$, which differs from $u(t)$, is the continuum velocity at the point occupied by a molecule and $q(t)$ is the random velocity. Here angle brackets denote an average over the random ensemble. The reason that the dispersion of a substance differs from that of elemental fluid particles is that molecules of the substance do not move with fluid particles at the local fluid velocity of the continuum, but rather with this velocity plus a random thermal component governed by the one-molecule probability distribution of molecular velocities relative to the mean continuum velocity.

Using a local solution of the passive-scalar advection-diffusion equation in the presence of the local straining and rotational motion relative to a material fluid element, and assuming that the random and the continuum motions are statistically uncorrelated, Saffman obtained

$$D^2 = \overline{Y^2} + 2\,\kappa\,(t - t_0) - \frac{1}{3}\kappa\,\overline{\omega^2}\,(t - t_0)^3 + \text{higher-order terms}, \tag{12.10}$$

a result that indicates that the effect of diffusivity is to *decelerate* the dispersion, in apparent contradiction to (12.8). The physical interpretation of this result is that the velocity of a molecule shows a smaller (in magnitude) correlation with its velocity at an earlier time in comparison with the material element velocity, which decreases the substance dispersion. This has the effect that the average velocity of a small blob of passive scalar decreases as the volume of the blob increases under molecular diffusion. Saffman attributed the discrepancy between (12.10) and (12.8) to the fact that Townsend had assumed that the instantaneous axis of the thermal wake was coincident with the direction of fluid particles passing through the source, whereas Saffman's analysis shows that, owing to the difference between the molecular and the macroscopic fluid velocity, this is not the case. Although (12.10) is valid only for small times, Saffman (1960) gave heuristic arguments that at least the sign of the interaction term would persist for longer. While experimental evidence (Mickelsen, 1960; Micheli, 1968) is inconclusive, Saffman's intriguing results remain interesting and counter-intuitive.

12.2.3 Marseilles meeting

The year 1961 saw two consecutive meetings on aspects of turbulence in Marseilles, France. The first was sponsored by the Centre Nationale de la Recherche Scientifique (CNRS) while the second was a joint International Union of Theoretical and Applied Mechanics (IUTAM) and International Union of Geodesy and Geophysics (IUGG) Symposium. This was a period when it was possible for almost all those active in a research discipline to attend a moderate-sized meeting. Participants included Taylor, Batchelor, Saffman, Moffatt, von Kármán, Liepmann, Roshko, Coles, Corrsin, Lumley, Kovasznay, Kraichnan, Kolmogorov, Obukhov, Yaglom, Millionshchikov, Favre, Hinze and others. Yaglom (1994) reports that

> this was the first and only time when A.N. (Kolmogorov) actively participated in an international meeting on turbulence.

It was also Saffman's first international meeting on turbulence. The first meeting at Marseilles is often remembered owing to Kolmogorov's famous presentation (Kolmogorov, 1962a) of his refined model of turbulent fluctuations later published as Kolmogorov (1962b) (see Chapter 6 on the Russian school). Saffman's contribution was an extension of his earlier work on turbulent diffusion (Saffman, 1962). His presentation was discussed in question time by Kraichnan, Yaglom and Liepmann.

12.2.4 Turbulent diffusion of passive vector fields

As discussed earlier, the statistical evolution of material surfaces and material lines within a turbulent velocity field was in the 1950s, and remains today, a topic of considerable interest in fluid dynamics and in magnetohydrodynamics (Moffatt, 2007). Saffman (1963) studied the small-scale statistical properties of vector fields F_i and G_i satisfying

$$\frac{\partial F_i}{\partial t} + u_j \frac{\partial F_i}{\partial x_j} = F_j \frac{\partial u_i}{\partial x_j} + \lambda \frac{\partial^2 F_i}{\partial x_j^2}, \qquad \frac{\partial F_i}{\partial x_i} = 0, \qquad (12.11)$$

$$\frac{\partial G_i}{\partial t} + u_j \frac{\partial G_i}{\partial x_j} = -G_j \frac{\partial u_i}{\partial x_j} + \lambda \frac{\partial^2 G_i}{\partial x_j^2}, \qquad \epsilon_{ijk} \frac{\partial G_j}{\partial x_k} = 0, \qquad (12.12)$$

where λ is a molecular or magnetic diffusivity. The magnetic field B_i in magnetohydrodynamics satisfies an equation similar to (12.11), as does the vorticity ω_i with $\lambda = \nu$. Under the conditions that neither F_i, nor G_i appear explicitly in the equations of motion for $u_i(x, t)$ nor do their initial conditions correspond to those of a dynamical quantity also governed by (12.11–12.12), then F_i, G_i

are considered passive vector fields. In particular, if the Lorentz force, proportional to $\epsilon_{ijk} J_j B_k$ (here J_i is the current density), is neglected in the Navier–Stokes momentum equations, then B_i behaves as a passive vector field. When $\lambda = 0$, (12.11) governs the evolution of infinitesimal material line elements. Equation (12.12) gives the evolution of $G_i = \partial\theta/\partial x_i$, where $\theta(x, t)$ is a passive scalar field whose evolution is described by the advection–(Fickian-)diffusion equation with scalar diffusivity $\lambda > 0$. The evolution of the gradient of $\phi(x, t)$ whose level sets correspond to Lagrangian material surfaces is given by (12.12) with $\lambda = 0$. An issue of interest is the long-time behavior of $\overline{F_i^2}$, $\overline{G_i^2}$ for small but finite diffusivity $\lambda \ll \nu$ when fluctuations are introduced at some scale. The significant example is the turbulent dynamo, where it is hypothesized that long-time fluctuations in B_i can be sustained by the turbulent motion.

Batchelor (1952) had considered the statistical properties of passive vector fields satisfying (12.11), concluding that, when $\lambda \ll \nu$, mean-square values can grow exponentially in time owing to the predominance of stretching of passive vectors by the rate-of-strain field of the turbulence. Focusing attention on fine-scale fluctuations at length scales less than the Kolmogorov length $(\nu^3/\epsilon)^{1/4}$, where ν is the kinematic viscosity and ϵ the mean kinetic energy dissipation, Saffman (1963) effectively extended to vector fields the earlier study of Batchelor (1959) for the sub-Kolmogorov-scale behavior of passive scalars. Starting from (12.11), (12.12), Saffman worked with equations for the covariance tensors $\overline{F_i F_j'}$ and $\overline{G_i G_j'}$, using their forms for isotropic distributions in terms of longitudinal and lateral scalar correlation functions. He utlized the idea that the local rate-of-strain is persistent, and that F_i and G_i tend to align with long-time (asymptotic) orientations of material lines and the normal to material surfaces respectively This was used to model the respective third-order correlations, or transfer terms, for small separation r, in terms containing the eigenvalues $0 < \alpha_1 \geq \alpha_2 \geq \alpha_3 < 0$, $\alpha_1 + \alpha_2 + \alpha_3 = 0$ of the local rate-of-strain tensor S_{ij}, and, in particular, a parameter $\sigma = -\overline{\alpha_1}/\overline{\alpha_3}$. He argued that the question of growth or decay appears to depend critically on the role played by the statistics of the eigenvalues of S_{ij} and on the spatial coherence of the local S_{ij} field.

Saffman concluded that, when perturbations of the vector fields are introduced at some scale, the intensification of the field through stretching and the production of smaller scales would ultimately be overcome by enhanced ohmic diffusion with subsequent long-term decay in both $\overline{F_i^2}$, $\overline{G_i^2}$. In particular, there is no sustained turbulent dynamo. Later, however, Moffatt (1970) showed that $\overline{F_i^2}$ will grow, with subsequent dynamo action, even when $\lambda \gg \nu$ provided only that the turbulence lacks reflection symmetry, which occurs when the helicity $\overline{\omega_i u_i}$ is nonzero. For the case $\lambda = \nu$ the issue has been further explored using modern computational methodology. When the turbulence is decaying, the behavior and ultimate decay of the F_i-field depends on the proximity of its initial condition to that of ω_i (Ohkitani, 2002). For statistically stationary turbulence, with both non-helical and strongly helical forcing, Tsinober and Galanti (2003) find exponential growth in $\overline{F_i^2}$ but decay in $\overline{G_i^2}$, so that these fields behave quite differently. Notably, the long-time behavior of F_i is qualitatively different to ω_i because the latter is constrained by $\omega_i = -\epsilon_{ijk}\,\partial u_j/\partial x_k$. This leads to long-time vorticity–rate-of-strain correlations which evidently limit vortex stretching but not passive-vector stretching. Saffman (1963) is correct for G_i but not for F_i.

12.3 Contributions to the theory of homogeneous turbulence

12.3.1 Homogeneous isotropic turbulence

In the mid 1960s a great deal of theoretical attention was being given to so-called homogeneous isotropic turbulence. Such an idealization is impossible to achieve experimentally, as even the best grid turbulence experiments do not approximate homogeneity well enough, but it was thought that the turbulence was of this character at scales intermediate between the largest and those associated with viscous dissipation (the so-called inertial range), and, indeed, there is evidence of this in simulations as well as experiments. But another motivation for making this simplification was that it made the theoretical developments more tractable. Saffman makes the following trenchant comment in this regard Saffman (1968):

> homogeneous turbulence is a theoretical idealization and it has been said ... with too much truth for comfort that the purpose of homogeneous turbulence is to provide full employment for mathematicians.

But the theory of homogeneous isotropic turbulence is useful in that it provides some important simplifications. Our main interest in a turbulent flow is to characterize the correlation tensors such as

$$R_{ij}(r) = \left\langle u_i(x,t) u_j(x+r,t) \right\rangle. \tag{12.13}$$

It is also often convenient to express results in terms of the 'spectral tensor' which is the Fourier transform of the velocity correlation tensor:

$$\Phi_{ij}(k) = \frac{1}{(2\pi)^3} \int R_{ij}(r) \exp(-ik \cdot r)\, dr. \tag{12.14}$$

One of the most important descriptors of the turbulent flow is the energy spectrum defined by

$$E(k) = 1/2 \int_{|k|=k} \Phi_{ii}(k) dA(k), \tag{12.15}$$

where $A(k)$ is the area of a shell in k-space of radius $|k| = k$.

Another spectral tensor which will play a role in the discussions below is the vorticity spectrum tensor defined by

$$\Omega_{ij}(k) = \frac{1}{(2\pi^3)} \int \langle \omega_i(x)\omega_j(x + r) \rangle \exp(-ik \cdot r) dr, \tag{12.16}$$

where ω_i is the fluid vorticity. The vorticity and velocity correlation tensors are related in Fourier space by

$$\Omega_{ij}(k) = \left(k^2\delta_{ij} - k_ik_j\right)\Phi_{kk}(k) - k^2\Phi_{ji}(k), \tag{12.17}$$

and, in particular,

$$\Omega_{ii}(k) = k^2\Phi_{ii}(k), \tag{12.18}$$

and

$$\overline{\omega^2} = 2 \int_0^\infty k^2 E(k) dk, \tag{12.19}$$

which is also related to the dissipation in the flow.

If we assume homogeneity and isotropy, then it can be shown (Batchelor, 1953) that the velocity correlation tensor takes the simple form

$$R_{ij}(r) = \overline{u_1^2} \left[\frac{f(r) - g(r)}{r^2} r_i r_j + g(r)\delta_{ij} \right], \tag{12.20}$$

where

$$g = f + rf'/2, \tag{12.21}$$

and u_1 is the x-component of the velocity. The functions $f(r)$ and $g(r)$ are, respectively, the longitudinal and lateral velocity correlations. The advantage of studying homogeneous isotropic turbulence is that one scalar function $f(r)$ is sufficient to describe the velocity correlation tensor.

The function $f(r)$ has a small distance expansion of the form

$$f(r) = 1 - \frac{r^2}{\lambda^2} + O(r^4), \tag{12.22}$$

where λ is the Taylor microscale and is given by

$$\lambda = \frac{\left[\overline{u_1^2}\right]^{1/2}}{\left[\overline{\left(\frac{\partial u_1}{\partial x_1}\right)^2}\right]^{1/2}} \tag{12.23}$$

For isotropic turbulence, the Taylor microscale is related to the dissipation rate of the flow via the relation

$$\epsilon = -\frac{d}{dt}\frac{\overline{\boldsymbol{u} \cdot \boldsymbol{u}}}{2} = \nu\overline{\left(\frac{\partial u_i}{\partial x_j}\right)^2} = \nu\overline{\omega^2} = 15\nu\frac{\overline{u_1^2}}{\lambda^2}. \tag{12.24}$$

This suggests that there may be a special role played by the Taylor microscale in that there may be some physical process that operates at this scale which is responsible for the bulk of the turbulent dissipation. We discuss this further below, but it is indicative of Saffman's attempt to understand turbulence in terms of the interactions of flow structures.

The third-order correlation tensor (associated with the skewness of the velocity field) is given by

$$S_{ijk}(\boldsymbol{r}) \equiv \langle u_i(\boldsymbol{x})u_j(\boldsymbol{x})u_k(\boldsymbol{x}+\boldsymbol{r})\rangle, \tag{12.25}$$

and can be shown to be of the form

$$S_{ijk}(\boldsymbol{r}) = (\overline{u_1^2})^{3/2}\left[\frac{k(r)-rk'(r)}{2r^3}r_ir_jr_k + \frac{2k+rk'}{4r}\left(r_i\delta_{jk} + r_j\delta_{ik}\right) - \frac{r}{2r}r_k\delta_{ij}\right], \tag{12.26}$$

where the function $k(r)$ is defined by

$$k(r) = \overline{(u_1'-u_1)^3}/[6(\overline{u_1^2})^{3/2}]. \tag{12.27}$$

Even using the simplification of homogeneous, isotropic turbulence the fundamental problem of closure of the equations for the correlation functions of course remains. If we take the Navier–Stokes equations

$$\frac{\partial u_i}{\partial t} + \frac{\partial}{\partial x_j}(u_iu_j) = -\frac{1}{\rho}\frac{\partial P}{\partial x_i} + \nu\nabla^2 u_i, \qquad \frac{\partial u_i}{\partial x_i} = 0, \tag{12.28}$$

and attempt to compute an evolution equation for the second-order correlation tensor $R_{ij}(\boldsymbol{r}, t)$ and further make the assumptions of homogeneity and isotropy, we derive the Kármán–Howarth equation in terms of the previously defined correlation functions $f(r)$ and $k(r)$:

$$\frac{\partial}{\partial t}\left[\overline{u_1^2}f(r)\right] = (\overline{u_1^2})^{3/2}\left(\frac{\partial}{\partial r} + \frac{4}{r}\right)k(r) + 2\nu\overline{u_1^2}\left(\frac{\partial^2}{\partial r^2} + \frac{4}{r}\frac{\partial}{\partial r}\right)f(r), \tag{12.29}$$

which relates a second-order correlation to a third-order correlation – the essence of the closure problem. This closure problem, as Saffman pointed out, is directly related to our lack of understanding of how to characterize turbulent dissipation. If we define

$$U(r,t) = 2\overline{u_1^2}\,[1 - f(r)], \qquad S(r,t) = \frac{6k(r)}{2^{3/2}(1 - f(r))^{3/2}}, \qquad (12.30)$$

then the Kármán–Howarth equation becomes

$$\frac{\partial U}{\partial t} + \frac{2^{1/2}}{3}\left(\frac{\partial}{\partial r} + \frac{4}{r}\right) S\,U^{3/2} - 2v\left(\frac{\partial^2}{\partial r^2} + \frac{4}{r}\right) U = \frac{\partial \overline{u_1^2}}{\partial t} = \frac{2\epsilon}{3}. \qquad (12.31)$$

If we understood how to model the dissipation, we would also then understand how to express the relation between the correlation functions. Further, the dissipation is exactly the mean square vorticity:

$$\epsilon = v\overline{\omega^2}. \qquad (12.32)$$

With this brief background, we next consider Saffman's contributions.

12.3.2 Decay of homogeneous isotropic turbulence

An important contribution was made by Saffman to the problem of the dynamics of the large eddies in homogeneous isotropic turbulence and, in particular, their rate of decay. The first estimate of this decay was made by Loitsyanski (1939) who examined what is now called *Loitsyanski's integral*

$$I = \int r^2 \langle \mathbf{u} \cdot \mathbf{u}' \rangle \, d\mathbf{r}. \qquad (12.33)$$

In the case of homogeneous isotropic turbulence this simplifies to

$$I = 8\pi u^2 \int r^4 f \, dr, \qquad (12.34)$$

where $u = \left(\overline{u_1^2}\right)^{1/2}$. Loitsyanski showed that, provided distant points in the turbulent region are statistically independent, the integral I was an invariant of the turbulent motion. He derived this result by integrating the Kármán–Howarth equation (12.29) over all r.

Landau (Landau and Lifshitz, 1987) showed that this result is related to the conservation of angular momentum

$$\mathbf{M} = \int \mathbf{x} \times \mathbf{u} \, dV. \qquad (12.35)$$

In particular, the angular momentum is related to I via

$$I = \langle M^2 \rangle / V + O(L/R) \tag{12.36}$$

where R is the radius of the assumed spherical turbulent domain, V is the volume of the sphere and L is the integral scale. In fact, the rate of change of angular momentum is related to the applied external torques, but the idea here is that these torques are oriented randomly and so, after averaging, the angular momentum should be conserved.

Landau considered next the rough value of the integral and argued that it should be of order

$$I \approx u^2 L^5 = \text{constant.} \tag{12.37}$$

By equating the rate of decay of kinetic energy of the flow to the dissipation we get

$$\rho u^2 / t = \rho u^3 / L, \tag{12.38}$$

and so we obtain a relation between the characteristic velocity, the time and the integral scale:

$$L \approx ut. \tag{12.39}$$

Using (12.37) we get an estimate of the decay rate of the turbulent velocity as well as the growth rate of the integral scale:

$$u \sim t^{-5/7}, \qquad L \sim t^{2/7}, \tag{12.40}$$

as derived by Kolmogorov (1941). There is another important consequence of this invariance which relates to an idea known as 'the permanence of the large eddies'. If I is indeed invariant, then at large distances (or small wavenumbers k), it can be shown that the energy spectrum $E(k)$ must vary as

$$E(k) = \left[\frac{I}{4\pi^2} \right] k^4 + \cdots , \tag{12.41}$$

and so the energy spectrum has a fixed shape at small wavenumbers.

Note that in order for any of this to make sense, the correlation function, f, must vanish sufficiently rapidly at large distances; this is implied in both Landau's and Loitsyanski's derivations. This assumption was not thought to be very crucial since at large distances the velocities at two distant points are assumed to be statistically independent. Proudman and Reid (1954) examined the consequences of the then-popular quasi-normal approximation closure scheme for the large distance behavior of the velocity correlations. They found that, for example, third-order correlations vanish only as fast as r^{-3}. Batchelor and Proudman (1956) considered anisotropic turbulence and showed that it was also possible there to have the third-order correlations vanish only as r^{-4}. Using the Kármán–Howarth result (12.29), this implies the second-order velocity correlations vanish only as r^{-5} and this makes the Loitsyanski integral at best conditionally convergent. Indeed, in the discussion of this issue in their text *Fluid Mechanics*, Landau and Lifshitz (1987) make the following uncharacteristic comment:

> Doubts have been recently expressed more than once concerning the applicability of the conservation law on account of the behavior of the velocity correlation at very large distances; for example, if this correlation does not decrease sufficiently rapidly, the integral may diverge. This whole subject seems to be as yet unclear.

Motivated by a discussion with H.W. Liepmann that cast doubt on the k^4 law, Saffman (1967) showed that, depending on the initial conditions, the Loitsyanski invariant might not exist, and that the large-scale energy spectrum could vary as

$$E(k) \sim Ck^2 + \cdots . \tag{12.42}$$

Batchelor and Proudman (1956) had assumed that at some initial time t_0 the turbulent velocity field was homogeneous and had convergent integral moments of cumulants of the velocity distribution. Saffman instead considered the dynamic evolution of a velocity field that had convergent integral moments of the cumulants of the vorticity. Such an initial condition is less restrictive because the convergence of velocity moments automatically implies the convergence of vorticity moments, but the converse is not true.

If all the integral moments of the vorticity distribution exist, then it follows that the spectral tensor of the vorticity, (12.16), has the following form

$$\Omega_{ij}(\boldsymbol{k}) = B_{ijmn}k_m k_n + O(k^3), \tag{12.43}$$

and further applications of the incompressibility constraint restrict the four-index tensor B_{ijkl} to the form

$$B_{ijkl} = \epsilon_{ik\alpha}\epsilon_{jl\beta}M_{\alpha\beta}, \tag{12.44}$$

where M is a symmetric tensor. As a result,

$$\Omega_{ij}(\boldsymbol{k}) = \left(\delta_{\alpha\beta} - \frac{k_\alpha k_\beta}{k^2}\right)k^2 M_{\alpha\beta} + O(k^3). \tag{12.45}$$

For small k the spectral velocity tensor, (12.14), can be recovered from Ω_{ij} via the relation

$$\Phi_{ij} = \left(\delta_{i\alpha} - \frac{k_i k_\alpha}{k^2}\right)\left(\delta_{i\beta} - \frac{k_i k_\beta}{k^2}\right)M_{\alpha\beta} + O(k), \tag{12.46}$$

from which follows the large-scale energy spectrum

$$E(k) = \frac{4\pi}{3} M_{\alpha\alpha} k^2 + O(k^3).$$ (12.47)

Unless the coefficient $M_{\alpha\alpha}$ vanishes, the energy spectrum will vary as k^2 rather than k^4. This also has implications for the decay rate of the turbulence in the final stages when the viscosity is important. At late times, Milliontschikov (see Monin and Yaglom, 1971) showed that the energy must decay as $t^{-5/2}$. If $M_{\alpha\alpha}$ does not vanish the energy decays like $t^{-3/2}$.

Saffman showed that there was no reason that $M_{\alpha\alpha}$ must vanish. He assumed that the turbulent motion was generated by uncorrelated impulsive forces f at the initial time. Then the impulsive approximation can be used to show that

$$\boldsymbol{u} = \boldsymbol{f} + \nabla\phi,$$ (12.48)

$$\nabla^2 \phi = -\nabla \cdot \boldsymbol{f},$$ (12.49)

$$\boldsymbol{\omega} = \nabla \times \boldsymbol{f}.$$ (12.50)

He further assumed that the spectral tensor of the force correlation exists and decreases exponentially with increasing distance. In that case, it is known as a result of analyticity that, as a function of wavenumber, the spectral correlation tensor of the force, $\mathcal{M}_{ij}$, will have the form

$$\mathcal{M}_{ij} = M_{ij} + O(k).$$ (12.51)

Using these results, Saffman derived the initial spectral tensors of the vorticity and velocity correlation functions:

$$\Omega_{ij}(k) = \epsilon_{ij\alpha}\epsilon_{jl\beta}k_k k_l \mathcal{M}_{\alpha\beta}(k),$$ (12.52)

$$\Phi_{ij}(\boldsymbol{k}) = \left[\delta_{i\alpha} - \frac{k_i k_\alpha}{k^2}\right]\left[\delta_{j\beta} - \frac{k_j k_\beta}{k^2}\right]\mathcal{M}_{\alpha\beta}(k).$$ (12.53)

We see immediately that there is a direct relationship between $\mathcal{M}$ and M as introduced in (12.44). Thus, turbulence generated by a solenoidal set of impulsive forces will lead to a velocity field in which moments of the vorticity correlation tensor will exist.

The existence of the Loitsyanski invariant now rests on whether or not M_{ij} vanishes. This in turn, as Saffman showed, depends on the way the flow is forced. A nonzero value of M_{ij} implies a force distribution whose correlation function grows linearly with increasing volume whereas the vanishing of M_{ij} implies a force–force correlation function that does not grow as rapidly. For example, a turbulent motion arising from some dynamic instability presumably imparts a bounded fluctuating momentum and we would then have $M_{ij} = 0$.

Saffman considered the case $M_{ij} \neq 0$ and examined the subsequent dynamics of the flow once this type of velocity distribution is set up. He argued that the subsequent redistribution of the momentum by the turbulence past the initial time will maintain the original force–force correlations. This maintains the near constant value of M_{ij} and leads to a large eddy structure which is a dynamical invariant of the motion, but with an energy spectrum that has the form

$$E(k) = \frac{4\pi}{3} M_{\alpha\alpha} k^2 + O(k^4).$$ (12.54)

He further considered the special case of isotropic turbulence and established that in this case

$$E(k, t) = 4\pi M k^2 + O(k^4),$$ (12.55)

where M is a dynamical invariant. This leads to the asymptotic form of the velocity correlation tensor

$$R_{ij} \sim 2\pi^3 M \left(\frac{3 r_i r_j}{r^5} - \frac{\delta_{ij}}{r^3} \right).$$ (12.56)

In terms of the correlation function $f(r)$ we have

$$R_{ij} = u^2 \left(f + \frac{r f'}{2} \right) \delta_{ij} - \frac{f'}{2r} r_i r_j,$$ (12.57)

which means the longitudinal correlation function decays as

$$u^2 f(r) \sim 4\pi^2 \frac{M}{r^3}.$$ (12.58)

But this means that the Loitsyanski invariant for homogeneous isotropic turbulence given by

$$I = \int_0^\infty r^4 f(r) dr,$$ (12.59)

does not even exist.

Saffman's own view of this work was typically self-deprecating. In his lectures he commented Saffman (1968):

> The work ... on the structure and invariance of the large eddies is believed to be both new and correct, but is of no real importance.

Presumably, he was of this opinion because the amount of energy in these scales is a small part of the turbulent energy budget. In addition, at the time of this work, it was not a simple matter to get experimental measurements which could settle the issue. The best measurements on grid turbulence at the time due to Batchelor and Townsend (1956) showed a $t^{-5/2}$ final decay of the flow which is consistent with the existence of the Loitsyanski invariant as opposed to a $t^{-3/2}$ behavior that would have arisen from the Saffman picture of the large scales.

Davidson (2009) shows that the issue is far from being of no real importance and in fact is not fully settled. It is also the case that the invariant can and does exist in the case of more complex turbulent flows such as those exhibiting anisotropy or originating from additional physical forcings such as those of magnetohydrodynamics. The invariant can also exist for some special kinematically consistent initial conditions, but for three-dimensional, fully dynamic, inhomogeneous isotropic turbulence, Saffman's original results are valid. Davidson discusses the distinction between what is now called 'Batchelor turbulence', in which Loitsyanski's invariant does exist with negligible variation, and what is termed 'Saffman turbulence', where the invariant does not exist and the decay rates are different. As Davidson shows (and as Saffman appreciated) the initial conditions play a decisive role.

There have been further computational and experimental investigations of this question since the original Saffman result. Experimentally, the decay of the large eddies was examined by several investigators using grid-generated turbulence generated in wind tunnels. These experiments have not been decisive in determining whether the grid turbulence is of 'Batchelor' or 'Saffman' type. Part of the difficulty here is the need to determine power-law behavior with limited data due to the small size of the wind-tunnel test sections. Recently, Krogstad and Davidson (2010) have performed new experiments in a much larger facility which provides a larger test section, and thus a greater range over which to measure power laws. Their results seem to indicate that grid turbulence is better described as being of Saffman type than being of Batchelor type.

The work remains important because it indicates the sometimes subtle role of initial conditions in turbulence and also illuminates the importance of the initial vorticity of the flow. This key role played by the vorticity is a theme that occurs repeatedly in Saffman's work as further discussed below.

12.3.3 Basis of the Kolmogorov law

One of the most important contributions to the theory of turbulence was the argument put forth originally by Kolmogorov and others that the small-scale structure of turbulence, which receives energy from the large-scale eddies, should be statistically independent of those large eddies. This is the concept of the energy cascade that in turn led to the development of the inertial range in the energy spectrum. Our purpose here is not to repeat these arguments, which are discussed in detail elsewhere in this volume, but to provide Saffman's perspective on the Kolmogorov theory. As the theory states, if an inertial range exists, a similarity argument indicates that the only relevant dimensional quantities are the turbulent dissipation ϵ, and the physical viscosity of the flow ν. From these a length scale, the Kolmogorov length, and a characteristic velocity u can be defined respectively as

$$\eta = \left(\frac{\nu^3}{\epsilon}\right)^{1/4}, \qquad u = (\nu\epsilon)^{1/4}. \tag{12.60}$$

The hypothesis further states that all statistical quantities in this inertial range at length scales $r \ll L_p$, where L_p is the integral scale associated with the energy spectrum, should be isotropic and should scale only in terms of η and u. This means for example that

$$\frac{\overline{(u_1 - u_1')^2}}{(\epsilon\nu)^{1/2}} = f_e(r/\eta), \qquad \frac{\overline{(u_1 - u_1')^3}}{\left[\overline{(u_1 - u_1')^2}\right]^{3/2}} = S_e(r/\eta), \qquad \frac{\overline{(u_1 - u_1')^4}}{\left[\overline{(u_1 - u_1')^2}\right]^2} = Q_e(r/\eta),$$
$$\tag{12.61}$$

and so forth, where f_e, S_e, and Q_e are universal functions. These quantities are simply related to the correlation functions discussed in section 12.3.1. Note that this also provides values of the skewness and flatness factors for the velocity distribution, and predicts that they are independent of the Reynolds number (a result that was later modified in (Kolmogorov, 1962b).

As long as the integral scale and the Kolmogorov length are well separated (i.e. $\eta \ll L_p$), and one assumes that between these scales energy is transferred at the rate ϵ, then one obtains the celebrated inertial sub-range spectrum

$$E(k) = K\epsilon^{2/3}k^{-5/3}, \tag{12.62}$$

where K is also meant to be a universal constant.

As is well known, the experimental evidence for this energy spectrum both supports and contradicts the hypothesis. On the one hand, the spectrum is seen experimentally in a number of settings. On the other hand, the behavior of higher order correlations such as the skewness and flatness factors show a clear dependence on Reynolds number.

In his 1966 lectures, Saffman argued that the form of the energy spectrum could be obtained under assumptions that were much less restrictive than those of an inertial cascade. In addition, he pointed out that it ought to be possible to derive the spectrum as a consequence of approximations made to the Navier–Stokes equations that could be connected to some sort of physical picture of the turbulent dissipation. While there have been many important contributions, this objective has yet to be fully realized.

Saffman observed that the Kolmogorov hypothesis could also be used to infer the energy spectrum associated with the Burgers equation,

$$\frac{\partial u}{\partial t} + u\frac{\partial u}{\partial x} = \nu\frac{\partial^2 u}{\partial x^2}, \tag{12.63}$$

that was put forward by Burgers in unpublished lectures at Caltech as a one dimensional model of turbulence. Saffman pointed out that there is a formal similarity between the Burgers equation and the Navier–Stokes equation as regards, for example, the Kármán–Howarth equation. Moreover, in one space dimension, cascade-like arguments can be applied to the Burgers equation, where fluid speed and hence kinetic energy is conserved on characteristics (particle paths). For smooth random solutions of (12.63) this reasoning seems plausible and suggests $E(k) \sim k^{-5/3}$, but in fact is wrong. Owing to the intersection of characteristics, which in one dimension brings into close proximity material particles with different speeds, jumps or shocks form whose thickness is $\delta = O(\nu/u')$ where u' is the root-mean-square speed. Burgers pointed out that solutions of the viscous Burgers equations comprised of shocks with random amplitude and spacing would give

$$E(k) \sim k^{-2}, \qquad L^{-1} \ll k \ll \delta^{-1}, \tag{12.64}$$

where L is proportional to the mean separation between shocks. Equation (12.64) is the power spectrum of a continuous stochastic function with random discontinuities. This is now well supported by numerical simulations. In one space dimension, there is no statistical transfer of energy to fine scales through a continuum of decreasing scales and therefore no physical cascade, but instead solutions are attracted to random superpositions of big and small scales, namely the spacing between shocks and the shocks themselves, the latter suitably smoothed by viscosity.

Of course, turbulence is far more complicated than the solutions of Burgers equation as a result of the inherent three-dimensionality of the flow and the constraints imposed by incompressibility which forbids the formation of shock-like structures. Saffman, however, felt that modal representations that lead to the Kolmogorov hypothesis were not the proper idiom for understanding turbulent dissipation, and remained interested for his entire career in identifying so-called 'vortical states', configurations of the vorticity field that were consistent with the dynamics of the Navier–Stokes equations and that could be used to understand from a physical space point of view the statistics of turbulent flow.

There are many exact solutions to the Navier–Stokes equations that represent simple vortical structures for which one can also obtain the response of the structure to the non-local straining flow induced by surrounding vorticity. Examples include vortex layers and tubes used to model turbulence by Townsend (1951). It has long been appreciated, however, that these are not fully consistent as descriptions of the structure of turbulence. First, the energy spectra of vortex sheets decays with increasing wave number like k^{-2} rather that $k^{-5/3}$. In addition, such simple structures are dynamically inconsistent in that sheets can be rolled up into tubes and tubes can be stretched into sheets by local straining flows.

Saffman proposed a heuristic hybrid structure which leads to a possible origin of the Kolmogorov scale. He assumed that the kinetic energy of the turbulent flow was associated with large eddies of length scale L and characteristic velocity

$$u = \left[\overline{u^2}/3\right]^{1/2}.$$ (12.65)

These eddies produce a straining field with rate of strain $\alpha \approx u/L$. The straining field will produce sheets and tubes of vorticity, and Saffman posited that it is the regions of enhanced dissipation arising from both sets of structures and their interactions that leads to the bulk of the dissipation. As the sheets are stretched by the straining flow, they will take on a characteristic length scale

$$\delta = \sqrt{\frac{\nu}{\alpha}} = \sqrt{\frac{\nu L}{u}}.$$ (12.66)

He further assumed the characteristic velocity of the vortices was also given by u.

He then considered as an example of subsequent motion the response of a weak vorticity field packed in a sphere of radius L to a simple velocity field of the form

$$\mathbf{u} = \alpha x \mathbf{i} + \beta y \mathbf{j} + \gamma z \mathbf{k}, \qquad \alpha + \beta + \gamma = 0.$$ (12.67)

Depending on the relative signs of the strains, such a sphere will either stretch into a sheet or will be pulled into an elliptic cylinder. Because of the stretching, the vorticity which was originally of size u/L will be amplified to $O(u/\delta)$. The sheet will bend back and forth about itself in the volume and Saffman estimated this would result in a net area within the box of size $(L/\eta)^{1/2}L^2 = (L^5/\delta)^{1/2}$. Therefore, the volume of fluid taken up by vortical sheets is

$$\sigma_{sheet} = \delta(L^5/\eta)^{1/2}/L^3 = (\delta/L)^{1/2}.$$ (12.68)

A similar argument for tubes leads to an estimate

$$\sigma_{tubes} = \delta/L.$$ (12.69)

This process was termed by Saffman the 'primary cascade'. He then estimated the amount of energy dissipation arising from the straining field, the collection of vortex sheets and the vortex tubes, and showed that the dissipation arising from sheets scales as

$$\epsilon_{sheet} \approx \nu \frac{u^2}{\delta^2} \sigma_{sheet} = \frac{1}{Re^{1/4}} \frac{u^3}{L},$$ (12.70)

and that from tubes as

$$\epsilon_{tubes} \approx \nu \frac{u^2}{\delta^2} \sigma_{tubes} = \frac{1}{Re^{1/2}} \frac{u^3}{L}.$$ (12.71)

Note that there is significant enhancement of dissipation due to sheets, but it is still dependent on the outer Reynolds number, Re. In addition, there is no obvious role for the Kolmogorov length.

Saffman hypothesized that secondary instabilities would arise on the vortex structures. For example, sheets could undergo Taylor–Gortler instabilities and tubes could undergo vortex breakdown or Couette-type instabilities. In this case, a secondary structure would arise with a characteristic length scale of δ such as that of the Gortler cells on the original sheets. These cells would be separated by internal layers with a characteristic thickness of size

$$\eta = \sqrt{\frac{\nu \delta}{u}} = \left(\frac{\nu^3 L}{u^3} \right)^{1/4}.$$ (12.72)

The characteristic vorticity in these internal layers is $O(u/\eta)$ and the fraction of volume occupied by these layers is $O((\eta/\delta)\sigma_{sheet})$ since most of the dissipation is taken up by sheets. This was termed the 'secondary cascade' by Saffman. Note that the total dissipation associated with the secondary cascade is

$$\epsilon_{secondary} \approx \nu \frac{u^2}{\eta^2} \frac{\eta}{\delta} \sigma_{sheet} = \frac{u^3}{L},$$ (12.73)

which is now independent of the Reynolds number. In addition we have the first equation of (12.60) for the Kolmogorov length. Note too, that the characteristic thickness of the sheets is the Taylor microscale. This is a quite different physical picture from that put forth by Kolmogorov. Saffman used a similar approach in his study of two-dimensional turbulence discussed below.

Saffman (1968) (see also Saffman, 1970b) also estimated the way in which the structure functions of the velocity of arbitrary order, in contrast to (12.61), depended on the Reynolds number based on the Taylor scale R_λ at small separation. This gave the prediction that the longitudinal structure functions at separation $r \ll \lambda$ were of the form (in Saffman's notation)

$$\left(\Delta u_p\right)_{2n} \equiv \left\langle \left(u'_p - u_p\right)^{2n} \right\rangle = u^{2n} f_{2n}\left(\frac{r}{\eta}\right) R_\lambda^{-1}, \qquad (12.74)$$

$$\left(\Delta u_p\right)_{2n+1} \equiv \left\langle \left(u'_p - u_p\right)^{2n+1} \right\rangle = u^{2n+1} f_{2n+1}\left(\frac{r}{\eta}\right) R_\lambda^{-1}, \qquad (12.75)$$

where the functions f_{2n} and f_{2n+1} are expected to be approximately the same for all kinds of turbulence, but may depend on the anisotropy of the large scales.

There is of course much that can be criticized in Saffman's picture. For example, the characteristic velocity is assumed to be u uniformly within the vortices. It is also possible that other scales can be generated beyond that of the secondary cascade. This approach can be criticized as imprecise in the sense that it is neither an exact consequence of Navier–Stokes dynamics, nor is it the result of an underlying, consistent statistical theory, involving many assumptions that have an unclear range of applicability. As expressed by the quote on the first page of this article, Saffman appreciated the limitations of this paradigm, but the results are included here to illustrate his attempt to understand the turbulent cascade and dissipation from a structural viewpoint, an approach he continued to pursue throughout much of his research career. This way of thinking continued to be valuable in later efforts to build models of the small scales based on vortex structures such as spirals (cf. Lundgren, 1982).

12.3.4 Wiener–Hermite expansions

A more formal approach that Saffman felt did not suffer these shortcomings is the Wiener–Hermite method based on expansions in terms of white noise polynomials. Since the Wiener–Hermite basis is complete, a realization of the turbulent velocity field can be expressed as a series with coefficients that are ordinary (non-random) functions whose evolution can be obtained from the Navier–Stokes equations. The equation set for the coefficients is of hierarchical form and there is then still a closure problem that can be handled by truncation or other techniques. Despite questions concerning convergence of the expansion, the method nonetheless has the attractive features that it is consistent with the accepted Gaussian form of the one-point probability distribution of the velocity in homogeneous turbulence and, importantly, truncation cannot lead to negative energy spectra which is a problem with some other turbulent closure hypotheses. Meecham and Siegel (1964) used the method for the Burgers equation while Saffman (1969) applied the Wiener–Hermite expansion to the diffusion of a passive scalar in homogeneous turbulence with the result that the Obukhov–Corrsin 'constant' $K_\theta \sim Re^{-1/4}$ where the Reynolds number Re is based on the integral scale. The prediction appears to be untested experimentally.

12.3.5 Two-dimensional turbulence

By the 1960s, it was well appreciated that the behavior of fluid-dynamical turbulence in one and two spatial dimensions was fundamentally different to that in three dimensions. For forced two-dimensional Navier–Stokes turbulence, the absence of vortex stretching implies that the mean dissipation must approach zero in the limit of vanishingly small viscosity, and so there cannot be a cascade of energy towards high wavenumbers. But vorticity is conserved on particle paths suggesting an enstrophy cascade, an observation that led Batchelor (1969) and Kraichnan (1967) to propose, based on enstrophy cascade arguments, an energy spectrum $E(k) \sim k^{-3}$ at wavenumbers exceeding some forcing wavenumber k_F. In the so-called dual cascade scenario, this is coupled with a reverse energy cascade to wavenumbers smaller than k_F with a corresponding $k^{-5/3}$ spectrum.

Saffman (1971), using an analogy with Burgers turbulence, obtained a quite different result. He argued that at large Reynolds numbers, a two-dimensional turbulent field would tend to develop near discontinuities in the vorticity field. Balancing convection and diffusion in regions of large vorticity gradients gives a thickness estimate for the vorticity jump of order $(\nu L/\omega')^{1/3}$ where L is the macroscale of the vorticity distribution and ω' the root-mean-square vorticity. This leads to a two-dimensional velocity spectrum

$$E(k) = \left(N\overline{J^2}/4\right)k^{-4}, \qquad N \ll k \ll \delta^{-1}, \qquad (12.76)$$

where N is the number of vorticity jumps per unit length along any straight line and $\overline{J^2}$ is the mean-square jump in vorticity across the discontinuity. Again, cascade and structure-based arguments lead to different results. The issue remains unresolved by numerical simulation, and although the weight of evidence leans perhaps in favor of k^{-3}, it seems clear that factors such as initial conditions and the degree of anisotropy are important to the extent that there may be no generic result (Kuznetsov et al., 2007). The dynamical issue appears to center on the extent to which the vorticity discontinuity is an attractor in two-dimensional turbulence over sufficiently long times and at large Reynolds numbers. In two-dimensional flow, characteristics do not intersect in a finite time although the distance between them may decrease exponentially with time, suggesting the formation of near discontinuities (Saffman, 1971).

Initial conditions may also be important. Saffman (1977) argues that a realization in which the vorticity is piecewise constant at $t = 0$ will have $E \sim k^{-4}$ initially. Since the piecewise constant state must persist when $v = 0$, then the spectrum will remain invariant.

12.3.6 Other work

An emerging trend in turbulence research in the 1970s was the interest in large-scale coherent structures as the dominant agent of turbulence transport, motivated in part by the perception that the statistical approach essentially discards almost all phase–amplitude correlations that characterize eddy structure. Coherent structures, such as the Kármán vortex street and turbulent spots, had been observed earlier, but were interpreted as special cases (see Liepmann, 1979). The galvanizing work was the seminal study by Brown and Roshko (1974) of the large-scale vortex structures in the turbulent mixing layer. Winant and Browand (1974) gave convincing evidence that the mechanism of linear stream-wise growth of the mixing layer was by the amalgamation, principally by pairing, of the large-scale vortices which form as a result of Kelvin–Helmholtz instability. Moore and Saffman (1975) proposed an alternative 'tearing' mechanism in which finite-sized vortices in a mixing layer will disintegrate if their spacing L relative to their local lateral scale, or mixing layer thickness δ, is too small. Their argument is based on the earlier analytical result of Moore and Saffman that a two-dimensional vortex of elliptical cross-section can exist within a uniform straining field only if its vorticity exceeded about 6.7 times the external maximum extensional rate of strain. Estimating the strain felt by a member vortex in a linear array of vortices of given circulation and separation, and using a simple elliptical model of a mixing layer vortex, they used this result to argue that if adjacent vortex centers are closer than about $2.8\,\delta$, their mutual interaction will lead to disintegration or tearing of the vortices, the remnants of which will then be captured or entrained by other neighboring vortices. This is an agile application of some simple ideas from vortex mechanics to a complex problem in turbulence. According to Moser and Rogers (1993), the Moore–Saffman tearing mechanism has been observed, but only for a small range of phase in combinations of the fundamental and subharmonic disturbances.

In the 1970s, Saffman's attention also turned to problems in turbulence of interest to engineering prediction and estimation, in particular, the construction of a closure model for the Navier–Stokes equations suitable for the calculation of complex turbulent flows. This work was motivated by earlier work of Kolmogorov (1942), not on statistical phenomenology, but on an eddy-viscosity-like model within the Reynolds-averaged Navier–Stokes (RANS) paradigm where Reynolds-stress closure is formulated in terms of model transport equations for the turbulent kinetic energy and the so-called 'pseudo-vorticity'. Saffman (1970a, 1974) introduced a Kolmogorov-type model predicated in part on the requirement that solutions should become asymptotic to the logarithmic version of the law of the wall near a solid boundary. The model contained several parameters, some obtained from accepted 'constants' such as the Kármán constant, and solutions for a variety of unidirectional flows were obtained by Knight (1975) and Wilcox (1975). Concerned that the basic eddy-viscosity concept was inadequate for flows with significant mean acceleration and flow curvature, Saffman (1976) and Saffman (1977) (see also Pope, 1975) proposed a general Reynolds-stress closure based on an equilibrium Reynolds stress $\overline{(u_i u_j)}_E$ expressed in terms of the rate-of-strain tensor S_{ij}, the rotation tensor Ω_{ij} and scalar quantities in a general way that satisfies invariance under axis rotations coupled to transport equations for the actual Reynolds stress $\overline{(u_i u_j)}$ that express diffusion of Reynolds stress and relaxation towards $\overline{(u_i u_j)}_E$. This relaxation stress model contains a large number of parameters that are supposed universal constants. Arguments based on canonical flows determines some but not all constants, and the latter are fixed by assumption. The model is shown to perform quite well against experiment (Saffman, 1976) for a class of flows where homogeneous turbulence is distorted by a uniform strain or uniform shear and for other applications (Knight and Saffman, 1978).

Saffman also focused in the 1970s on interesting problems related to turbulent trailing vortices shed by aircraft. Saffman (1973) considers the structure of turbulent line vortices. It is assumed that after some initial time the internal vortex structure depends on only the circulation Γ_0, the kinematic viscosity v and time t. Using a self-similar ansatz coupled to a Reynolds stress closure and a structural model, Saffman found an outer region with radius greater than the radius at which the tangential velocity is a maximum and in which there is a logarithmic circulation profile, an inner region where the azimuthal velocity decreases towards zero and a viscous region of radius $O(v t)^{1/2}$. A simplified model of the inner region is proposed and in both the inner and viscous regions there is close to solid body rotation. Saffman claims that his arguments for the log profile differ from those proposed earlier by Hoffmann and Joubert (1963). An estimate for the time for an overshoot in the radial distribution of circulation predicted earlier by Govindaraju and Saffman (1971) using an inviscid closure is obtained as are estimates of the magnitude of induced axial velocities (towards the wing in aircraft trailing vortices) using ideas applied to axial flow in laminar trailing vortices (Moore and Saffman, 1973).

12.4 Saffman as critic

The late 1970s saw the advent of several new techniques and approaches to theoretical treatments of turbulence flow phenomena. Taylor's statistical paradigm (Taylor, 1935) was well established for descriptive kinematics, in the techniques of experimental diagnostics and in theoretical analysis and modeling (see Leslie, 1973). Other theoretical formulations had also been proposed based on vortex dynamics (Townsend, 1951; Synge and Lin, 1943), dynamical systems theory (Ruelle, 1976), renormalization (Forster et al., 1977), the Wiener–Hermite method (Meecham and Siegel, 1964; Hogge and Meecham, 1978) and the proper-orthogonal decomposition (see Berkooz et al., 1993). Lundgren (1967) and Monin (see Monin and Yaglom, 1971) had independently developed the probability-density formulation of turbulence with proposed closure schemes. Computational fluid dynamics, still in it infancy, had nevertheless progressed to the point where the concepts of direct-numerical simulation, where all kinematically and dynamically relevant scales are resolved, was well established. Spectral simulations at moderate resolution (Orszag and Patterson, 1972) for box-turbulence had been performed and the first channel-flow and boundary-layer simulations had appeared. Both two-dimensional and three-dimensional vortex methods (Leonard, 1974b) had been applied to simple but realistic turbulence-flow simulations and the ideas and implementation of large-eddy-simulation (LES) were in place (Leonard, 1974a; Schumann, 1977). The role of large structures as the principal agent of transport in free turbulent flow was recognized (Brown and Roshko, 1974) and earlier observations of the presence of streaks in the near-wall region of turbulent boundary layers were confirmed (Kline et al., 1967).

These and other developments represented progress from time of publication of Batchelor's and Townsend's monographs of the 1950s. In something of a heroic effort, Saffman (1978) extended a critical survey first outlined in the 1966 lectures, conducting a wide-ranging review of the state of turbulence theory, including numerical simulation, dating roughly from the mid 1950s to the late 1970s. Saffman begins by defining contact with experiment and intellectual challenge as the principal hallmarks of 'progress'. Of six desirable properties of a theory he values clear physical purpose and intelligibility above rigor, remarking that

> A theory which cannot be understood without enormous effort would seem to lack value.

Predictive capability is ranked above rigor but below intelligibility. For accurate DNS of homogeneous turbulence Saffman estimates resolution in each of three directions as $N \sim Re_\lambda^{3/4}$ and with operation count $\sim R_\lambda^{11/2}[\log R_\lambda + K]$

where R_λ is the Reynolds number based on the Taylor scale. DNS for real-flow geometries are "awaited with interest". Jiménez (2003) estimates 2015 for this, signaling future progress.

Saffman states that

> practically everything that is useful in turbulence theory is a scaling law

but that the scaling approach is rarely useful for most complex flows as described by Bradshaw (1977). Vortex-based numerical methods, the structural-vortex approach and Lumley's POD approach seem to Saffman to have potential while the main utility of Reynolds-averaged (RANS) methods seems

> to lie in the interpolation of experimental data ...

Subgrid and vortex modeling are assessed as

> combining heavy computing with approximations of unknown validity ... [but they] ... may provide insight.

If LES can handle walls and

> produce the logarithmic layer and the Kármán constant, then a practical method of calculating gross details of turbulent flows of engineering interest would perhaps be available.

Saffman is less than sanguine when assessing the dynamical statistical theories of Kraichnan (but see the accompanying article on R.H. Kraichnan, Chapter 10) remarking that the

> absence of a physical basis is unfortunately usually combined with obscurity of the details.

Eight questions are posed to which

> answers are not really known.

These include the proposed independence of the dissipation rate (homogeneous turbulence) on the viscosity ν, the Reynolds number dependence of the inertial range, small eddies and intermittency and long-time existence of solutions to the Euler and Navier–Stokes equations.

The Springer article is notable for its scope combined with a forthright and critical tone. Perhaps disturbingly, many problems discussed seem just as relevant at the end of the first decade of the twenty-first century as in the late 1970s. In the interim, DNS has moved forward to produce fascinating simulations which nonetheless remain, in 2010, limited to moderate Reynolds numbers for free turbulent flows (e.g. Livescu et al., 2009), and for the most part to wall-bounded flows at Reynolds numbers somewhat beyond transition (see Schlatter and Örlü, 2010). Nonetheless there is now overlap between DNS and experiment, and it is predicted that DNS at mid-range laboratory Reynolds numbers will be reached within a decade (Jiménez, 2003). The processing of the ocean of numerical data, perhaps under-anticipated in the 1970s, will remain a challenge. Long-time existence for the Euler and Navier–Stokes equations remains open and while there is some experimental support for the independence of the rate of energy dissipation on viscosity for grid turbulence at moderate Reynolds numbers (see Sreenivasan, 1984), the issue remains unresolved. While RANS-type methods are still the principal workhorse of industry, to the present authors perhaps the most compelling advance made within the topics covered by Saffman have been in SGS modeling and LES where now quite realistic unions of complex transport, mixing and reaction physics/chemistry models have been combined with advances in numerical algorithms and the treatment of complex bounding geometry to produce an era of increasingly high-fidelity numerical simulations.

Saffman concludes the written account of his 1966 lectures with

> Finally, it would seem that turbulence theory to date can summarized by the quotation from *Macbeth*, "full of sound and fury, signifying nothing." It is to be hoped that future work will render this quotation inappropriate.

Twelve years later, in his discussion of general principles, Saffman (1978) tempers this as

> In searching for a theory of turbulence, perhaps we are looking for a chimera ... So perhaps there is no 'real turbulence problem', but a large number of turbulent flows and our problem is the self imposed and possibly impossible task of fitting many phenomena into the Procrustean bed of a universal turbulence theory.

The reader of this volume, three decades later, may have a response to these comments.

References

Batchelor, G.K. 1951. Pressure fluctuations in isotropic turbulence. *Proceedings of the Cambridge Philosophical Society*, **47**, 359.

Batchelor, G.K. 1952. The effect of homogeneous turbulence on material lines and surfaces. *Proceedings of the Royal Society of London. Series A, Mathematical and Physical Sciences*, **213**, 349–366.

Batchelor, G.K. 1953. *The Theory of Homogeneous Turbulence*. Cambridge University Press.

Batchelor, G.K. 1959. Small-scale variation of convected quantities like temperature in turbulent fluid Part 1. General discussion and the case of small conductivity. *Journal of Fluid Mechanics*, **5**(01), 113–133.

Batchelor, G.K. 1969. Computation of the energy spectrum in homogeneous two-dimensional turbulence. *Physics of Fluids (Supp. 2)*, **12**, 233.

Batchelor, G.K. and Proudman, I. 1956. The large-scale structure of homogeneous turbulence. *Philosophical Transactions of the Royal Society of London. Series A, Mathematical and Physical Sciences*, **248**(949), 369–405.

Batchelor, G.K. and Townsend, A.A. 1956. Turbulent diffusion. In *Surveys in Mechanics*, G.K. Batchelor, ed., Cambridge University Press, 352–399.

Berkooz, G., Holmes, P. and Lumley, J.L. 1993. The proper orthogonal decomposition in the analysis of turbulent flows. *Annual Review of Fluid Mechanics*, **25**, 539–575.

Bradshaw, P. 1977. Complex turbulent flows. In *Theoretical and Applied Mechanics; Proceedings of the Fourteenth International Congress, Delft, Netherlands, August 30–September 4, 1976. (A78-13990 03–31) Amsterdam*, North-Holland Publishing Co., 1977, 103–113.

Brown, G.L. and Roshko, A. 1974. Density effects and large structure in turbulent mixing layers. *Journal of Fluid Mechanics*, **64**, 775–816.

Cocke, W.J. 1969. Turbulent hydrodynamic line stretching. Consequences of isotropy. *Physics of Fluids*, **12**, 2488.

Cohen, S. 1999. Philip Saffman, a memoir. *Caltech Archives*, **1**, 1–91.

Davidson, P.A. 2009. The role of angular momentum conservation in homogeneous turbulence. *Journal of Fluid Mechanics*, **32**, 329–358.

Etemadi, N. 1990. On curve and surface stretching in isotropic turbulent flow. *Journal of Fluid Mechanics*, **221**, 685–692.

Forster, D., Nelson, D.R. and Stephen, M.J. 1977. Large-distance and long-time properties of a randomly stirred fluid. *Physical Review A*, **16**(2), 732–749.

Govindaraju, S.P. and Saffman, P.G. 1971. Flow in a turbulent trailing vortex. *Physics of Fluids*, **14**, 2074.

Hoffmann, E.R. and Joubert, P.N. 1963. Turbulent line vortices. *Journal of Fluid Mechanics*, **16**, 395–411.

Hogge, H.D. and Meecham, W.C. 1978. The Wiener–Hermite expansion applied to decaying isotropic turbulence using a renormalized time-dependent base. *Journal of Fluid Mechanics*, **85**, 325–347.

Jiménez, J. 2003. Computing high-Reynolds-number turbulence: will simulations ever replace experiments? *Journal of Turbulence*, **4**, 1–13.

Kline, S.J., Reynolds, W.C., Schraub, F.A. and Runstadler, P.W. 1967. The structure of turbulent boundary layers. *Journal of Fluid Mechanics*, **30**, 741–773.

Knight, D. 1975. Turbulence-model predictions for a flat plate boundary layer. *AIAA Journal*, **13**, 945–947.

Knight, D.D. and Saffman, P.G. 1978. Turbulence model predictions for flows with significant mean streamline curvature. In *AIAA, Aerospace Sciences Meeting*.

Kolmogorov, A.N. 1941. Dissipation of energy in locally isotropic turbulence. *Izv. Akad. Nauk. SSR Seria fizichka*, **32**, 16–18.

Kolmogorov, A.N. 1942. Equations of turbulent motion in an incompressible liquid. *Izv. Akad. Nauk. SSR Seria fizichka*, **VI**, 56.

Kolmogorov, A.N. 1962a. Precisions sur la structure locale de la turbulence dans un fluide visqueux aux nombres de Reynolds élevés. In *Mecanique de la*

Turbulence; Colloques Internationaux du Centre National de la Recherche Scientifique. CNRS, 447–458.

Kolmogorov, A.N. 1962b. A refinement of previous hypotheses concerning the local structure of turbulence in a viscous incompressible fluid at high Reynolds numbers. *Journal of Fluid Mechanics*, **13**, 82–85.

Kraichnan, R.H. 1967. Inertial ranges in two-dimensional turbulence. *Physics of Fluids*, **10**, 1417.

Krogstad, P.A., and Davidson, P.A. 2010. Is grid turbulence Saffman turbulence? *Journal of Fluid Mechanics*, **642**, 373–394.

Kuznetsov, E.A., Naulin, V., Nielsen, A.H. and Rasmussen, J.J. 2007. Effects of sharp vorticity gradients in two-dimensional hydrodynamic turbulence. *Physics of Fluids*, **19**, 105110.

Landau, L., and Lifshitz, E. 1987. *Fluid Mechanics.* Butterworth-Heinemann.

Leonard, A. 1974a. Energy cascade in large-eddy simulations of turbulent fluid flows. *Advance in Geophysics*, **18A**, 237–248.

Leonard, A. 1974b. Numerical studies of turbulence using vortex filaments. *Bulletin of the Americal Physical Society*, **19**, 1163–1164.

Leslie, D.C. 1973. *Developments in the Theory of Turbulence.* Oxford University Press.

Liepmann, H.W. 1979. The rise and fall of ideas in turbulence. *American Scientist*, **67**, 221–228.

Livescu, D., Ristorcelli, J.R., Gore, R.A., Dean, S.H., Cabot, W.H. and Cook, A.W. 2009. High-Reynolds number Rayleigh–Taylor turbulence. *Journal of Turbulence*, **10**(13), 1–32.

Loitsyanski, L.G. 1939. Some basic laws for isotropic turbulent flow. *Trudy Tsentr. Aero.-Gidrodyn*, **3**, 33.

Lundgren, T.S. 1967. Distribution functions in the statistical theory of turbulence. *Physics of Fluids*, **10**, 969.

Lundgren, T.S. 1982. Strained spiral vortex model for turbulent fine structure. *Physics of Fluids*, **25**, 2193.

Meecham, W.C. and Siegel, A. 1964. Wiener–Hermite expansion in model turbulence at large Reynolds number. *Physics of Fluids*, **7**, 1178.

Micheli, P.L. 1968. Dispersion in a turbulent field. PhD Thesis, Stanford University.

Mickelsen, W.R. 1960. Measurements of the effect of molecular diffusivity in turbulent diffusion. *Journal of Fluid Mechanics*, **7**, 397–400.

Moffatt, H.K. 1970. Turbulent dynamo action at low magnetic Reynolds number. *Journal of Fluid Mechanics*, **41**, 435–452.

Moffatt, H.K. 2007. The birth and adolescence of MHD turbulence. In *Magnetohydrodynamics – Historical Evolution and Trends*, S. Molokov, R. Moreau and H.K. Moffatt, eds., 213–222.

Monin, A.S. and Yaglom, A.M. 1971. *Statistical Fluid Mechanics*, vols 1&2. MIT Press.

Moore, D.W. and Saffman, P.G. 1973. Axial flow in laminar trailing vortices. *Proceedings of the Royal Society of London. Series A, Mathematical and Physical Sciences*, **333**(1595), 491–508.

Moore, D.W. and Saffman, P.G. 1975. The density of organized vortices in a turbulent mixing layer. *Journal of Fluid Mechanics*, **69**, 465–473.

Moser, R.D., and Rogers, M.M. 1993. The three-dimensional evolution of a plane mixing layer: pairing and transition to turbulence. *Journal of Fluid Mechanics*, **247**, 275–320.

Ohkitani, K. 2002. Numerical study of comparison of vorticity and passive vectors in turbulence and inviscid flows. *Physical Review E*, **65**(4), 046304.

Orszag, S. and Patterson, G. 1972. The numerical simulation of 3-dimensional homogeneous isotropic turbulence. *Phys. Rev. Letters*, 76.

Orszag, S.A. 1970. Comments on turbulent hydrodynamic line stretching. Consequences of isotropy. *Physics of Fluids*, **13**, 2203.

Pope, S.B. 1975. A more general effective-viscosity hypothesis. *Journal of Fluid Mechanics*, **72**, 331–340.

Proudman, I. and Reid, W.H. 1954. On the decay of a normally distributed and homogeneous turbulent velocity field. *Philosophical Transactions of the Royal Society of London. Series A, Mathematical and Physical Sciences*, **247**, 163–189.

Ruelle, D. 1976. Statistical mechanics and dynamical systems. Chapter I of *Statistical Mechanics and Dynamical Systems by David Ruelle and papers from the 1976 Duke Turbulence Conference*, Duke University Mathematics Series III.

Saffman, P.G. 1960. On the effect of the molecular diffusivity in turbulent diffusion. *Journal of Fluid Mechanics*, **8**, 273–283.

Saffman, P.G. 1962. Some aspects of the effects of the molecular diffusivity in turbulent diffusion. *Colloques Internationaux du Centre National de la Recherche Scientifique*, 53.

Saffman, P.G. 1963. On the fine-scale structure of vector fields convected by a turbulent fluid. *Journal of Fluid Mechanics*, **16**, 545–572.

Saffman, P.G. 1967. The large-scale structure of homogeneous turbulence. *Journal of Fluid Mechanics*, **27**(03), 581–593.

Saffman, P.G. 1968. Lectures on homogeneous turbulence. *Topics in Nonlinear Physics*, 485–614.

Saffman, P.G. 1969. Application of Wiener–Hermite expansion to diffusion of a passive scalar in a homogeneous turbulent field. *Physics of Fluids*, **12**, 1786–1789.

Saffman, P.G. 1970a. A model for inhomogeneous turbulent flow. *Proceedings of the Royal Society of London. Series A, Mathematical and Physical Sciences*, **317**, 417–433.

Saffman, P.G. 1970b. Dependence on Reynolds number of high-order moments of velocity derivatives in issotropic turbulence. *Physics of Fluids*, **13**, 2193.

Saffman, P.G. 1971. On the spectrum and decay of random two-dimensional vorticity distributions at large Reynolds number. *Studies in Applied Mathematics*, **50**, 377–383.

Saffman, P.G. 1973. Structure of turbulent line vortices. *Physics of Fluids*, **16**, 1181.

Saffman, P.G. 1974. Model equations for turbulent shear flow. *Studies in Applied Mathematics*, **53**, 17–34.

Saffman, P.G. 1976. Development of a complete model for the calculation of turbulent shear flows. Chapter II of *Statistical mechanics and Dynamical Systems by David Ruelle and papers from the 1976 Duke Turbulence Conference*, Duke University Mathematics Series III.

Saffman, P.G. 1977. Results of a two equation model for turbulent flows and development of a relaxation stress model for application to straining and rotating flows. In

Turbulence in Internal Flows: Turbomachinery and Other Engineering Applications; Proceedings of the SQUID Workshop, Warrenton, VA., June 14, 15, 1976. (A78-34826 14–34) Washington, DC. Hemisphere Publishing Corp., 1977, 191–226; Discussion, 226–231.

Saffman, P.G. 1978. Problems and progress in the theory of turbulence. In *Structure and Mechanisms of Turbulence II*, 273–306.

Saffman, P.G. and Turner, J.S. 1956. On the collision of drops in turbulent clouds. *Journal of Fluid Mechanics*, **1**, 16–30.

Schlatter, P. and Örlü, R. 2010. Assessment of direct numerical simulation data of turbulent boundary layers. *Journal of Fluid Mechanics*, **659**, 116–126.

Schumann, U. 1977. Realizability of Reynolds-stress turbulence models. *Physics of Fluids*, **20**, 721.

Sreenivasan, K.R. 1984. On the scaling of the turbulent energy dissipation rate. *Physics of Fluids*, **5**, 1048.

Synge, J.L. and Lin, C.C. 1943. On a statistical model of isotropic turbulence. *Trans. Roy. Soc. Canada*, **37**, 45–63.

Taylor, G.I. 1922. Diffusion by continuous movements. *Proc. London Math. Soc*, 2(20), 196–212.

Taylor, G.I. 1935. Statistical theory of turbulence. *Proceedings of the Royal Society of London. Series A, Mathematical and Physical Sciences*, **151**(873), 421–444.

Townsend, A.A. 1951. On the fine-scale structure of turbulence. *Proceedings of the Royal Society of London. Series A, Mathematical and Physical Sciences*, **208**, 534–542.

Townsend, A.A. 1954. The diffusion behind a line source in homogeneous turbulence. *Proceedings of the Royal Society of London. Series A, Mathematical and Physical Sciences*, **224**(1159), 487–512.

Tsinober, A. and Galanti, B. 2003. Exploratory numerical experiments on the difference between genuine and 'passive' turbulence. *Physics of Fluids*, **15**, 3514–3531.

Wilcox, D.C. 1975. Turbulence-model transition predictions. *AIAA Journal*, **13**, 241–243.

Winant, C.D. and Browand, F.K. 1974. Vortex pairing: the mechanism of turbulent mixing-layer growth at moderate Reynolds number. *Journal of Fluid Mechanics*, **63**(02), 237–255.

Yaglom, A.M. 1994. A.N. Kolmogorov as a fluid mechanician and founder of a school in turbulence research. *Annual Review of Fluid Mechanics*, **26**(1), 1–23.

13

Epilogue: a turbulence timeline

The Editors

To supplement the foregoing chapters, we offer below a table listing some key developments in turbulence research over the period covered by this book, i.e. roughly up to mid-1970s. Later developments involving massive computations, low-dimensional dynamics, the renormalization group, turbulence control, modern instrumentation, and so on, are not included; nor do we include such closely related areas as turbulent thermal convection, combustion, wave turbulence, or the vast field of applications in geophysics, astrophysics and plasma physics. Moreover, the table is 'internal' to the subject, in that we make no attempt to relate the events to developments in other scientific fields or to the wider historical context. Despite these limitations, it is our hope that the table, necessarily subjective to some extent, will provide a useful point of reference for the reader. We thank the authors of this book for their comments on the table, especially Professor R. Narasimha for the inspiration he provided.

Table 13.1 *Some major events in the history of turbulence*

Reference	Brief description of the event
Leonardo da Vinci (*c.* 1500)	Used the word 'turbolenza' and sketched a variety of turbulent flows
Katsushika Hokusai (*c.* 1831)	Sketched the "Great wave off Kanagawa" depicting turbulent broken waves
Hagen (1839)	Formally recognized two states of fluid motion
Saint-Venant (1851), Boussinesq (1870)	Postulated eddy viscosity
Reynolds (1874)	Analogy between eddy motion of fluid and heat transport
Reynolds (1883)	Direct and sinuous motion in pipe flows; Reynolds number
Kelvin (1887)	Used 'turbulence' in modern scientific literature
Rayleigh (1892)	Inviscid instability
Reynolds (1895)	Reynolds decomposition; Reynolds stresses
Prandtl (1904)	The concept of the boundary layer; its separation and control
Orr (1907), Sommerfeld (1908)	Equation for viscous stability
Eiffel (1912)	Demonstration that turbulence reduces drag on sphere
Blasius (1913)	1/4-power law for friction in pipe flows
Prandtl (1914)	Correct explanation for Eiffel's observation
King (1914)	Hot-wire
Taylor (1915)	Eddy motion; introduction of vorticity transport theory, modified in Taylor (1937)
Taylor (1921)	Diffusion by continuous movements; Lagrangian autocorrelation function
Richardson (1922)	Cascade of scales; possibility of numerical weather prediction
Taylor (1923)	Stability of Couette flow
Keller & Friedman (1924)	Eulerian correlation function; moment equations and an *ad hoc* closure scheme
Prandtl (1925)	Mixing length theory
Richardson (1926)	4/3 dependence of turbulent diffusion on scale size
Tollmien (1929)	Viscous stability solutions
Kármán (1930)	Log-law and the outer law for wall flows
Prandtl (1932)	Rederivation of the log-law
Nikuradse (1932, 1933)	High-Reynolds-number pipe flow measurements
Leray (1934)	Existence of weak solutions of the Navier–Stokes equations
Taylor (1935)	Isotropic turbulence; statistical theory
Taylor (1938)	Introduction of spectral analysis

(*cont.*)

Table 13.1 *Some major events in the history of turbulence (continued)*

Reference	Brief description of the event
Kármán & Howarth (1938)	Self-similarity; Kármán–Howarth equation
Millionshchikov (1939), Proudman & Reid (1956), Tatsumi (1957), Orszag (1970)	Quasi-normal closures, EDQNM models
von Neumann (1940s)	Possibility of electronic computing
Kolmogorov (1941)	Local isotropy; universality of small scale; inertial-range scaling of structure functions
Obuhkov (1941)	Inertial-range scaling of power spectrum
Kolmogorov (1942), Prandtl (1945)	Model transport equations for computing turbulent flows
Landau (1944), Hopf (1948)	Successive bifurcations leading to turbulence
Landau & Lifschitz (1944)	Criticism of small-scale universality
Schubauer & Skramstad (1947)	Observation of Tollmien–Schlichting waves
Burgers (1948)	One-dimensional model-equation
Obukhov (1948), Yaglom (1949), Corrsin (1951)	Kolmogorov's ideas extended to passive scalars
Onsager (1949)	Statistical equilibria of point vortices in two dimensions
Emmons (1951)	Turbulent 'spots'
Lighthill (1952)	Aerodynamically generated noise
Batchelor & Townsend (1951)	Dissipation-scale intermittency
Batchelor (1953)	*The Theory of Homogeneous Turbulence*
Dhawan (1953)	Direct measurement of skin friction
Kolmogorov (1954), Arnold (1963), Moser (1962)	KAM theory
Feynman (1955)	Quantum turbulence
Corrsin & Kistler (1955)	Outer intermittency
Batchelor & Proudman (1956), Saffman (1967)	Low wavenumber spectrum
Townsend (1956)	*Structure of Turbulent Shear Flows*
Hinze (1959)	*Turbulence: An Introduction to Its Mechanisms and Theory*
Batchelor (1959)	Passive scalar theory for high Schmidt number mixing
Kraichnan (1959, 1965)	Field theoretic methods (DIA and LHDIA)
Grant et al. (1962)	Experimental verification of inertial-range scaling
Obukhov (1962), Kolmogorov (1962)	Intermittency; local averaging; log-normality; refined similarity hypotheses
Smagorinsky (1962), Lilly (1967)	LES models
Lorenz (1963), Ueda (1960s)	Deterministic chaos
Yeh & Cummins (1964)	Laser Doppler velocimetry
Favre (1965)	Variable density averaging
Steenbeck et al. (1966)	Mean field electrodynamics

Table 13.1 *Some major events in the history of turbulence (continued)*

Reference	brief description of the event
Kraichnan (1967), Batchelor (1969)	Two-dimensional turbulence
Kline et al. (1967), Rao et al. (1971)	Bursting phenomena
Moreau (1961), Moffatt (1969)	Helicity an inviscid invariant
Kovasznay et al. (1971)	Conditional sampling
Ruelle & Takens (1971)	Strange attractors
Monin & Yaglom (1971, 1975)	*Statistical Fluid Mechanics*, vols 1 and 2
Barenblatt, Zeldovich (1970s)	Intermediate asymptotics, incomplete similarity
Brown & Roshko (1974)	Resurgence of coherent structures
Mandelbrot (1974)	Application of fractals

Bibliography and comments

Arnold, V.I. 1963. Proof of a theorem by A.N. Kolmogorov on the invariance of quasi-periodic motions under small perturbations of the Hamiltonian. *Usp. Math. Nauk* **18**, 13–40.

Barenblatt, G.I. 2003. *Scaling.* Cambridge University Press. This book is a summary and accessible account of many years of work of the author with Ya.B. Zeldovich.

Batchelor, G.K. 1953. *The Theory of Homogeneous Turbulence.* Cambridge University Press. Besides systematizing the then-available statistical theory of turbulence, the book brought Kolmogorov's work to the attention of the Western world. For a fuller account of Batchelor's contributions, see the accompanying article by H.K. Moffatt.

Batchelor, G.K. 1959. Small-scale variation of convected quantities like temperature in turbulent fluid. Part 1. General discussion and the case of small conductivity. *J. Fluid Mech.* **5**, 113–33.

Batchelor, G.K. 1969. Computation of the energy spectrum in homogeneous two-dimensional turbulence. *Phys. Fluids Suppl.* **11** 233–239.

Batchelor, G.K. & Proudman, I. 1956. The large-scale structure of homogeneous turbulence. *Phil. Trans. Roy. Soc. Lond. A* **248**, 369–405.

Batchelor, G.K. & Townsend, A.A. 1949. The nature of turbulent motion at large wavenumbers. *Proc. Roy. Soc. Lond. A* **199**, 238–55.

Blasius, H. 1913. Das Ähnlichkeitsgesetz bei Reibúngsvorgängen in Flüssigkeiten. *Forschungsarbeiten auf dem Gebiete des Ingenieurwesens* no. 131, Berlin.

Boussinesq, J. 1870. Essai théorique sur les lois trouvées expérimentalement par M. Bazin pour l'écoulment unifrome de l'eau dans les canaux découverts. *C.R. Acad. Sci. Paris* **71**, 389–393.

Brown, G.L. & Roshko, A. 1974. Density effect and large structure in turbulent mixing layers. *J. Fluid Mech.* **64**, 775–816.

Burgers, J.M. 1948. A mathematical model illustrating the theory of turbulence. *Adv. Appl. Mech.* **1**, 171–199. For a brief description of Burgers' work, see the accompanying article by K.R. Sreenivasan.

Corrsin, S. 1951. On the spectrum of isotropic temperature fluctuations in isotropic turbulence. *J. Appl. Phys.* **22**, 469–473. For a fuller account of Corrsin's contributions, see the accompanying article by C. Meneveau & J.J. Riley.

Corrsin, S. & Kistler, A.L. 1955. The free-stream boundaries of turbulent flows. NASA Tech. Rep. 1244.

Dhawan, S. 1952. Direct measurement of skin friction. NASA Tech. Note 2567. For a fuller account of Dhawan's contributions, see the accompanying article by R. Narasimha.

Eiffel, G. 1912. Sur la résistance des sphéres dans l'air en mouvement. *Compt. Rend.* **155**, 1587–1599.

Emmons, H.W. 1951. The laminar turbulent transition in a boundary layer. *J. Aero Sci.* **18**, 490–498.

Favre, A., 1965. Equations des gaz turbulents compressibles. *J. de Mécanique* **4**, 361–390.

Feynman, R.P. 1955. Application of quantum mechanics to liquid helium. *Prog. Low Temp. Phys.* **1**, 17–53.

Grant, H.L., Stewart, R.W. & Moilliet, A. 1962. Turbulence spectra from a tidal channel. *J. Fluid Mech.*, **12**, 263–272. Obukhov verified the equivalent result of the 2/3 power for the second-order structure function in 1949, using the data of K. Gödecke, obtained in 1935.

Hagen, G. 1939. Über die Bewegnung des Wassers in engen zylindrichen Röhren. *Pogg. Ann.* **46**, 423–442.

Hinze, J.O. 1959. *Turbulence. An Introduction to its Mechanisms and Theory.* McGraw Hill Co. New York.

Hopf, E. 1948. A mathematical example displaying the features of turbulence. *Commun. Pure Appl. Math.* **1**, 303–322.

Kármán, Th. von 1930. Mechanische Ahnlichkeit und Turbulenz. *Nach. Ges. Wiss. Göttingen, Math.-Phys.* **Kl**, 58–76. For various nuances of shared credit for the log-law between Kármán and Prandtl, see the accompanying article by A. Leonard & N. Peters on Kármán and that on Prandtl by E. Bodenschatz & M. Eckert in this volume.

Kármán, Th. von & Howarth, L. 1938. On the statistical theory of isotropic turbulence. *Proc. Roy. Soc. Lond.* **A164**, 192–215. For scientific exchanges between Kármán and Taylor on this problem, see the accompanying article by A. Leonard & N. Peters on Kármán and that on Taylor by K.R. Sreenivasan in this volume.

Keller, L.V. & Friedman, A.A. 1924. Differentialgleichung für die turbulente Bewegung einer kompressiblen Flüssigkeit. *Proc. 1st Intern. Cong. Appl. Mech.* Delft, pp. 395–405.

Kelvin, Lord. 1887. On the propagation of laminar motion through a turbulently moving inviscid liquid. *Phil. Mag.* **24**, 342–353.

King, L.V. 1914. On the convection of heat from small cylinders in a stream of fluid: determination of the convection constants of small platinum wires, with applications to hot-wire anemometry. *Proc. Roy. Soc.* **90**, 563–570.

Kline, S.J., Reynolds, W.C., Schraub, F.A. & Runstadler, P.W. 1967. The structure of turbulent boundary layers. *J. Fluid Mech.* **30**, 741–773.

Kolmogorov, A.N. 1941. The local structure of turbulence in incompressible viscous fluid for very large Reynolds numbers. *Dokl. Akad. Nauk SSSR* **30**, 9–13. (reprinted in *Proc. Roy. Soc. Lond.* **A434**, 9–13). The main results were rederived independently by Onsager: 'The distribution of energy in turbulence'. *Phys. Rev.* **68**, 286 (1945); by Heisenberg: 'Zur statistichen Theorie der Turbulenz'. *Z. Phys.* **124**, 628–657 (1948) and *Proc. Roy. Soc. Lond. A.* **195**, 402–406 (1948); and by von Weizsäcker: 'Das Spektrum der Turbulenz bei grossen Reynoldschen Zahlen'. *Zeit. f. Phys.* **124**, 614–627 (1948). Kolmogorov followed up this seminal paper of his by two others on different aspects of the same topic: they appeared in the same journal. For a more complete list of references, and for a discussion of further contributions by Kolmogorov, see the accompanying article by G. Falkovich.

Kolmogorov, A.N. 1942. Equations of turbulent motion of an incompressible fluid. *Izv. AN SSSR. Ser. Fiz.* **6**, 56–58.

Kolmogorov, A.N. 1954. On the conservation of conditionally periodic motions for a small change in Hamilton's function. *Dokl. Akad. Nauk SSSR* **98**, 525–530.

Kolmogorov, A.N. 1962. A refinement of previous hypotheses concerning the local structure of turbulence in a viscous incompressible fluid at high Reynolds number. *J. Fluid Mech.* **13**, 82–85.

Kovasznay, L.S.G., Kibens, V. & Blackwelder, R.F. 1970. Large scale motion in the intermittent region of a turbulent boundary layer. *J. Fluid Mech.* **41**, 283–325.

Kraichnan, R.H. 1959. The structure of isotropic turbulence at very high Reynolds numbers. *J. Fluid Mech.* **5**, 497–543. For a fuller account of Kraichnan's work, including his passive scalar work not included here, see the accompanying article by G.L. Eynik & U. Frisch.

Kraichnan, R.H. 1965. Lagrangian-history closure approximation for turbulence. *Phys. Fluids* **8**, 575–598.

Kraichnan, R.H. 1967. Inertial ranges in two-dimensional turbulence. *Phys. Fluids* **10**, 1417–1423.

Landau, L.D. 1944. On the problem of turbulence. *Akad. Nauk.* **44** 339–342.

Landau, L.D. & Lifschitz, E.M. 1944. *Fluid Mechanics* (published in English by Pergamon Press in 1963). The book contains other important contributions to turbulence, not touched upon here.

Leray, J. 1934. Sur le mouvement d'un liquide visqueux emplissant l'espace. *Acta Math.* **63**, 193–248. The work on weak solutions was extended by W. Hopf in 1951 and O.A. Ladyzhenskaya in late 1950s. The latter has summarized the essentials in 2003 as: 'Sixth problem of the millennium: Navier–Stokes equations, existence and smoothness', *Usp. Mat. Nauk* **58**, 45–78.

Lighthill, M.J. 1952. On sound generated aerodynamically. *Proc. Roy. Soc. Lond. A* **211**, 564–587. Subsequent papers of Lighthill on this topic followed in the same journal.

Lilly, D.K., 1967. The representation of small-scale turbulence in numerical simulation experiments. *Proc. of the IBM Sci. Comp. Symp. on Env. Sci.*, IBM-Form No. 320–1951.

Lorenz, E.N. 1963. Deterministic nonperiodic flow. *J. Atmos. Sci.* **20**, 130–141.

Mandelbrot, B.B. 1974. Intermittent turbulence in self-similar cascades; divergence of high moments and dimension of the carrier. *J. Fluid Mech.* **62**, 331–358. Mandelbrot (1983) contains a vivid description of his work on turbulence.

Mandelbrot, B.B. 1983. *The Fractal Geometry of Nature*. W.H. Freeman and Co. New York.

Millionshchikov, M.D. 1939. Decay of homogeneous isotropic turbulence in viscous incompressible fluids. *Dokl. AN SSSR*, **22**, 236–240.

Moffatt, H.K. 1969. The degree of knottedness of tangled vortex lines. *J. Fluid Mech.* **35**, 117–129.

Monin, A.S. & Yaglom, A.M. 1971. *Statistical Fluid Mech.*, vol. I. MIT Press (Russian edition 1965)

Monin, A.S. & Yaglom, A.M. 1975. *Statistical Fluid Mech.*, vol. II. MIT Press (Russian edition 1965). The two volumes made a valiant effort to bring together much of the knowledge available at that time.

Moreau, J.-J. 1961. Constants d'un îlot tourbillonaire en fluide parfait barotrope. *Comptes Rendus, Acad. des Sciences* **252**, 2810–2813.

Moser, J.K. 1962. On invariant curves of area-preserving mappings of an annulus. *Nachr. Akad. Wiss. Göttingen Math.-Phys. Kl. II.* **1**, 1–20.

Nikuradse, J. 1932. Gesetzmässigkeiten der turbulenten Strömung in glatten Röhren. *VDI-Forschungsheft* no. 356. The work on rough pipes appeared in 1933 as: Strömungs gesetze in rauhen Röhren, in *VDI-Forschungsheft* no. 361.

Obukhov, A.M. 1941. Energy distribution in the spectrum of a turbulent flow. *Izv. AN SSSR Ser. Geogr. Geofiz.* **5**, 453–466.

Obukhov, A.M. 1949. Structure of temperature fields in a turbulent flow. *Izv. AN SSSR Ser. Geogr. Geofiz.* **13**, 58–69.

Obukhov, A.M. 1962. Some specific features of atmospheric turbulence. *J. Fluid Mech.* **13**, 77–81.

Onsager, L. 1949. Statistical hydrodynamics. *Neuvo Cimento* **6**, Suppl. no. 2, 279–287. This article is about both the statistical equilibria of point vortices in two dimensions and the energy spectrum in three-dimensional turbulence. For a fuller account of Onsager's turbulence work, see G.L. Eyink, & K.R. Sreenivasan 'Onsager and the theory of hydrodynamic turbulence'. *Rev. Mod. Phys.* **78**, 87–135 (2006).

Orr, W.M. 1907. The stability or instability of the steady motions of a perfect liquid and of a viscous liquid. *Proc. Roy. Irish Acad. A* **27**, 9–68; 69–138.

Orszag, S.A. 1970. Analytical theories of turbulence. *J. Fluid Mech.* **41**, 363–386.

Prandtl, L. 1904. Über Flüssigkeitsbewegung bei sehr kleiner Reibung. In *Verhandlungen des dritten Internationalen Mathematiker-Kongresses in Heidelberg 1904*, edited by A. Krazer, Teubner, Leipzig (1905), 574–584. (English translation in *Early Developments of Modern Aerodynamics*, edited by J.A.K. Ackroyd, B.P. Axcell & A.I. Ruban, Butterworth–Heinemann, Oxford, UK (2001), pp. 77–87.) For several other lasting contributions to turbulence by Prandtl and his school, see the accompanying article by E. Bodenschatz & M. Eckert, this volume.

Prandtl, L. 1914. Der Luftwiderstand von Kugelin. *Nachrichten der Gesselschaft der Wissenschaften zu Göttingen, Math.-Phys. Klasse*, 177–190.

Prandtl, L., 1925. Bericht uber Untersuchungen zur ausgebildeten Turbulenz. *ZAMM* **5**, 136–139.

Prandtl, L. 1932. Zur turbulenten Strömung in Rohren und längs Platten. *Ergebnisse der Aerodynamischen Versuchsanstalt zu Göttingen*. **4**, 18–29.

Prandtl, L. 1945 Über die Rolle der Zähigkeit im Mechanismus der ausgebildete Turbulenz (The role of viscosity in the mechanism of developed turbulence). *Göttinger Archiv des DLR, Göttingen* 3712.

Proudman, I. & Reid, W.H. 1954. On the decay of a normally distributed and homogeneous turbulent velocity field. *Phil. Trans. Roy. Soc. Lond. A* **247**, 163–189.

Rao, K.N., Narasimha, R. & Badri Narayanan, M.A. 1971. The 'bursting' phenomenon in a turbulent boundary layer. *J. Fluid Mech.* **48**, 339–352.

Rayleigh, Lord. 1892. On the question of stability of the flow of fluids. *Phil. Mag.* **34**, 59–70.

Reynolds, O. 1874. On the extent and action of the heating surface for steam boilers. *Proc. Manchester Lit. Phil. Soc.* **14**, 7–12. For Reynolds' contributions to turbulence and his place in history, see the article by B.E. Launder & J.D. Jackson, this volume.

Reynolds, O. 1883. An experimental investigation of the circumstances which determine whether the motion of water shall be direct or sinuous, and of the law of resistance in parallel channels. *Phil. Trans. Roy. Soc. Lond.* **174**, 935–982.

Reynolds, O. 1895. On the dynamical theory of incompressible viscous fluids and the determination of the criterion. *Phil. Tran. Roy. Soc. Lond.* **86**, 123–164.

Richardson, L.F. 1922. *Weather Prediction by Numerical Methods*. Cambridge University Press. For Richardson's other contributions to turbulence and his eclectic work, see the article by R. Benzi, this volume.

Richardson, L.F. 1926, Atmospheric diffusion shown on a distance–neighbour graph. *Proc. Roy. Soc. Lond. A* **110**, 709–737.

Ruelle, D. & Takens, F. 1971. On the nature of turbulence. *Commun. Math. Phys.* **20**, 167–192.

Saffman, P.G. 1967. The large-scale structure of homogeneous turbulence. *J. Fluid Mech.* **27**, 581–593. For Saffman's other contributions to turbulence, see the article by D.I. Pullin & D.I. Meiron, this volume.

Saint-Venant, A.J.C. 1850. Mémoire sur des formulaes nouvelles pour la solution des problémes relatifs aux eaux courantes. *C. R. Acad. Sci. Paris* **31**, 283–286.

Schubauer, G.B. & Skramstad, H.K. 1947. Laminar boundary-layer oscillations and stability of laminar flow. *J. Aero. Sci.* **14**, 69–76.

Smagorinsky, J. 1963. General circulation experiments with the primitive equations, I. The basic experiment. *Monthly Weather Rev.* **91**, 99–164.

Sommerfeld, A. 1908. Ein Beitrag zur hydrodynamischen Erklärung der turbulenten Flüssigkeitsbewegungen. *Proc. 4th Internat. Cong. Math.* Rome, **3**, 116–124.

Steenbeck, M., Krause, F. & Radler, K.-H. 1966. Berechnung der mittleren Lorentz-Feldstarke fur ein elektrisch leitendes Medium in turbulenter, durch Coriolis-Krafte beeinflusster Bewegung. 2. *Naturf.* **21a**, 369–376.

Tatsumi, T. 1957. The theory of decay process of incompressible isotropic turbulence. *Proc. Roy. Soc. Lond. A* **239**, 16–45.

Taylor, G.I. 1915. Eddy motion in the atmosphere. *Phil. Trans. Roy. Soc. Lond. A* **215**, 1–26.

Taylor, G.I. 1921. Diffusion by continuous movements. *Proc. Lond. Math. Soc.* **20**, 196–212.

Taylor, G.I. 1923. Stability of a viscous liquid contained between two rotating cylinders. *Phil. Trans. Roy. Soc. Lond. A* **223**, 289–343.

Taylor, G.I. 1935. Statistical theory of turbulence. I. *Proc. Roy. Soc. Lond. A* **151**, 421–444. Subsequent parts II–V on this topic appeared in the same journal. For a full list of references and a more complete description of Taylor's contributions, see the article in this volume by K.R. Sreenivasan.

Taylor, G.I. 1937. Flow in pipes and between parallel planes. *Proc. Roy. Soc. Lond. A* **159**, 496–506.

Taylor, G.I. 1938. The spectrum of turbulence. *Proc. Roy. Soc. Lond. A* **164**, 476–481.

Townsend, A.A. 1956. *The Structure of Turbulent Shear Flow*. Cambridge University Press. Townsend's book emphasized the presence of structure within statistical description. See the article by I. Marusic & T. Nichols, this volume, for an elaboration of this aspect and the other work of Townsend. Similar recognitions of the importance of flow structures were made by others, e.g., T. Theodorsen, 'Mechanism of turbulence', in *Proc. Second Midwestern Conf. on Fluid Mech.* Ohio State University, Columbus, Ohio, pp. 1–19 (1952). Townsend initiated the modeling of small scales through vortex sheets and tubes in 'On the fine-scale structure of turbulence', *Proc. Roy. Soc. A*, **208**, 534–642 (1951).

Tollmien, W. 1929. Über die Entstehung der Turbulenz. *Nachr. Ges. Wiss. Göttingen Math-Phys.* **Kl, II**, 21–44

Ueda, Y. 1970. In 1961, Ueda posed a mathematical model on an analog computer that displayed chaotic dynamics. However, this work was not published until 1970; see Y. Ueda, C. Hayashi, N. Akamatsu, & H. Itakura, On the behavior of self-oscillatory systems with external force. *Electronics & Communication in Japan* **53**, 31–39 (1970).

Yaglom, A.M. 1949. Local structure of the temperature field in a turbulent flow. *Dokl. Akad. Nauk. SSSR* **69**, 743–746.

Yeh, Y. & Cummins, H.Z. 1964. Localized fluid flow measurements with an He–Ne laser spectrometer. *Appl. Phys. Lett.* **4**, 176–178.

Printed in the United States
By Bookmasters